# Lecture Notes in Computer Scie

T0238405

*Commenced Publication in 1973*
Founding and Former Series Editors:
Gerhard Goos, Juris Hartmanis, and Jan van Leeuwen

Jin Akiyama   William Y.C. Chen
Mikio Kano   Xueliang Li   Qinglin Yu (Eds.)

# Discrete Geometry, Combinatorics and Graph Theory

7th China-Japan Conference, CJCDGCGT 2005
Tianjin, China, November 18-20, 2005
Xi'an, China, November 22-24, 2005
Revised Selected Papers

 Springer

Volume Editors

Jin Akiyama
Tokai University
Research Institute of Education
Shibuya-ku, Tokyo, Japan
E-mail: fwjb5117@mb.infoweb.ne.jp

William Y.C. Chen
Xueliang Li
Qinglin Yu
Nankai University
Center for Combinatorics
Tianjin, China
E-mail: {chen,lxl,yu}@nankai.edu.cn

Mikio Kano
Ibaraki University
Department of Computer and Information Sciences
Hitachi, Ibaraki, Japan
E-mail: kano@mx.ibaraki.ac.jp

Library of Congress Control Number: 2006940628

CR Subject Classification (1998): I.3.5, G.2, F.2.2, E.1

LNCS Sublibrary: SL 1 – Theoretical Computer Science and General Issues

ISSN        0302-9743
ISBN-10     3-540-70665-8 Springer Berlin Heidelberg New York
ISBN-13     978-3-540-70665-6 Springer Berlin Heidelberg New York

Springer is a part of Springer Science+Business Media

springer.com

© Springer-Verlag Berlin Heidelberg 2007
Printed in Germany

Typesetting: Camera-ready by author, data conversion by Scientific Publishing Services, Chennai, India
Printed on acid-free paper      SPIN: 11980766      06/3142      5 4 3 2 1 0

# Preface

This volume consists of the peer-reviewed papers of the China-Japan Conference on Discrete Geometry, Combinatorics and Graph Theory (CJCDGCGT 2005). The conference was held in two places in China—Nankai University, Tianjin (November 18-20, 2005) and Northwestern Polytechnical University, Xi'an, Shaanxi (November 22-24, 2005). The conference sponsors include the Center for Combinatorics of Nankai University, the Department of Applied Mathematics of Northwestern Polytechnical University and the Institute for Education Development of Tokai University of Japan. More than 200 participants from China, Japan, USA, Malaysia, Mexico, Thailand, Iran, etc. attended. The opening ceremony was chaired by the Vice-President of Nankai University, Bill Chen. Yuan Wang, a distinguished academician of the Chinese Academy of Sciences, and the Co-chair of the conference, Jin Akiyama, gave welcome speeches.

There are many established graph theorists in China but few discrete geometers. It is hoped that the conference inspired Chinese mathematicians to pursue some of the many challenging problems in discrete geometry.

The conference was the seventh in a series of conferences held on Discrete Geometry and Graph Theory since 1997. The earlier ones were held in Tokyo, Manila (Philippines) and Bandung (Indonesia). The proceedings of five of these conferences were published by Springer as part of the series *Lecture Notes in Computer Science* (LNCS) volumes 1763, 2098, 2866, 3330 and 3742. The sixth was also published by Springer as a special issue of the journal *Graphs and Combinatorics*, Vol.18, No.4, 2002. The organizers of CJCDGCGT 2005 would like to express sincere thanks to the sponsors, the conference secretariat and the invited plenary speakers: Kiyoshi Ando, Guizhen Liu, Zhiming Ma, János Pach, Jorge Urrutia, Fuji Zhang and Chuanming Zong.

November 2006

Jin Akiyama
William Y.C. Chen
Mikio Kano
Xueliang Li
Qinglin R. Yu

The Organizing Committee
Co-chairs: Jin Akiyama and William Y.C. Chen
Members: Mikio Kano, Xueliang Li, Chie Nara, Guohua Peng, Toshinori Sakai, Xuehou Tan, Masatsugu Urabe, Qinglin R. Yu, Shenggui Zhang
Secretary: Matthew Q.H. Guo

# Table of Contents

# Infinite Series of Generalized Gosper Space Filling Curves

Jin Akiyama[1], Hiroshi Fukuda[2], Hiro Ito[3], and Gisaku Nakamura[1]

[1] Research Institute of Education, Tokai University, 2-28-4 Tomigaya,
Shibuya-ku Tokio, Japan
[2] School of Administration and Informatics, University of Shizuoka,
52-1 Yada, Shizuoka, 422-8526 Japan
[3] Department of Communications and Computer Engineering, School of Informatics,
Kyoto University, Kyoto 606-8501, Japan

**Abstract.** We report on computer search for generalized Gosper curve for $37 < N < 61$, where $N$ is the degree of the generalized Gosper curve. From the results of the computer search and some geometrical insight, we conjecture that the degree $N$ satisfies $N = 6n + 1$. We investigate the existence of infinite series of generalized Gosper curves. We show how to generate these series and introduce two new methods, the 'decomposition method' and the 'modified layer method'.

## 1 Generalized Gosper Space Filling Curves

The Gosper curve is a space filling curve discovered by William Gosper, an American computer scientist, in 1973, and was introduced by Martin Gardner in 1976 [1,2]. The curve is constructed by recursively replacing a bold arrow, called *initiator*, by seven arrows, called *generators*, Fig. 1(a), Fig. 1(b) and Fig. 1(c) illustrate the curves obtained by replacing the initiator by generators once and twice respectively.

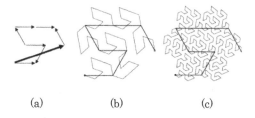

(a)          (b)          (c)

**Fig. 1.** Gosper curve

The Gosper curve is said to be a monster curve since it is a path from the root to the tip of the initiator visiting all interior lattice points on a regular triangular lattice. In 2001 [3], we found many such monster curves, called *generalized Gosper curves*, by computer search. Their shapes are similar to the original Gosper curve as shown in Fig. 2.

J. Akiyama et al. (Eds.): CJCDGCGT 2005, LNCS 4381, pp. 1–9, 2007.

**Fig. 2.** Generalized Gosper curve with $N = 13, 43, 91$

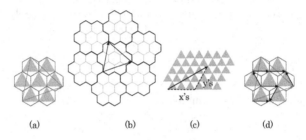

(a)              (b)              (c)              (d)

**Fig. 3.** B-figure and initiator triangle

The generalized Gosper curves are defined by the following procedure used in the computer search in [3].

1. Choose the degree of the generalized Gosper curve $N$, the number of arrows included in the generator.
2. Assemble $N$ hexagons having equilateral triangles inside as shown in Fig. 3 (a). We call this assembly of hexagons *B-figure*.
3. Choose two vertices from the triangles in B-figure as root and tip of the initiator.
4. Create a tiling by B-figures as shown in Fig. 3 (b), by using edges of the equilateral triangle having the initiator as its one side, as translation vectors. Then

$$N = x^2 + y^2 + xy \tag{1}$$

   where $x$ and $y$ $(= 0, 1, 2, \ldots)$ are the numbers of horizontal and vertical triangles between the end points of the initiator as shown in Fig. 3 (c).
5. The B-figure has to have three-fold rotational axis. Therefore,

$$N = 3m \text{ or } 3m + 1, \ m = 0, 1, 2, \ldots. \tag{2}$$

6. Both the root and tip of the initiator have to be three-fold rotational axes of the tiling pattern of the B-figure.
7. By taking one edge from each equilateral triangle in the B-figure, construct a path from the root to the tip of the initiator. Confirm that the path visits all lattice points in the B-figure, except the points on the boundary, where the points are covered by only one triangle.

8. Convert the edges in the path to the arrow so that the direction of the arrow is counterclockwise in the equilateral triangle. Then obtain the generator arrows as shown in Fig. 3 (d).

In [3], computer search for generalized Gosper curves for $N \leq 37$ was performed and the existence of infinite series of generalized Gosper curves was confirmed. In this paper, we report on two new results. In section 2, results of computer search for $37 < N \leq 61$ are reported. In section 3, many infinite series of generalized Gosper curves are shown.

## 2   Degree and Number of Curves

Computer search was performed for the generalized Gosper curves for $N \leq 61$. We were able to determine the number of generalized Gosper curves $C_N$ and the number of corresponding B-figures $B_N$ for given $N$. In Table 1, we tabulate $C_N$, $B_N$ and also $C_N(i)$, the number of curves on the $i$'th B-figure, for $N$ given by Eqs. (1) and (2).

**Table 1.** Number of generalized Gosper curves $C_N$. $N$'s with $C_N = 0$ are omitted.

| $N$ x y $C_N = C_N(1) + C_N(2) + \cdots + C_N(B_N)$ |
| --- |
| 7 1 2 1 |
| 13 1 3 1 |
| 19 2 3 1 |
| 31 1 5 7 $= 1 + 5 + 1$ |
| 37 3 4 8 $= 1 + 2 + 3 + 2$ |
| 43 1 6 24 $= 3 + 1 + 19 + 1$ |
| 49 0 7 134 $= 58 + 5 + 3 + 60 + 4 + 4$ |
| 49 3 5 75 $= 41 + 18 + 1 + 6 + 2 + 1 + 5 + 1$ |
| 61 4 5 (486) $= 1 + 13 + 14 + 153 + 19 + 166 + 24 + 45 + 10 + 5 + 1 + 34 + 1$ |

From Table 1, we observe that among $N = x^2 + y^2 + xy$ only $N = 6n + 1$ seem to be degrees allowed for the curves. Therefore,

*Conjecture.* The degree $N$ of the generalized Gosper curves has to satisfy

$$N = 6n + 1, \tag{3}$$

where $n$ is positive integers.

We have a rough scheme of the proof of this conjecture, but it is not complete and not simple. We hope to produce the proof in a future paper.

From Table 1 we observe that the number of B-figures, $B_N$, increases gradually with $N$. On the other hand, the computational time for $B_N$ increases exponentially with $N$ and this is the reason why we stopped calculation at $N = 61$. We show all B-figures with $43 \leq N < 61$ in Fig. 4–6. The B-figures for $N < 43$ are given in [3]. We note that the calculation for $N = 61$ is not completed, and that the number $C_N$ given in parenthesis in Table 1 is a lower bound.

**Fig. 4.** B-figures for $N = 43$. $i = 1, 2, 3, 4$ from the left.

**Fig. 5.** B-figures for $N = 49$, $(x, y) = (7, 0)$. $i = 1, 2, \ldots, 6$ from the left top to right bottom.

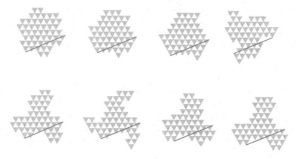

**Fig. 6.** B-figures for $N = 49$, $(x, y) = (5, 3)$. $i = 1, 2, \ldots, 8$ from the left top to right bottom.

## 3  Infinite Series of Curves

We discussed the natural extension of the Gosper curve in [3]. It is a simple procedure to generate infinitely many curves with

$$N = 3k^2 + 3k + 1, \ k = 1, 2, 3, \ldots \tag{4}$$

from the Gosper curve as shown in Fig. 7, where we can find a simple relation between successive generators in $k$. Since the B-figure for $k + 1$ is obtained by adding a layer to the B-figure for $k$ as shown in Fig. 8, we call this procedure *layer method*. In this section, we introduce two new methods to generate infinite series and show that there exist several other infinite series of curves.

**Fig. 7.** Infinite series of generalized Gosper curves with $N = 3k^2 + 3k + 1$. $N = 7, 19, 37$.

**Fig. 8.** Layer method

(a)          (b)

**Fig. 9.** Wrapped B-figures for $N = 13$ and generalized Gosper curve on the B-figure

**Fig. 10.** T-figure for $N = 13$. The filled circle is the three fold rotational axis.

## 3.1 Decomposition Method

We consider the infinite series from $N = 13$. From the case $N = 7$, if we wrap the B-figure by the layer method, we obtain 31 triangles shown in Fig. 9 (a). By comparing the results for $N = 31$ in [3], we find that this is a B-figure having one generalized Gosper curve shown in Fig. 9 (b). However, it is difficult to find any simple relation between the curve with $N = 13$ and Fig. 9 (b).

We introduce an assembly of hexagons which are obtained by replacing the nodes of the generator by hexagons except two end points as shown in Fig. 10. We call this assembly of $N - 1$ hexagons *T-figure*. From the definitions of the B-figure, the T-figure has three-fold rotational axis. Therefore, the T-figure can be decomposed into three congruent pieces $\alpha$, $\beta$ and $\gamma$ as shown in Fig. 10.

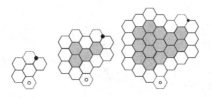

**Fig. 11.** Decomposed T-figure for $N = 13$ and wrapping process

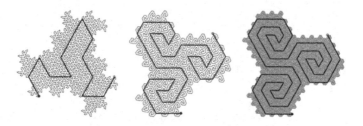

**Fig. 12.** Infinite series of generalized Gosper curves from $N = 13$ by decomposition method. $N = 13, 43, 91$.

Then we find that we can obtain the B-figure by adding layers of hexagons onto each decomposed T-figure, as shown in Fig. 11, and can construct a path from $\alpha$ to $\beta$ then $\gamma$. In Fig. 11, the dashed hexagon with a circle inside is an endpoint of the generator, and the filled circle is a three-fold rotational axis of the T-figure.

As can be seen from Fig. 10, for $N = 13$, the path in $\alpha$ and $\gamma$ starts from the dashed hexagon with a circle inside and goes toward the hexagon with three-fold rotational axis. The path in $\beta$ starts from the dashed hexagon without a circle and goes toward the hexagon with three-fold rotational axis. The path in successive bigger pieces are obtained similarly. In this way, we get an infinite series from $N = 13$ shown in Fig. 12 with

$$N = 9k^2 + 3k + 1, k = 1, 2, 3, \ldots. \tag{5}$$

We call this method *decomposition method*.

Up to now, we have not found another series generated by the layer method other than the series found in [3]. However, the decomposition method generates several infinite series. For example, the decomposition method works for a curve with $N = 31$ and yields an infinite series shown in Fig. 13 with

$$N = 9k^2 + 15k + 7, \ k = 1, 2, 3, \ldots. \tag{6}$$

The decomposed piece in this case is shown in Fig. 14.

For the generalized Gosper curve with $N = 31$ shown in Fig. 15 (a), a layer is added to the decomposed T-figure in different a way as shown in Fig. 16. In

**Fig. 13.** Infinite series of generalized Gosper curves from $N = 31$ by decomposition method. $N = 31, 73$.

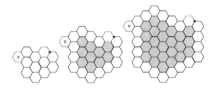

**Fig. 14.** Decomposed T-figure for $N = 31$ and wrapping process

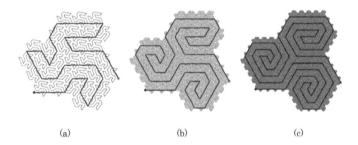

(a)                              (b)                              (c)

**Fig. 15.** Infinite series of generalized Gosper curves from $N = 31$ by decomposition method. $N = 31, 73, 133$.

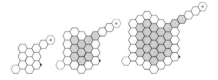

**Fig. 16.** Decomposed T-figure for $N = 31$ and wrapping process

this case, the starting points and terminal points of the path in the decomposed T-figures are also different from the previous two examples.

The path in $\alpha$ and $\beta$ starts from the dashed hexagon with a circle inside and goes toward the hexagon with a three-fold rotational axis. The path in $\gamma$ starts from the dashed hexagon without a circle and goes toward the hexagon with a circle. This procedure yields the infinite series shown in Fig. 15 with

$$N = 9k^2 + 15k + 7, \ k = 1, 2, 3, \ldots \tag{7}$$

from $N = 31$.

### 3.2 Modified Layer Method

In Fig. 17, a generalized Gosper curve for $N = 37$ and the corresponding B-figure are shown. From this curve, we can generate an infinite series by modifying the layer method as follows.

**Fig. 17.** A generalized Gosper curve for $N = 37$ and its B-figure

1. It is impossible to wrap the B-figure in Fig. 17 by a single layer as shown in Fig. 18 (a).

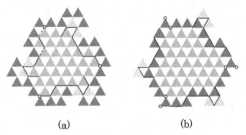

(a)                    (b)

**Fig. 18.** Modified layer for $N = 37$

2. We modify by removing the striped triangles from the B-figure and wrap it by a single layer as shown in Fig. 18 (b).
3. After wrapping the modified B-figure, we add the removed triangles (striped triangles) as shown in Fig. 18 (b).
4. Then we can connect a starting point of the new initiator and a terminal point of the old initiator by a path, and a terminal point of the new initiator and a starting point of the old initiator by a path.
5. It is possible to continue this procedure to obtain a new infinite series with

$$N = 3k^2 + 15k + 19, \ k = 1, 2, 3, \ldots \tag{8}$$

starting from $N = 37$ as shown in Fig. 19.

**Fig. 19.** Infinite series of generalized Gosper curve from $N = 37$ by modified layer method. $N = 37, 61, 91$.

## 4 Summary

In this paper, we reported on new results of computer search for generalized Gosper curves for $37 < N < 61$. We then observed that the empirically allowed degree $N$ for generalized Gosper curves seems to be $N = 6n + 1$. The proof of this conjecture looks complex and lengthy. It is left for a future work.

Finally, we pointed out that there exist several infinite series of generalized Gosper curves generated by three methods we introduced: the layer method, the decomposition method, and the modified layer method. However, so far as our investigation is concerned, several other series and methods exist and more than three quarters of generalized Gosper curves can generate infinitely many curves. For example, an infinite series with

$$N = 9k^2 + 3k + 1 \qquad (9)$$

shown in Fig. 2 can not be generated by the above three method. Thus more systematic investigation of infinite series for every generalized Gosper curve is desired.

## References

1. M. Gardner, "In which 'monster' curves force redefinition of the word 'curve', Scientific American 235 (December issue), 124–133(1976).
2. B. B. Mandelbrot, "The fractal geometry", W. H. Freeman and Company, New York (1977).
3. H. Fukuda, M. Shimizu and G. Nakamura, "New Gosper Space Filling Curves", Proceedings of the International Conference on Computer Graphics and Imaging (CGIM2001), 34–38 (2001).

# Contractible Edges in a $k$-Connected Graph

Kiyoshi Ando

The University of Electro-Communications, 1-5-1 Chofugaoka,
Chofu City, Tokyo, 182-8585, Japan

**Abstract.** An edge of a $k$-connected graph is said to be $k$-contractible if
the contraction of the edge results in a $k$-connected graph. Some results
concerning $k$-contractible edges in a $k$-connected graph are presented.

## 1  Introduction

We present some recent results concerning contractible edges in $k$-connected
graphs which are given by the author with Egawa, Kaneko, Kawarabayashi and
others.

We deal with finite undirected graphs with neither loops nor multiple edges.
For a graph $G$, let $V(G)$ and $E(G)$ denote the set of vertices of $G$ and the set
of edges of $G$, respectively. For a vertex $x \in V(G)$, we denote by $N_G(x)$ the
neighborhood of $x$ in $G$ and let $N_G[x] = N_G(x) \cup \{x\}$. Moreover, for a subset
$S \subset V(G)$, let $N_G(S) = \cup_{x \in S} N_G(x) - S$ and let $N_G[S] = \cup_{x \in S} N_G(x) \cup S$.
We denote the degree of $x \in V(G)$ by $\deg_G(x)$, namely $\deg_G(x) = |N_G(x)|$.
Let $V_i(G)$ be the set of vertices of degree $i$. We denote the minimum degree of
$G$ by $\delta(G)$. Let $G[S]$ denote the subgraph induced by $S \subset V(G)$. The *square*
of a graph $G$ is the graph obtained from $G$ by adding edges joining each pair
of vertices whose distance in $G$ is 2. For two graphs $G$ and $H$, we denote the
join of $G$ and $H$ by $G + H$. Let $K_k^-$ denote the graph obtained from $K_k$ by
deleting one edge. A bowtie is the graph consisted of two triangles with exactly
one common vertex. Hence $K_4^- \cong 2K_1 + K_2$ and a bowtie can be written by
$K_1 + 2K_2$. For a graph $H$, a graph $G$ is said to be $H$-free if $G$ has no $H$ as
a subgraph. Let $G$ be a connected graph. A subset $S \subset V(G)$ is said to be a
*cutset* of $G$, if $G - S$ is not connected. A cutset $S$ is said to be a $k$-*cutset* if
$|S| = k$. A $k$-cutset $S$ of a $k$-connected graph $G$ is to be *trivial* if there is a
vertex $x \in V_k(G)$ such that $S = N_G(x)$. Let $k$ be an integer such that $k \geq 2$
and let $G$ be a $k$-connected graph. An edge $e$ of $G$ is said to be $k$-*contractible*
if the contraction of the edge results in a $k$-connected graph. An edge which
is not $k$-contractible is called a *noncontractible* edge. We denote the set of $k$-
contractible edges of $G$ by $E_c(G)$. If the contraction of $e \in E(G)$ results in a
graph with minimum degree $k - 1$, then $e$ is said to be *trivially* noncontractible.
In other words, $e$ is trivially noncontractible if and only if the end vertices of
$e$ have a common neighbor of degree $k$, or (equivalently) they are contained in
some trivial cutset. A $k$-connected graph with no $k$-contractible edge is said to
be *contraction critically $k$-connected*.

J. Akiyama et al. (Eds.): CJCDGCGT 2005, LNCS 4381, pp. 10–20, 2007.

It is known that every 3-connected graph of order 5 or more contains a 3-contractible edge (Tutte [28]).

On the distribution of 3-contractible edges in a 3-connected graph, the followings are known.

**Theorem A. (Ando, Enomoto and Saito [5])** *Every 3-connected graph $G$ with at least 5 vertices has $\frac{1}{2}|V(G)|$ contractible edges.*

Theorem A was also independently proved by W. McCuaig ([22]).

**Theorem B. (Ota [25])** *Every 3-connected graph $G$ with at least 19 vertices has at least $(2|E(G)| + 12)/7$ contractible edges.*

The classification of contraction critically 4-connected graphs was obtained by Fonted and, independently, by Martinov.

**Theorem C. (Fonted [15], Martinov [21])** *If $G$ is a 4-connected graph with no 4-contractible edge, then $G$ is either the square of a cycle or the line graph of a cyclically 4-connected 3-regular graph.*

Egawa proved the following minimum degree condition for a $k$-connected graph to have a $k$-contractible edge.

**Theorem D. (Egawa [13])** *Let $k \geq 2$ be an integer, and let $G$ be a $k$-connected graph with $\delta(G) \geq \lfloor \frac{5k}{4} \rfloor$. Then $G$ has a $k$-contractible edge, unless $2 \leq k \leq 3$ and $G$ is isomorphic to $K_{k+1}$.*

Kriesell extended Egawa's Theorem and proved the following degree sum condition for a $k$-connected graph to have a $k$-contractible edge.

**Theorem E. (Kriesell [17])** *Let $k \geq 2$ be an integer, and let $G$ be a noncomplete $k$-connected graph. If $\deg_G(x) + \deg_G(y) \geq 2\lfloor \frac{5k}{4} \rfloor - 1$ for any pair of distinct vertices $x$, $y$ of $G$, then $G$ has a $k$-contractible edge.*

Mader proved the following theorem which states that each contraction critically $k$-connected graph has many triangles.

**Theorem F. (Mader [20])** *Let $G$ be a $k$-connected graph of order $n$ with no contractible edges. Then $G$ contains at least $n/3$ triangles.*

Thomassen proved that "triangle-free" is one of the sufficient conditions for a $k$-connected graph to have a $k$-contractible edge.

**Theorem G. (Thomassen [27])** Every $k$-connected triangle-free graph has a $k$-contractible edge.

Egawa et al. studied the distribution of $k$-contractible edges in a $k$-connected triangle-free graph and proved the following.

**Theorem H. (Egawa, Enomoto and Saito [14])** Each $k$-connected triangle-free graph with $|V(G)| \geq 3k$ contains $\min\{|V(G)| + \frac{3}{2}k^2 - 3k, |E(G)|\}$ $k$-contractible edges.

## 2  Conditions for a $k$-Connected Graph to Have a $k$-Contractible Edge

By Theorem C, we know that each contraction critically 4-connected graph is 4-regular and each edge of it is contained in some triangle. Hence, each contraction critically 4-connected graph has very restricted induced subgraphs. However, for $k \geq 5$, there is no restriction on induced subgraphs of a contraction critically $k$-connected graph.

**Theorem 1. (Ando, Kaneko and Kawarabayashi [7])**  *For a given graph $G$ and a given integer $k$ ($k \geq 5$), there exists a contraction critically $k$-connected graph which has $G$ as its induced subgraph.*

In view of Theorem H, a $k$-connected "triangle-free" graph has many $k$-contractible edges, and this situation indicates the possibility of the existence of a weaker forbidden subgraph condition for a $k$-connected graph to have a $k$-contractible edge. In this direction we get the following results.

Since $K_4^-$ contains a triangle, the following is an extension of Theorem G in $k$ being odd case.

**Theorem 2. (Kawarabayashi [16])**  *Let $K_4^-$ be the graph obtained from $K_4$ by removing one edge. Let $k \geq 3$ be an odd integer, and let $G$ be a $k$-connected graph which does not contain $K_4^-$. Then $G$ has a $k$-contractible edge.*

Recall bowtie is isomorphic to $K_1 + 2K_2$. Since bowtie contains a triangle, the following is also an extension of Theorem G.

**Theorem 3. (Ando, Kaneko, Kawarabayashi and Yoshimoto[11])**  *A $k$-connected bowtie free graph has a $k$-contractible edge.*

The following Theorem is an extension of Theorem 3 because $(K_1 + P_4)$ contains a bowtie.

**Theorem 4. (Ando, Kaneko and Kawarabayashi [10])**  *Let $k \geq 5$ be an integer and let $G$ be a $k$-connected $(K_1 + P_4)$-free graph. If $G[V_k(G)]$ is bowtie-free, then $G$ has a $k$-contractible edge.*

The following is another but more general extension of Theorem G. Note that if $s = t = 1$, then Theorem 5 is equivalent to Theorem G. Also note that "$K_4^- \cong K_2 + 2K_1$" and "bowtie $\cong K_1 + 2K_2$". Hence $K_2 + sK_1$ and $K_1 + tK_2$ may regard a generalized $K_4^-$ and a generalized bowtie, respectively.

**Theorem 5. (Ando and Kawarabayashi [12])**  *Let $k$ be a positive integer such that $k \geq 5$. Take two positive integers $s$ and $t$ such that $s(t - 1) < k$. If a $k$-connected graph $G$ has neither $K_2 + sK_1$ nor $K_1 + tK_2$, then $G$ has a $k$-contractible edge.*

We can not replace the condition $s(t - 1) < k$ with $s(t - 1) \leq k$ in Theorem 5. In this sense Theorem 5 is sharp.

Theorems G and Theorem 2 through 5 deal with forbidden subgraph conditions for a $k$-connected graph to have a $k$-contractible edge. On the other hand, Theorem D gives a minimum degree condition for a $k$-connected graph to have a $k$-contractible edge. In [13], contraction critically $k$-connected graphs with $\delta(G) = \lfloor \frac{5k}{4} \rfloor - 1$ are displayed, hence the bound of the minimum degree in Theorem D is sharp in this sense. However, if we restrict ourselves to a class of graphs which satisfy some forbidden subgraph conditions, then we may relax the minimum degree bound in Theorem D. In this direction, we get a forbidden subgraph condition which relaxes the minimum degree bound to $k+1$. Note that if $k \geq 5$, then $\lfloor \frac{5k}{4} \rfloor \geq k + 1$.

**Theorem 6. (Ando and Kawarabayashi [12])** *Let $G$ be a $k$-connected graph which contains neither $K_5^-$ nor $5K_1 + P_3$ with $k \geq 5$. If $\delta(G) \geq k + 1$, then $G$ has a $k$-contractible edge.*

Since there is a $k$-regular contraction critically $k$-connected graph which has neither $K_5^-$ nor $5K_1 + P_3$, we can not replace $\delta(G) \geq k + 1$ with $\delta(G) \geq k$ in Theorem 6. In this sense, the minimum degree bound of Theorem 6 is sharp.

# 3   Contractible Edges in a 4-Connected Graph

By Theorem C, we know that if a 4-connected graph $G$ has either a vertex whose degree is greater than 4, or an edge whose end vertices have no common neighbor of degree 4, then $G$ has a contractible edge.

Hence, it seems to be natural to expect that there is a contractible edge near each vertex with degree greater than 4 in a 4-connected graph. In this direction we get the following.

**Theorem 7. (Ando, Egawa [3])** *Let $G$ be a 4-connected graph with $V(G) - V_4(G) \neq \emptyset$, and let $u \in V(G) - V_4(G)$. Then there exists a 4-contractible edge $e$ such that either $e$ is incident with $u$, or at least one of the endvertices of $e$ is adjacent to $u$. Further if $G[N_G(u) \cap V_4(G)]$ is not a path of order 4, then there are two such 4-contractible edges.*

The following graph illustrated in Fig. 1 shows that, in Theorem 7, the necessity of the condition $G[N_G(u) \cap V_4(G)] \not\cong P_4$ for a vertex $u$ has two contractible edges within distance of 1 from $u$. The graph has only two 4-contractible edges, $y_1 z_1$ and $y_2 z_2$. The graph has $x_1$ which is a vertex of degree five such that there is only one 4-contractible edge $y_1 z_1$ within distance of 1 from it.

There is a 4-connected graph which has a vertex of degree greater than 4 such that there are exactly two 4-contractible edges within distance of 1 from the vertex. Hence, in Theorem 7, "two" is sharp. The graph in Fig. 2 has four 4-contractible edges, $y_1 z_1$, $y_2 z_2$, $y_3 z_3$ and $y_4 z_4$. It has $x_1$ which is a vertex of degree five such that there are exactly two 4-contractible edges $y_1 z_1$ and $y_4 z_4$ with in distance of 1 from $x_1$.

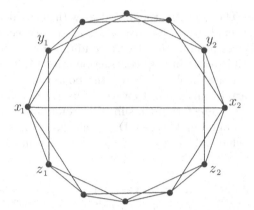

**Fig. 1.** A 4-connected graph $G$ which shows that the necessity of the condition $G[N_G(u) \cap V_4(G)] \not\cong P_4$ in Theorem 7

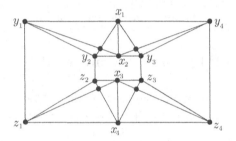

**Fig. 2.** A 4-connected graph which shows sharpness of Theorem 7

As a consequence of Theorem 7, we get the following result.

**Theorem 8. (Ando, Egawa [3])** *Every 4-connected graph $G$ has $(|V(G)| - |V_4(G)|)$ 4-contractible edges.*

There are infinitely many 4-connected graphs $G$ such that $V(G) - V_4(G) \neq \emptyset$ and which has exactly $(|V(G)| - |V_4(G)|)$ 4-contractible edges. Note that the graph in Fig. 1 is one of such graphs.

Let $\tilde{E}(G)$ denote the set of edges whose end vertices has no common neighbor. In view of Theorem C, for a 4-connected graph $G$, we may also expect that there is a contractible edge near each edge in $\tilde{E}(G)$. In this direction we get the following.

**Theorem 9. (Ando, Egawa [4])** *Let $G$ be a 4-connected graph with $\tilde{E}(G) \neq \emptyset$, and let $uv \in \tilde{E}(G)$. Suppose that $uv$ is noncontractible and let $S$ be a 4-cutset with $u, v \in S$, and let $A$ be the vertex set of a component of $G - S$. Then there exists a 4-contractible edge $e$ such that either $e$ is incident with $u$ or there exists $a \in N_G(u) \cap (S \cup A) \cap V_4(G)$ such that $e$ is incident with $a$.*

In Theorem 9, we need the condition "$uv$ is noncontractible".

If $G$ is 4-regular, then we can say a stronger statement as follows.

**Theorem 10. (Ando, Egawa [4])** *Let $G$ be 4-regular, 4-connected and let $e$ be an edge of $G$ whose end vertices have no common neighbor. Then either $e$ is 4-contractible or there is a 4-contractible edge adjacent to $e$.*

There is a 4-connected graph which has an edge $e$ such that neither $e$ is 4-contractible nor there is a 4-contractible edge adjacent to $e$. Hence in Theorem 10, the condition that $G$ is 4-regular can not be dropped.

For 4-regular graphs we have the following.

**Theorem 11. (Ando, Egawa [4])** *Every 4-regular, 4-connected graph $G$ has at least $\frac{|\tilde{E}(G)|}{2}$ 4-contractible edges. Moreover, there is a 4-regular, 4-connected graph which has exactly $\frac{|\tilde{E}(G)|}{2}$ 4-contractible edges.*

## 4   Contractible Edges in a 5-Connected Graph

When $k$ is greater than 4, there is a contraction critically $k$-connected graph which is not $k$-regular. However, from Theorem D we know that the minimum degree of a contraction critically $k$-connected graph is $k$ for $k = 5, 6, 7$. So we pose the following problem.

*Problem 1.* Let $k$ be an integer such that $5 \leq k \leq 7$. Determine the largest value of $c_k$ such that $|V_k(G)| \geq c_k|V(G)|$ holds for each contraction critically $k$-connected graph $G$.

The following theorem says that each contraction critically 5-connected graph has many vertices of degree 5.

**Theorem 12. (Ando, Kaneko and Kawarabayashi [7])** *Let $G$ be a 5-connected graph on $n$ vertices which does not have a 5-contractible edge. Then each vertex of $G$ has a neighbor of degree 5 and hence $G$ has at least $n/5$ vertices of degree 5.*

There exists a contraction critically 5-connected graph on $n$ vertices which has exactly $(8n)/13$ vertices of degree 5. Hence we know $8/13 \geq c_5 \geq 1/5$.

*Conjecture 1.* $c_5 = 8/13$.

For $k = 6$ and $k = 7$, concerning the distribution of vertices of degree $k$ in a contraction critically $k$-connected graph, we get the following results.

**Theorem 13. (Ando, Kaneko and Kawarabayashi [8])** *Every 6-connected graph with no 6-contractible edge of order $n$ has at least $n/7$ vertices of degree 6.*

**Theorem 14. (Ando, Kaneko and Kawarabayashi [9])** *Every 7-connected graph with no 7-contractible edge of order $n$ has at least $n/64$ vertices of degree 7.*

Note that there is a contraction critically 6-connected graph on $n$ vertices which has exactly $n/2$ vertices of degree 6 and there is a contraction critically 7-connected graph on $n$ vertices which has exactly $(6n)/13$ vertices of degree 7. Hence we know $1/2 \geq c_6 \geq 1/7$ and $6/13 \geq c_7 \geq 1/64$.

An edge $e = xy$ of a $k$-connected graph is said to be trivially noncontractible if $x$ and $y$ have a common neighbor of degree $k$. Theorem C says that each edge in a contraction critically 4-connected graph is trivially noncontractible.

From Theorem 12 and Theorem F, it seems to be natural to expect that each contraction critically 5-connected graph has many trivially noncontractible edges.

**Theorem 15. (Ando [1])** *Each contraction critically 5-connected graph of order $n$ has at least $\frac{n}{2}$ trivially noncontractible edges.*

The bound $\frac{n}{2}$ is improved to $n$ by Li and Yuan ([23]).

There is a contraction critically 5-connected graph on $n$ vertices which has exactly $2n$ trivially noncontractible edges.

*Conjecture 2.* Each contraction critically 5-connected graph of order $n$ has $2n$ trivially noncontractible edges.

Using Theorem 15, we can improve the lower bound in Theorem 12.

**Theorem 16. (Ando [1])** *Each contraction critically 5-connected graph of order $n$ has at least $(2n)/9$ vertices of degree 5.*

Very recently the lower bound of Theorems 13, 14 and 16 are improved by a group of Chinese graph theorists as follows.

**Theorem 17. (C. Qin, X. Yuan and J. Su [26])** *Every 5-connected graph with no 5-contractible edge of order $n$ has at least $(4n)/9$ vertices of degree 5.*

**Theorem 18. (Q. Zhao, C. Qin, X, Yuan and M. Li [30])** *Every 6-connected graph with no 6-contractible edge of order $n$ has at least $n/5$ vertices of degree 6.*

**Theorem 19. (M. Li, X. Yuan, J. Su [24])** *Every 7-connected graph with no 7-contractible edge of order $n$ has at least $n/22$ vertices of degree 7.*

Hence we now know that $8/13 \geq c_5 \geq 4/9$, $1/2 \geq c_6 \geq 1/5$ and $6/13 \geq c_7 \geq 1/22$.

## 5   A Local Structure Theorem of 5-Connected Graphs

In this section we present a theorem of 5-connected graphs concerning the local structure around a vertex near which there are no 5-contractible edges, which is an actual extension of Theorem 12. Let $Edge^{(i)}(x)$ denote the set of edges whose distance from $x$ is $i$ or less. Before we state the result, we need to introduce some specified configurations in $k$-connected graphs. Let $x$ be a vertex of a $k$-connected graph $G$. A triangle which contains $x$ and also contains a vertex other than $x$ of degree $k$ is called an $x^*$-triangle. Namely, a triangle $T$ is an $x^*$-triangle if $x \in V(T)$ and $(V(T) - \{x\}) \cap V_k(G) \neq \emptyset$. A configuration which consists of two triangles with nothing in common but $x$ is called an $x$-bowtie, hence an $x$-bowtie is isomorphic to $2K_2 + K_1$, and whose vertex of degree 4 is $x$. A $K_4^-$ is called a reduced $x$-bowtie if one of the vertices of degree 3 is $x$. If, in each triangle of an $x$-bowtie, there is a vertex of degree $k$ other than $x$, then the $x$-bowtie is said to be an $x^*$-bowtie, in other words, an $x$-bowtie is an $x^*$-bowtie if whose two triangles are both $x^*$-triangles. If a reduced $x$-bowtie has at least two vertices of degree $k$ other than $x$, then it is called a reduced $x^*$-bowtie. Hence, In Fig.3, (1) is an $x^*$-bowtie if neither $\{y_1, y_2\} \cap V_k(G)$ nor $\{z_1, z_2\} \cap V_k(G)$ is empty, and (2) is a reduced $x^*$-bowtie if $|\{w_1, w_2, w_3\} \cap V_k(G)| \geq 2$.

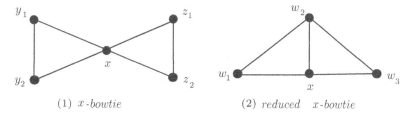

(1)  $x$-bowtie          (2)  reduced   $x$-bowtie

**Fig. 3.** An $x$-bowtie and a reduced $x$-bowtie

We need introduce one more specified configuration in 5-connected graphs. Let $S = \{a_1, a_2, x, b_1, b_2\}$ be a 5-cutset of a 5-connected graph $G$ and let $A$ be a component of $G - S$ such that $V(A) \subset V_5(G)$, $|V(A)| = 4$ and $G[A] \cong K_4^-$, say $A = \{u_1, u_2, v_1, v_2\}$, with edges within $A$ and between $A$ and $S$ exactly as in Fig.4; there may be edges between vertices of $S$. We call this configuration, $G[V(A) \cup S]$, a $K_4^-$-configuration with center $x$. Note that $\{u_1, u_2, v_1, v_2\} \subset V_5(G)$ and edges in Fig.4 other than $xu_1$ and $xv_1$ are all trivially noncontractible. Moreover, we can find two nontrivial 5-cutsets, $\{u_1, u_2, x, b_1, b_2\}$ and $\{v_1, v_2, x, a_1, a_2\}$ which contain $V(xu_1)$ and $V(xv_1)$, respectively. Hence all edges in Fig.4 are noncontractible. Finally we observe that if there is an edge between vertices of $S$, then it clearly is noncontractible since $S$ is a 5-cutset of $G$.

To prove Theorem A, we needed to prove the following Lemma which is regarded a weak local structure theorem of 3-connected graph.

**Lemma 1. (Ando, Enomoto and Saito [5])** Let $x$ be a vertex of a 3-connected graph $G$. If $x$ is incident to no contractible edge, then there is an $x^*$-triangle.

Kriesell proved a more strong local structure theorem of 3-connected graphs.

**Lemma 2. (Kriesell [18])** Let $G$ be a 3-connected graph $G$ such that $G \not\cong K_4$ and let $x \in V(G)$. If $x$ is incident to no contractible edge, then $x$ is adjacent

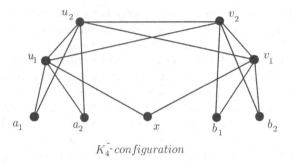

$K_4^-$-$configuration$

**Fig. 4.** A $K_4^-$-configuration with center $x$

to at least four degree-three vertices each of which is incident to exactly two contractible edges.

This result was also independently proved by Wu([29]).

To prove Theorem 7, we needed to prove the following Lemma which is a local structure theorem of 4-connected graph.

**Lemma 3. (Ando, Egawa [3])** *Let $x$ be a vertex of 4-connected graph $G$. If $x$ is incident to no contractible edge, then $G$ has an $x^*$-bowtie.*

In order to get a 5-connected version analogue of Lemma 3, we investigate carefully around a vertex in 5-conncted graph which has no contractible edge near it and we get the following.

**Theorem 20. (Ando [2])** *Let $x$ be a vertex of 5-connected graph $G$ such that $Edge^{(2)}(x) \cap E_c(G) = \emptyset$. If $G$ has neither an $x^*$-bowtie nor a reduced $x^*$-bowtie, then $G$ has a $K_4^-$-configuration with center $x$.*

The following graph shows that the necessity of the condition $Edge^{(2)}(x) \cap E_c(G) = \emptyset$ in Theorem 20.

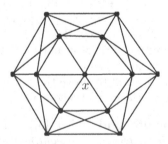

**Fig. 5.** A 5-connected graph which have none of an $x^*$-bowtie nor a reduced $x^*$-bowtie, a $K_4^-$-configuration with center $x$

The two graphs illustrated in Fig. 6 show that the necessity of all three configurations, an $x^*$-bowtie, a reduced $x^*$-bowtie and a $K_4^-$-configuration with center

$x$, in Theorem 20. We observe that the two graphs are both contraction critically 5-connected. The left graph has a reduced $x^*$-bowtie but it has neither an $x^*$-bowtie nor a $K_4^-$-configuration with center $x$. Also we observe that the right graph has a $K_4^-$-configuration with center $y$ but it has neither a $y^*$-bowtie nor a reduced $y^*$-bowtie. Finally we observe that the right graph has a $z^*$-bowtie but it has neither a reduced $z^*$-bowtie nor a $K_4^-$-configuration with center $z$.

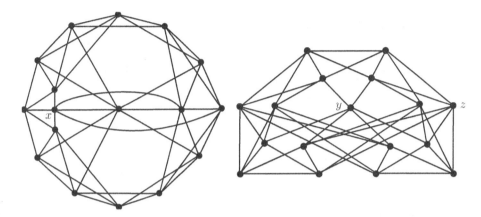

**Fig. 6.** Two graphs which show the necessity of all three configurations

## Acknowledgements

The author would like to thank professors Egawa, Kaneko and Kawarabayashi for the fruitful discussions with them. The author would also thank professors Kriesell and Yuan for their information of the recent progress in this field. The author thank referees for their useful comments.

## References

1. Ando, K.: Trivially Noncontractible Edges in a Contraction Critically 5-Connected Graph. Disc. Math. **293** (2005) 61–72
2. Ando, K.: A Local Structure Theorem of 5-Connected Graphs. Preprint
3. Ando, K., Egawa, Y.: Contractible Edges in a 4-Connected Graph with Vertices of Degree Greater than 4. Preprint
4. Ando, K., Egawa, Y.: Edges not Contained in Triangles and the Distribution of Contractible Edges in a 4-Connected Graph. Preprint
5. Ando, K., Enomoto, H., Saito, A.: Contractible Edges in 3-Connected Graphs. J. Combin. Theory Ser. B **42** (1) (1987) 87–93
6. Ando, K., Kawarabayashi, K., Kaneko, A.: Contractible Edges in Minimally $k$-Connected Graphs. SUT J. Math. **36** (1) (2000) 99–103
7. Ando, K., Kaneko, A., Kawarabayashi, K.: Vertices of Degree 5 in a Contraction Critically 5-Connected Graphs. Graphs Combin. **21** (2005) 27–37

8. Ando, K., Kaneko, A., Kawarabayashi, K.: Vertices of Degree 6 in a Contraction Critically 6-Connected Graph. Discrete Math. **273** (2003) 55–69
9. Ando, K., Kaneko A., Kawarabayashi, K.: Vertices of Degree 7 in a Contraction Critically 7-Connected Graph. Preprint
10. Ando, K., Kaneko, A., Kawarabayashi, K.: Some Sufficient Conditions for a $k$-Connected Graph to Have a $k$-Contractible Edge. Preprint
11. Ando, K., Kaneko, A., Kawarabayashi, K., Yoshimoto, K.: Contractible Edges and Bowties in a $k$-Connected Graph. Ars Combin. **64** (2002) 239–247
12. Ando, K., Kawarabayashi, K.: Some Forbidden Subgraph Conditions for a Graph to Have a $k$-Contractible Edge. Discrete Math. **267** (2003) 3–11
13. Egawa, Y.: Contractible Edges in $n$-Connected Graphs with Minimum Degree Greater than or Equal to $[\frac{5n}{4}]$. Graphs Combin. **7** (1991) 15-21
14. Egawa, Y., Enomoto, H., Saito A. Contractible Edges in Triangle-Free Graphs. Combinatorica. **6** (1986) 269–274
15. Fontet, M.: Graphes 4-Essentiels. C. R. Acad. Sci. Paris **287** (1978) 289–290
16. Kawarabayashi, K.: Note on $k$-Contractible Edges in $k$-Connected Graphs. Australas. J. Combin. Combin. **24** (2001) 165–168
17. Kriesell, M.: A Degree Sum Condition for the Existence of a Contractible Edge in a $\kappa$-Connected Graph. J. Combin. Theory Ser. B **82** (1) (2001) 81–101
18. Kriesell, M.: Contractible Subgraphs in 3-Connected graphs. J. Combin. Theory Ser. B **80** (1) (2000) 32–48
19. Mader, W. Disjunkte Fragmente in kritisch $n$-Fach Zusammenhängende Graphen. European J. Combin. **6** (1985) 353–359
20. Mader, W.: Generalizations of Critical Connectivity of Graphs. Discrete Math. **72** (1988) 267–283
21. Martinov, N.: Uncontractible 4-Connected Graphs. J. Graph Theory **6** (1982) 343–344
22. McCuaig, W.: Edge Contractions in 3-Connected Graphs. Ars Combin. **29** (1990) 299–308
23. Li, X., Yuan, X.: private communication.
24. Li, M., Yuan, X., Su, J.: Vertex of Degree 7 in a Contraction Critical 7-Connected Graph. Preprint
25. Ota, K.: The Number of Contractible Edges in 3-Connected Graphs. Graphs Combin. **4** (4) (1988) 333–354
26. Qin, C., Yuan, X., Su, J.: Some Properties of Contraction-critical 5-cnnected Graphs. Preprint
27. Thomassen, C.: Non-Separating Cycles in $k$-Connected Graphs. J. Graph Theory **5** (1981) 351-354
28. Tutte, W.T.: A Theory of 3-Connected Graphs. Indag. Math. **23** (1961) 441–455
29. Wu, H.: Contractible Elements in Graphs and Matroids. Combin. Probab. Comput. **12** (4) (2003) 457–465
30. Zhao, Q., Qin, C., Yuan, X., Li, M.: Vertices of Degree 6 in a Contraction-Critical 6 Connected Graph. Journal of Guangxin Normal University **25** (2) (2005) 38–43

# An Implicit Weighted Degree Condition for Heavy Cycles in Weighted Graphs

Bing Chen[1], Shenggui Zhang[1,2], and T.C. Edwin Cheng[2]

[1] Department of Applied Mathematics, Northwestern Polytechnical University,
Xi'an, Shaanxi 710072, P.R. China
chenbingnwpu@hotmail.com, sgzhang@nwpu.edu.cn
[2] Department of Logistics, The Hong Kong Polytechnic University,
Hung Hom, Kowloon, Hong Kong, P.R. China

**Abstract.** Let $G$ be a 2-connected weighted graph and $m$ a nonnegative number. As introduced by Li as the weighted analogue of a concept due to Zhu et al, we use $id^w(v)$ to denote the implicit weighted degree of a vertex $v$ in $G$. In this paper we prove that $G$ contains either a Hamilton cycle or a cycle of weight at least $m$, if the following two conditions are satisfied: (1) $\max\{id^w(u), id^w(v)\} \geq m/2$ for each pair of nonadjacent vertices $u$ and $v$ that are vertices of an induced claw or an induced modified claw of $G$; (2) In each induced claw, each induced modified claw and each induced $P_4$ of $G$, all the edges have the same weight. This is a common generalization of several previous results on the existence of long cycles in unweighted graphs and heavy cycles in weighted graphs.

## 1 Terminology and Notations

We use Bondy and Murty [2] for terminology and notations not defined here and consider finite simple connected graphs only.

Let $G = (V, E)$ be a simple graph. For a vertex $v \in V$, its *neighborhood*, denoted by $N(v)$, and its *degree*, denoted by $d(v)$, are defined as the set and the number of vertices in $G$ that are adjacent to $v$, respectively. The *distance* between two vertices $u$ and $v$, denoted by $d(u, v)$, is the length of a shortest path between them. We use $N_2(v)$ to denote the set of vertices of $G$ that are at distance 2 from $v$.

In [12], Zhu *et al.* proposed the concept of implicit degrees of vertices as follows:

**Definition 1. (Zhu *et al.* [12])** *Let $v$ be a vertex of a graph $G$. If $N_2(v) \neq \emptyset$ and $d(v) \geq 2$, then set $k = d(v) - 1$, $m_2 = \min\{d(u) \,|\, u \in N_2(v)\}$ and $M_2 = \max\{d(u) \,|\, u \in N_2(v)\}$. Suppose $d_1 \leq d_2 \leq \cdots \leq d_{k+2} \leq \cdots$ is the degree sequence of the vertices of $N(v) \cup N_2(v)$. Let*

$$d^*(v) = \begin{cases} m_2, & \text{if } d_k \leq m_2; \\ d_{k+1}, & \text{if } d_{k+1} > M_2; \\ d_k, & \text{if } d_k \geq m_2 \text{ and } d_{k+1} \leq M_2. \end{cases}$$

J. Akiyama et al. (Eds.): CJCDGCGT 2005, LNCS 4381, pp. 21–29, 2007.
© Springer-Verlag Berlin Heidelberg 2007

*Then the* implicit degree *of* $v$, *denoted by* $id(v)$, *is defined as* $id(v) = \max\{d(v), d^*(v)\}$. *If* $N_2(v) = \emptyset$ *or* $d(v) \leq 1$, *then we define* $id(v) = d(v)$.

It is clear that $id(v) \geq d(v)$ for every vertex $v$ in a graph $G$.

A graph $G$ is called a *weighted graph* if each edge $e$ is assigned a nonnegative number $w(e)$, called the *weight* of $e$. We define the *weighted degree* of $v$ in $G$ by

$$d^w(v) = \sum_{h \in N(v)} w(vh).$$

For a subgraph $H$ of $G$, the *weight* of $H$ is defined by

$$w(H) = \sum_{e \in E(H)} w(e).$$

Motivated by the concept of implicit degrees, Li [10] introduced the concept of implicit weighted degrees of vertices in weighted graphs as follows:

**Definition 2. (Li [10])** *Let* $v$ *be a vertex of a graph* $G$. *If* $N_2(v) \neq \emptyset$ *and* $d(v) \geq 2$, *then set* $k = d(v) - 1$, $m_2 = \min\{d^w(u) | u \in N_2(v)\}$ *and* $M_2 = \max\{d^w(u) | u \in N_2(v)\}$. *Suppose* $d_1^w \leq d_2^w \leq \cdots \leq d_{k+2}^w \leq \cdots$ *is the weighted degree sequence of the vertices of* $N(v) \cup N_2(v)$. *Let*

$$d^{*w}(v) = \begin{cases} m_2, & \text{if } d_k^w < m_2; \\ d_{k+1}^w, & \text{if } d_{k+1}^w > M_2; \\ d_k^w, & \text{if } d_k^w \geq m_2 \text{ and } d_{k+1}^w \leq M_2. \end{cases}$$

*Then the* implicit weighted degree *of* $v$, *denoted by* $id^w(v)$, *is defined as* $id^w(v) = \max\{d^w(v), d^{*w}(v)\}$. *If* $N_2(v) = \emptyset$ *or* $d(v) \leq 1$, *then we define* $id^w(v) = d^w(v)$.

Clearly, $id^w(v) \geq d^w(v)$ for every vertex $v$ in a graph $G$.

An unweighted graph can be regarded as a weighted graph in which each edge $e$ is assigned a weight $w(e) = 1$. Thus, in an unweighted graph, $d^w(v) = d(v)$ and $id^w(v) = id(v)$ for every vertex $v$, and the weight of a subgraph is simply the number of its edges.

We call the graph $K_{1,3}$ a *claw*, and the graph $K_{1,3} + e$ ($e$ is an edge between two nonadjacent vertices in the claw) a *modified claw*.

## 2    Results

There are many results on the existence of long cycles in unweighted graphs in terms of degrees of vertices. Among them, the following is well-known.

**Theorem A. (Fan [8])** *Let* $G$ *be a 2-connected graph and* $c$ *an integer. If* $\max\{d(u), d(v) | d(u, v) = 2\} \geq c/2$, *then* $G$ *contains either a Hamilton cycle or a cycle of length at least* $c$.

Chen [7] showed that the degree condition in Theorem A can be replaced by the implicit degree condition.

**Theorem B. (Chen [7])** *Let $G$ be a 2-connected graph and $c$ an integer. If* $\max\{id(u),\ id(v)\mid d(u,\ v)=2\}\geq c/2$, *then $G$ contains either a Hamilton cycle or a cycle of length at least $c$.*

On the other hand, Bedrossian *et al.* [1] gave a further generalization of Fan's theorem. They imposed one more restriction on the pair of vertices $u$ and $v$: they must be vertices of an induced claw or an induced modified claw.

**Theorem C. ([Bedrossian** *et al.* **[1])** *Let $G$ be a 2-connected graph and $c$ an integer. If $\max\{d(u),\ d(v)\}\geq c/2$ for each pair of nonadjacent vertices $u$ and $v$ that are vertices of an induced claw or an induced modified claw of $G$, then $G$ contains either a Hamilton cycle or a cycle of length at least $c$.*

Recently, Chen and Zhang [6] gave an extension of Theorem C.

**Theorem D. (Chen and Zhang [6])** *Let $G$ be a 2-connected graph and $c$ an integer. If $\max\{id(u),\ id(v)\}\geq c/2$ for each pair of nonadjacent vertices $u$ and $v$ that are vertices of an induced claw or an induced modified claw of $G$, then $G$ contains either a Hamilton cycle or a cycle of length at least $c$.*

Bondy and Fan [3,4] began the study of the existence of heavy paths and cycles in weighted graphs. They generalized several classical theorems of Dirac, and of Erdös and Gallai, on paths and cycles to weighted graphs. Later, Bondy *et al.* [5] gave a weighted generalization of Ore's theorem. In [11], it was shown that if one wants to generalize Theorem A to weighted graphs, some extra conditions cannot be avoided. By adding two extra conditions, Zhang *et al.* [11] gave a weighted generalization of Theorem A.

**Theorem 1. (Zhang** *et al.* **[11])** *Let $G$ be a 2-connected weighted graph and $m$ a nonnegative number. Then $G$ contains either a Hamilton cycle or a cycle of weight at least $m$ if it satisfies the following conditions:*

(1) $\max\{d^w(u),\ d^w(v)\mid d(u,\ v)=2\}\geq m/2$;
(2) $w(ux)=w(vx)$ *for every vertex $x\in N(u)\cap N(v)$ with $d(u,\ v)=2$;*
(3) *In every triangle $T$ of $G$, either all the edges of $T$ have different weights or all the edges of $T$ have the same weight.*

Theorem B also admits a weighted generalization similar to Theorem 1.

**Theorem 2. (Li [10])** *Let $G$ be a 2-connected weighted graph and $m$ a nonnegative number. Then $G$ contains either a Hamilton cycle or a cycle of weight at least $m$ if it satisfies the following conditions:*

(1) $\max\{id^w(u),\ id^w(v)\mid d(u,\ v)=2\}\geq m/2$;
(2) $w(ux)=w(vx)$ *for every vertex $x\in N(u)\cap N(v)$ with $d(u,\ v)=2$;*
(3) *In every triangle $T$ of $G$, either all the edges of $T$ have different weights or all the edges of $T$ have the same weight.*

Fujisawa [9] gave the so-called claw conditions for the existence of heavy cycles in weighted graphs, generalizing Theorem C.

**Theorem 3. (Fujisawa [9])** *Let $G$ be a 2-connected weighted graph and $m$ a nonnegative number. Then $G$ contains either a Hamilton cycle or a cycle of weight at least $m$ if it satisfies the following conditions:*

**(1)** $\max\{d^w(u), d^w(v)\} \geq m/2$ *for each pair of nonadjacent vertices $u$ and $v$ that are vertices of an induced claw or an induced modified claw of $G$;*
**(2)** *For each induced claw and each induced modified claw of $G$, all of its edges have the same weight.*

Now, we have the following question: Can $\max\{d^w(u), d^w(v)\} \geq m/2$ in Theorem 3 be replaced by $\max\{id^w(u), id^w(v)\} \geq m/2$? If the answer is affirmative, then we can give a generalization of Theorem 3. In this paper we give a partial answer to this problem. Our result is a generalization of Theorems D and 2.

**Theorem 4.** *Let $G$ be a 2-connected weighted graph and $m$ a nonnegative number. Then $G$ contains either a Hamilton cycle or a cycle of weight at least $m$ if it satisfies the following conditions:*

**(1)** $\max\{id^w(u), id^w(v)\} \geq m/2$ *for each pair of nonadjacent vertices $u$ and $v$ that are vertices of an induced claw or an induced modified claw of $G$;*
**(2)** *In each induced claw, each induced modified claw and each induced $P_4$ of $G$, all of its edges have the same weight.*

We postpone the proof of this theorem to the following sections.

## 3   Proof of Theorem 4

We call a path $P$ a *heaviest longest path* if $P$ has the following properties

- $P$ is a longest path of $G$, and
- $w(P)$ is the maximum among all the longest paths in $G$.

A *y-path* is a path that has $y$ as one of its end-vertices. A *y-heaviest longest path* is a heaviest longest path that has $y$ as one of its end-vertices.

Our proof of Theorem 4 is based on the following two lemmas.

**Lemma 1. (Bondy *et al.* [5])** *Let $G$ be a non-hamiltonian 2-connected weighted graph and $P = v_1 v_2 \cdots v_p$ be a heaviest longest path in $G$. Then there is a cycle $C$ in $G$ with $w(C) \geq d^w(v_1) + d^w(v_p)$.*

**Lemma 2.** *Let $G$ be a non-hamiltonian 2-connected weighted graph that satisfies the conditions of Theorem 4. Suppose that $y$ is an end-vertex of a heaviest longest path in $G$ and there exists $e_1, e_2 \in E(G)$ such that $w(e_1) \neq w(e_2)$. Then, there exists a y-heaviest longest path such that the other end-vertex has weighted degree at least $m/2$.*

We prove this lemma in the next section.

*Proof of Theorem 4*

Suppose that all the edges of $G$ have the same weight $t$. If $t = 0$, then the assertion is obvious. If $t \neq 0$, then $id^w(v) = t * id(v)$ for all $v \in V(G)$. Hence, $\max\{id(x), id(y)\} = \max\{id^w(x), id^w(y)\}/t \geq m/(2t)$. It follows from Theorem D that $G$ contains either a Hamilton cycle or a cycle $C$ of length at least $m/t$, which implies $w(C) \geq m$.

If there exist two edges $e_1$ and $e_2$ in $E(G)$ with $w(e_1) \neq w(e_2)$ and $G$ does not contain a Hamilton cycle, then, by using Lemma 2 twice, we obtain a heaviest longest path with both end-vertices having weighted degree at least $m/2$. By Lemma 1, we can find a cycle of weight at least $m$. The proof of Theorem 4 is complete.                                                               □

## 4   Proof of Lemma 2

Let $P = v_1 v_2 \cdots v_p$ be a path of a graph $G$. In the following we denote $N^-(v_1) = \{v_i \mid v_1 v_{i+1} \in E(G)\}$, and use $k(P)$ to denote the maximum index $i$ with $v_1 v_i \in E(G)$. To prove Lemma 2, we need the following two lemmas.

**Lemma 3.** **(Li [10])** *Let $G$ be a 2-connected weighted graph and let $P = v_1 v_2 \cdots v_p$ be a heaviest longest path of $G$. If $d^w(v_1) < id^w(v_1)$ and $v_1 v_p \notin E(G)$, then either*

(1) *there is some $v_j \in N^-(v_1)$ such that $d^w(v_j) \geq id^w(v_1)$; or*
(2) $N(v_1) = \{v_2, v_3, \ldots, v_{d(v_1)+1}\}$ *and* $id^w(v_1) = \min\{d^w(u) \mid u \in N_2(v_1)\}$.

**Lemma 4.** *Let $G$ be a connected weighted graph with at least 4 vertices that satisfies Condition (2) of Theorem 4. If there exist two edges with different weights, then each pair of vertices of $G$ are at distance at most 2.*

*Proof.* Since $G$ is connected, we can choose a vertex $x \in V(G)$ so that there exist $u, v \in N(x)$ with $w(ux) \neq w(vx)$.

Let $\bigcup_{i=1}^{l} V_i$ be a partition of $N(x)$ such that for $u \in V_i$ and $v \in V_j$, $w(ux) = w(vx)$ if and only if $i = j$. Now we denote the weight of the edges joining $x$ and the vertices of $V_i$ by $w_i$ for $1 \leq i \leq l$.

*Claim 1. If there exists a vertex $y \in V(G)$ such that $d(x, y) = 2$, then $vy \in E(G)$ for all $v \in N(x)$.*

*Proof.* The fact $d(x, y) = 2$ shows that there is a vertex $v_1 \in N(x)$ such that $yv_1 \in E(G)$. Without loss of generality, we may assume $v_1 \in V_1$. Now suppose that there exists a vertex $v \in \bigcup_{i=2}^{l} V_i$ with $yv \notin E(G)$. Then $\{v, x, v_1, y\}$ induces a $P_4$ or a modified claw. From the condition of the lemma, we have $w(v_1 x) = w(vx)$. This contradicts the definition of the partition $\bigcup_{i=1}^{l} V_i$. So we have $yv \in E(G)$ for all $v \in \bigcup_{i=2}^{l} V_i$.

Applying the same argument to $v_2 \in V_2$ and $v \in V_1$, we have $yv \in E(G)$ for every $v \in V_1$.                                                               □

*Claim 2. There exists no vertex $z$ such that $d(x, z) = 3$.*

*Proof.* Suppose that there exists a vertex $z \in V(G)$ such that $d(x, z) = 3$. Then $z$ has a neighbor $y$ such that $d(x, y) = 2$. By Claim 1, there exist $v_1 \in V_1$ and $v_2 \in V_2$ such that $v_1 y, v_2 y \in E(G)$. Now each of $\{x, v_1, y, z\}$ and $\{x, v_2, y, z\}$ induces a $P_4$. By the condition of Lemma 4, we have $w(v_1 x) = w(yz) = w(v_2 x)$. This contradicts the definition of the partition $\bigcup_{i=1}^{l} V_i$.  $\square$

The lemma follows from Claims 1 and 2 immediately.  $\square$

*Proof of Lemma 2*

*Claim 1. There exists a $y$-heaviest longest path such that the other end-vertex has implicit weighted degree at least $m/2$.*

*Proof.* By contradiction. Let $P = v_1 v_2 \cdots v_p$ ($v_p = y$) be a $y$-heaviest longest path such that $k(P)$ is as large as possible. Clearly, $N(v_1) \cup N(v_p) \subseteq V(P)$, and there exists no cycle of length $p$. Since $G$ is 2-connected, $v_1$ is adjacent to at least one vertex on $P$ other than $v_2$. So, $3 \le k(P) \le p - 1$. Now let $k = k(P)$.

*Claim 1.1. $N(v_1) \ne \{v_2, \ldots, v_k\}$.*

*Proof.* Suppose $N(v_1) = \{v_2, \ldots, v_k\}$. Since $P$ is a longest path, $N(v_i) \subset V(P)$ for $i = 2, \ldots, k - 1$. By the fact that $G$ is non-hamiltonian and 2-connected, we have $k + 2 \le p$ and $v_p \notin N(v_i)$ for $i = 1, \ldots, k - 1$. Furthermore, since $G - v_k$ is connected, there must be an edge $v_j v_s \in E(G)$ with $j < k < s$.

We assume that such an edge $v_j v_s$ was chosen so that

$(i)$   $s$ is as large as possible;
$(ii)$   $j$ is as large as possible, subject to $(i)$.

Clearly we have $s \le p - 1$.

Suppose $j \le k - 2$. By the choices of $v_k$, $v_j$ and $v_s$, we have $v_1 v_s \notin E(G)$ and $v_{j+1} v_s \notin E(G)$. Then $\{v_j, v_{j+1}, v_1, v_s\}$ induces a modified claw, which implies that $w(v_1 v_{j+1}) = w(v_j v_{j+1})$.

Suppose $j = k - 1$. By Lemma 4, we have $d(v_1, v_p) = 2$. So $v_k v_p \in E(G)$. Now $\{v_k, v_{k-1}, v_1, v_p\}$ induces a modified claw, which implies that $w(v_1 v_k) = w(v_{k-1} v_k)$.

From the above discussion we can see that $P' = v_j v_{j-1} \cdots v_1 v_{j+1} v_{j+2} \cdots v_p$ is a $y$-heaviest longest path with $k(P') = s > k$, a contradiction.  $\square$

Therefore, we may confine our attention to the case $N(v_1) \ne \{v_2, \ldots, v_k\}$.

Choose $v_r \notin N(v_1)$ with $2 < r < k$ such that $r$ is as large as possible. Then $v_1 v_i \in E(G)$ for every $i$ with $r < i \le k$. Let $j$ be the smallest index such that $j > r$ and $v_j \notin N(v_1) \cap N(v_r)$. Since $v_{r+1} \in N(v_1) \cap N(v_r)$, we have $j \ge r + 2$. On the other hand, it is obvious that $j \le k + 1$.

*Claim 1.2.* $id^w(v_r) \geq m/2$.

*Proof.* By the choice of $v_r$, $v_1v_r \notin E(G)$. Then it follows from the choice of $v_j$ that $\{v_{j-1}, v_j, v_1, v_r\}$ induces a claw or a modified claw. Since $id^w(v_1) < m/2$, we have $id^w(v_r) \geq m/2$. □

*Claim 1.3.* $w(v_1v_{r+1}) \neq w(v_rv_{r+1})$.

*Proof.* Suppose $w(v_1v_{r+1}) = w(v_rv_{r+1})$. Now, $v_rv_{r-1}\cdots v_1v_{r+1}v_{r+2}\cdots v_p$ is a $y$-heaviest longest path different from $P$. So $id^w(v_r) < m/2$, contradicting Claim 1.2. □

*Claim 1.4.* $v_{r+1}v_j \notin E(G)$.

*Proof.* By the choice of $v_r$, $v_1v_r \notin E(G)$. So, if $v_{r+1}v_j \in E(G)$, then from the choice of $v_j$, we know that $\{v_{r+1}, v_1, v_r, v_j\}$ induces a claw or a modified claw. Thus $w(v_1v_{r+1}) = w(v_rv_{r+1})$, contradicting Claim 1.3. □

Suppose $j \leq k$. Now $\{v_r, v_{r+1}, v_1, v_j\}$ induces a $P_4$. So $w(v_1v_{r+1}) = w(v_rv_{r+1})$, contradicting Claim 1.3.

Suppose $j = k + 1$. If $v_rv_j \in E(G)$, then $\{v_j, v_r, v_{r+1}, v_1\}$ induces a $P_4$. So $w(v_1v_{r+1}) = w(v_rv_{r+1})$, contradicting Claim 1.3. If $v_rv_j \notin E(G)$, then both $\{v_{r+1}, v_1, v_k, v_j\}$ and $\{v_{r+1}, v_r, v_k, v_j\}$ induce $P_4$s or modified claws. So $w(v_1v_{r+1}) = w(v_rv_{r+1})$, again contradicting Claim 1.3.

The proof of Claim 1 is complete. □

By Claim 1, we can choose a heaviest longest path $P = v_1v_2\cdots v_p$ in $G$ such that $id^w(v_1) \geq m/2$.

We distinguish two cases.

*Case 1.* There is some $v_j \in N^-(v_1)$ such that $d^w(v_j) \geq id^w(v_1)$.

If $j = 1$, then there is nothing to prove. So we assume $j \geq 2$.

*Claim 2.* $w(v_1v_{j+1}) = w(v_jv_{j+1})$.

*Proof.* Suppose $v_{j+1}v_p \in E(G)$. Then $\{v_{j+1}, v_1, v_j, v_p\}$ induces a claw or a modified claw, which implies that $w(v_1v_{j+1}) = w(v_jv_{j+1})$.

Suppose $v_{j+1}v_p \notin E(G)$. Then by Lemma 4, $d(v_{j+1}, v_p) = 2$. So, there exists a vertex $v_s$ such that $v_sv_{j+1} \in E(G)$ and $v_sv_p \in E(G)$. Now $\{v_1, v_{j+1}, v_s, v_p\}$ induces a $P_4$ or a modified claw. At the same time, $\{v_j, v_{j+1}, v_s, v_p\}$ induces a $P_4$ or a modified claw. So $w(v_1v_{j+1}) = w(v_jv_{j+1})$. □

It follows from Claim 2 that $v_jv_{j-1}\cdots v_1v_{j+1}v_{j+2}\cdots v_p$ is a $y$-heaviest longest path with $d^w(v_j) \geq id^w(v_1) \geq m/2$.

*Case 2.* $d^w(v_i) < id^w(v_1)$ for every vertex $v_i \in N^-(v_1)$.

Specifically, we have $d^w(v_1) < id^w(v_1)$. Choose $v_k \in N(v_1)$ such that $k$ is as large as possible. By Lemma 3, we know that $N(v_1) = \{v_2, v_3, \ldots, v_k\}$, and for any $u \in N_2(v_1)$, $d^w(u) \geq id^w(v_1) \geq m/2$. So, with similar arguments, we can choose an edge $v_j v_s$ as in Claim 1.1 and prove the following claim.

*Claim 3.* $w(v_1 v_{j+1}) = w(v_j v_{j+1})$.

*Claim 4.* $s \geq k + 2$.

*Proof.* Suppose $s = k + 1$. Since $d(v_1, v_p) = 2$, we have $v_k v_p \in E(G)$. Now $v_1 v_2 \cdots v_j v_s v_{s+1} \cdots v_p v_k v_{k-1} \cdots v_{j+1} v_1$ is a cycle of length $p$, a contradiction.    $\square$

*Claim 5.* $w(v_{s-1} v_s) = w(v_j v_s)$.

*Proof.* If $v_j v_{s-1} \in E(G)$, then $\{v_j, v_{s-1}, v_s, v_1\}$ induces a modified claw, which implies that $w(v_{s-1} v_s) = w(v_j v_s)$. If $v_j v_{s-1} \notin E(G)$, then $\{v_s, v_{s-1}, v_{s+1}, v_j\}$ induces a claw or a modified claw, so $w(v_{s-1} v_s) = w(v_j v_s)$.    $\square$

It follows from Claims 3 and 5 that $v_{s-1} v_{s-2} \cdots v_{j+1} v_1 \ v_2 \cdots v_j v_s v_{s+1} \cdots v_p$ is a $y$-heaviest longest path different from $P$. By Claim 4 and Lemma 4, we have $d(v_1, v_{s-1}) = 2$, so $d^w(v_{s-1}) \geq m/2$.

The proof of the lemma is complete.    $\square$

## Acknowledgements

This work was supported by NSFC (No. 10101021) and SRF for ROCS of SEM. The second and the third author were supported by The Hong Kong Polytechnic University under grant number G-YX42.

## References

1. Bedrossian, P., Chen, G., Schelp, R.H.: A Generalization of Fan's Condition for Hamiltonicity, Pancyclicity, and Hamiltonian Connectedness. Discrete Math. **115** (1993) 39–50
2. Bondy, J.A., Murty, U.S.R.: Graph Theory with Applications. Macmillan London and Elsevier. New York (1976)
3. Bondy, J.A., Fan, G.: Optimal Paths and Cycles in Weighted Graphs. Ann. Discrete Math. **41** (1989) 53–69
4. Bondy, J.A., Fan, G.: Cycles in Weighted Graphs. Combinatorica **11** (1991) 191–205
5. Bondy, J.A., Broersma, H.J., van den Heuvel, J., Veldman, H.J.: Heavy Cycles in Weighted Graphs. Discuss. Math. Graph Theory **22** (2002) 7–15
6. Chen, B., Zhang, S.: An Implicit Degree Condition for Long Cycles in 2-Connected Graphs. Applied Mathematics Letters. **19** (11) (2006) 1148–1151
7. Chen, G.: Longest Cycles in 2-Connected Graphs. Journal of Central China Normal University (Natural Science) **20** (3) (1986) 39–42
8. Fan, G.: New Sufficient Conditions for Cycles in Graphs. J. Combin. Theory Ser. B **37** (1984) 221–227
9. Fujisawa, J.: Claw Conditions for Heavy Cycles in Weighted Graphs. Graphs & Combin. **21** (2) (2005) 217–229

10. Li, P.: Implicit Weighted Degree Condition for Heavy Paths in Weighted Graphs. Journal of Shandong Normal University (Natural Science) **18** (1) (2003) 11–13

11. Zhang, S., Broersma H.J., Li, X., Wang, L.: A Fan Type Condition for Heavy Cycles in Weighted Graphs. Graphs & Combin. **18** (2002) 193–200

12. Zhu, Y., Li, H., Deng, X.: Implicit-Degrees and Circumferences. Graphs & Combin. **5** (1989) 283–290

# On the Choice Numbers of Some Complete Multipartite Graphs

G.L. Chia and V.L. Chia

Institute of Mathematical Sciences, University of Malaya,
50603 Kuala Lumpur, Malaysia
glchia@um.edu.my,
chiavuileong@yahoo.co.uk

**Abstract.** For every vertex $v$ in a graph $G$, let $L(v)$ denote a list of colors assigned to $v$. A *list coloring* is a proper coloring $f$ such that $f(v) \in L(v)$ for all $v$. A graph is *k-choosable* if it admits a list coloring for every list assignment $L$ with $|L(v)| = k$. The *choice number* of $G$ is the minimum $k$ such that $G$ is $k$-choosable. We generalize a result (of [4]) concerning the choice numbers of complete bipartite graphs and prove some uniqueness results concerning the list colorability of the complete $(k + 1)$-partite graph $K_{n,\dots,n,m}$. Using these, we determine the choice numbers for some complete multipartite graphs $K_{n,\dots,n,m}$. As a byproduct, we classify (i) completely those complete tripartite graphs $K_{2,2,m}$ and (ii) almost completely those complete bipartite graphs $K_{n,m}$ (for $n \leq 6$) according to their choice numbers.

## 1 Introduction

Let $G$ denote a graph and let $V(G)$ denote its vertex set. If $v \in V(G)$, let $L(v)$ denote a list of colors assigned to $v$. Then $G$ is said to be *L-colorable* if there is a proper coloring $f$ of $G$ such that $f(v) \in L(v)$ for all $v \in V(G)$. In the event that $G$ is $L$-colorable and $|L(v)| = k$ for all $v \in V(G)$, then $G$ is said to be *k-choosable*. The *choice number* of $G$, denoted $ch(G)$, is the smallest integer $k$ for which $G$ is $k$-choosable.

The problem of coloring the vertices of graphs with given lists of colors was studied by Vizing [10], and independently by Erdös, Rubin and Taylor [2] where all 2-choosable graphs were completely characterized. In [2], it was also proved that the choice number of a complete bipartite graph could be arbitrarily large.

The characterization of 3-choosable complete bipartite graphs was subsequently carried out by Mahadev, Roberts and Santhanakrishnan [6], Shende and Tesman [8] and finally completed with a result of O'Donnell [7].

In [2], it was proved that the complete $k$-partite graph $K_{2,2,\dots,2}$ has choice number equals to $k$. The choice numbers for $K_{2,\dots,2,m}$ for $m = 3$ and for $4 \leq m \leq 5$ were determined in [3] and [1] respectively while the case for $K_{3,3,\dots,3}$ was determined in [5]. Motivated by these, we consider the choice number for the complete multipartite graph $K_{n,\dots,n,m}$.

J. Akiyama et al. (Eds.): CJCDGCGT 2005, LNCS 4381, pp. 30–37, 2007.
© Springer-Verlag Berlin Heidelberg 2007

In [2], it was also observed that the complete bipartite graph $K_{n,m}$ is always $(n+1)$-choosable for all $m$ and that it is $n$-choosable if and only if $m < n^n$. The exact range for which the choice number of $K_{n,m}$ is $n+1$ or $n$ was determined by Hoffman and Johnson Jr. in [4] (see Corollary 2). This result, together with those mentioned in the preceding paragraph enable one to classify completely all complete bipartite graphs $K_{n,m}$ according to their choice numbers for $n \leq 5$ (see Section 3). In Section 3, we show that $K_{6,m}$ is not 4-choosable if $m \geq 125$. This leads to an almost-complete classification of $K_{6,m}$ according to their choice numbers.

In Section 2, we generalize the result of Hoffman and Johnson Jr. to complete multipartite graphs $K_{n,...,n,m}$. In the final section, we classify all complete tripartite graphs $K_{2,2,m}$ according to their choice numbers.

## 2    Complete Multipartite Graphs

Suppose $n$ and $k$ are positive integers and let $G$ denote the complete $(k+1)$-partite graph $K_{n,...,n,m}$ with partite sets $V_j = \{u_{1,j}, u_{2,j}, \ldots, u_{n,j}\}$, $j = 1, \ldots, k$ and $V_{k+1}$. Let $A_1, A_2, \ldots, A_n$ be pairwise disjoint sets each of size $nk$. That is, $|A_i| = nk$, for each $i = 1, 2, \ldots, n$ and $A_i \cap A_j \neq \emptyset$ for $i \neq j$. Let

$$\mathcal{C}_{n,k} = \{B_1 \cup B_2 \cup \cdots \cup B_n \mid B_i \subseteq A_i, \ |B_i| = k, \ i = 1, 2, \ldots, n\}.$$

Then $|\mathcal{C}_{n,k}| = \binom{nk}{k}^n$.

Suppose $m = \binom{nk}{k}^n$ and let $G(n, k)$ denote the complete $(k+1)$-partite graph $K_{n,...,n,m}$. Let $L_{n,k}$ denote a list assignment on $G(n, k)$ such that (i) $L_{n,k}(u_{i,j}) = A_i$ for each $i = 1, 2, \ldots, n$ and for each $j = 1, \ldots, k$, and that (ii) each element in $\mathcal{C}_{n,k}$ is assigned to precisely one vertex in $V_{k+1}$.

It is not hard to see that if $G$ is the complete $(k+1)$-partite graph $G(n, k)$, then $G$ is not $L_{n,k}$-colorable. This is because, if $G$ is $L_{n,k}$-colorable, then any $L_{n,k}$-coloring of $G$ must color the $nk$ vertices in $V_1 \cup \cdots \cup V_k$ with $nk$ colors which takes the form $B_1 \cup B_2 \cup \cdots \cup B_n$ where $B_i \subseteq A_i$ and $|B_i| = k$. This being so, since the $k$ vertices $u_{i,1}, \ldots, u_{i,k}$ must be colored with $k$ distinct colors from $A_i$, say $B_i$, where $i = 1, 2, \ldots, n$. However this causes the vertex $v \in V_{k+1}$ to be not colorable from the list if $L_{n,k}(v) = B_1 \cup B_2 \cup \cdots \cup B_n$.

**Theorem 1.** *Let $G$ denote the complete $(k+1)$-partite graph $K_{n,...,n,m}$. Then*

**(i)** $ch(G) = nk + 1$, *if* $m \geq \binom{nk}{k}^n$ *and*

**(ii)** $ch(G) \geq nk - k + 1$, *if* $m \geq \binom{(n-1)k}{k}^{n-1} - \binom{(n-2)k}{k}^{n-1}$, *where* $n \geq 3$.

*Proof.* (i) Suppose $m \geq \binom{nk}{k}^n$. Then $ch(G) \geq nk+1$ because $G$ contains $G(n, k)$ as a subgraph and $G(n, k)$ is not $L_{n,k}$-colorable and hence not $nk$-choosable.

If $L$ is a list assignment on $G$ such that $|L(v)| \geq nk+1$, then $G$ is $L$-colorable because any coloring of $V_1, \ldots, V_k$ (where $|V_j| = n$ for each $j = 1, \ldots, k$) uses up at most $nk$ colors (say $c_1, c_2, \ldots, c_{nk}$) and as such each vertex $w$ in $V_{k+1}$ can be colored from the list (because $|L(w) - \{c_1, \ldots, c_{nk}\}| \geq 1$). Hence $ch(G) \leq nk+1$ so that $ch(G) = nk + 1$.

(ii) Let $A$ be a set of size $(n - 1)k$ and let $A$ be partitioned into $n - 1$ subsets $C_2, C_3, \ldots, C_n$ each of size $k$. Let $D_2, D_3, \ldots, D_n$ be pairwise disjoint sets each of size $(n - 1)k$ such that $D_i \cap A = C_i$ for each $i = 2, 3, \ldots, n$. Let

$$\mathcal{D}_{n,k} = \{ B = B_2 \cup B_3 \cup \cdots \cup B_n \mid B_i \subseteq D_i,$$

$$|B_i| = k, \; i = 2, 3, \ldots, n, \; B \cap A \neq \emptyset \}.$$

Then $|\mathcal{D}_{n,k}| = \binom{(n-1)k}{k}^{n-1} - \binom{(n-2)k}{k}^{n-1}$.

Note that $\binom{(n-1)k}{k}^{n-1}$ (respectively $\binom{(n-2)k}{k}^{n-1}$) counts the number of those lists $B_2 \cup B_3 \cup \cdots \cup B_n$ where $B_i \subseteq D_i$ (respectively $B_i \subseteq D_i - C_i$) and $|B_i| = k$, $i = 2, 3, \ldots, n$.

Suppose $m = \binom{(n-1)k}{k}^{n-1} - \binom{(n-2)k}{k}^{n-1}$. Let $M_{n,k}$ denote a list assignment on $G$ such that (i) $M_{n,k}(u_{1,j}) = A$ and $M_{n,k}(u_{i,j}) = D_i$ for each $i = 2, 3, \ldots, n$ and for each $j = 1, \ldots, k$, and that (ii) each element in $\mathcal{D}_{n,k}$ is assigned to precisely one vertex in $V_{k+1}$.

Then, just as in the case of $L_{n,k}$, it is not hard to see that $G$ is not $M_{n,k}$-colorable and hence not $(n - 1)k$-choosable.    □

**Problem 1.** Is $L_{n,k}$ the only list assignment $L$ with $|L(v)| \geq nk$ for all vertices $v \in V(G(n, k))$ that causes $G(n, k)$ not $L$-colorable?

In the sequel, we shall show that the above question is true for three cases: (i) $n = 1$ (Section 2), (ii) $k = 1$ (Section 3), and (iii) $n = 2 = k$ (Section 4).

**Definition 1.** *Let $L$ and $L'$ be two list assignments of a graph $G$. Then $L$ is said to be equivalent to $L'$ if there exist a bijection $\tau$ of colors and an automorphism $\varphi$ of $G$ satisfying $\tau \circ L \circ \varphi = L'$.*

When $n = 1$, $G(n, k)$ is the complete graph $K_{k+1}$. In this case, recall that $L_{1,k}$ is the list assignment on $G(1, k)$ such that $L_{1,k}(v) = A_1$ for all vertices $v$ in $G(1, k)$ and $|A_1| = k$.

**Theorem 2.** *Let $G$ denote the complete graph $K_{k+1}$. Let $L$ be a list assignment on $G$ such that $|L(v)| \geq k$ for all vertices $v$ in $G$. Then $G$ is not $L$-colorable if and only if $L$ is equivalent to $L_{1,k}$.*

*Proof.* We need only prove the necessity. Suppose $G$ is not equivalent to $L_{1,k}$. Then there exist two vertices $u$ and $v$ such that $L(u) \neq L(v)$. We can give $u$ a color not in $L(v)$, then color the remaining vertices successively ending with $v$ thereby establishing an $L$-coloring for $G$.    □

**Corollary 1.** *Let $G$ denote the complete $(k + 1)$-partite graph $K_{1,\ldots,1,m}$. Then $ch(G) = k + 1$.*

*Proof.* Direct consequence of Theorem 1.    □

## 3   Bipartite Case

Note that when $k = 1$, $G(n, 1)$ is the complete bipartite graph $K_{n, n^n}$. In [4], it was shown that the list assignment $L_{n, 1}$ is essentially unique. For completeness and for uniformity of presentation, the proof is also included below.

**Theorem 3.** [4] *For $n \geq 2$, let $G$ denote the complete bipartite graph $K_{n, n^n}$. Let $L$ be a list assignment on $G$ such that $|L(v)| \geq n$ for all $v \in V(G)$. Then $G$ is not $L$-colorable if and only if $L$ is equivalent to $L_{n, 1}$.*

*Proof.* We need only prove the necessity. Suppose $G$ is not $L$-colorable.

We first assert that (i) $L(u) \cap L(v) = \emptyset$ for any distinct vertices $u$, $v \in V_1$.

Suppose $c \in L(u) \cap L(v)$. Then an $L$-coloring of $G$ is possible: color $u$ and $v$ with $c$ and the rest of the vertices in $V_1$ with $c_1, \ldots, c_{n-2}$. Then each vertex $w$ in $V_2$ can be colored from the list because $|L(w) - \{c, c_1, \ldots, c_{n-2}\}| \geq 1$. This contradiction proves the assertion.

Next, we assert that (ii) for any coloring of $V_1$ which uses $n$ colors $\alpha_1, \alpha_2, \ldots, \alpha_n$, there exists a vertex $w \in V_2$ such that $L(w) = \{\alpha_1, \alpha_2, \ldots, \alpha_n\}$.

If the assertion is not true, then there exists a coloring of $V_1$ with $n$ colors $\alpha_1, \alpha_2, \ldots, \alpha_n$ such that $|L(z) - \{\alpha_1, \alpha_2, \ldots, \alpha_n\}| \geq 1$ for any vertex $z \in V_2$ and this means that each vertex in $V_2$ can be colored from the list, a contradiction.

Since $|V_2| = n^n$, assertion (i) and (ii) imply that each vertex in $V_1$ is of the same list size $n$ and so is every vertex in $V_2$. This finishes the proof.   □

**Corollary 2.** [4] *For $n \geq 2$, let $G$ denote the complete bipartite graph $K_{n, m}$. Then the choice number of $G$ is given by*

$$ch(G) = \begin{cases} n + 1, & \text{if } m \geq n^n, \\ n, & \text{if } (n-1)^{n-1} - (n-2)^{n-1} \leq m < n^n, \text{ where } n \geq 3. \end{cases}$$

*Moreover $ch(G) < n$, if $m < (n-1)^{n-1} - (n-2)^{n-1}$.*

*Proof.* The first part of the corollary follows from Theorems 1 and 3. The second part of the corollary was established in [4].   □

Obviously, $ch(K_{1, m}) = 2$ and it follows as an immediate consequence of Corollary 2 that $ch(K_{2, m}) = 3$ if $m \geq 4$ and that $ch(K_{2, m}) = 2$ otherwise. Furthermore,

$$ch(K_{3, m}) = \begin{cases} 4, & \text{if } m \geq 3^3, \\ 3, & \text{if } 3 \leq m < 3^3, \\ 2, & \text{if } m < 3. \end{cases}$$

In [6], the authors studied the 3-choosability of complete bipartite graphs. In particular, they proved that $K_{4, m}$ is 3-choosable if and only if $m \leq 18$ while in [8], it was shown that $K_{5,m}$ is 3-choosable if and only if $m \leq 12$. These results together with Corollary 2 lead to the following.

$$ch(K_{4,m}) = \begin{cases} 5, & \text{if } m \geq 4^4, \\ 4, & \text{if } 19 \leq m < 4^4, \\ 3, & \text{if } 2 \leq m < 19, \\ 2, & \text{if } m = 1, \end{cases}$$

$$ch(K_{5,m}) = \begin{cases} 6, & \text{if } m \geq 5^5, \\ 5, & \text{if } 175 \leq m < 5^5, \\ 4, & \text{if } 13 \leq m < 175, \\ 3, & \text{if } 2 \leq m < 13, \\ 2, & \text{if } m = 1. \end{cases}$$

In [7], it was shown that $K_{6,m}$ is 3-choosable if and only if $m \leq 10$. Here we shall show that $K_{6,m}$ is not 4-choosable if $m \geq 125$ and this leads to the following.

$$ch(K_{6,m}) = \begin{cases} 7, & \text{if } m \geq 6^6, \\ 6, & \text{if } 2101 \leq m < 6^6, \\ 5, & \text{if } 125-? \leq m < 2101, \\ 4, & \text{if } 11 \leq m < 125-?, \\ 3, & \text{if } 2 \leq m < 11, \\ 2, & \text{if } m = 1. \end{cases}$$

In [9], it was mentioned that N. Eaton gave a construction of list assignment that shows that $K_{6,125}$ is not 4-choosable. Since this construction is not available, we offer a construction here which is interesting in its own in that the construction is described using the Petersen graph.

It is most probably the case that $ch(K_{6,124}) = 4$ but a proof, if available, will take much space and has to appear elsewhere.

**Proposition 1.** *If $m \geq 125$, then $K_{6,m}$ is not 4-choosable.*

*Proof.* It suffices to show that $K_{6,125}$ is not 4-choosable. Let $V_1 = \{u_1, u_2, \ldots, u_6\}$ and $V_2$ be the partite sets of $K_{6,125}$.

Let $L$ be a list assignment on $K_{6,125}$, where

$$L(u_1) = \{1, 2, 3, 4\} \quad L(u_2) = \{1, 5, 6, 7\} \quad L(u_3) = \{2, 5, 8, 9\}$$
$$L(u_4) = \{3, 6, 8, 10\} \quad L(u_5) = \{4, 7, 9, 10\} \quad L(u_6) = \{11, 12, 13, 14\}$$

Now, let $X$ denote the graph (associated with the above lists) whose vertex set is $V(X) = \{1, 2, \ldots, 10\}$ and whose edge set $E(X)$ is defined by $uv \in E(X)$ if and only if there exist four distinct integers $i, j, r, s \in \{1, 2, \ldots, 5\}$ such that $u \in L(u_i) \cap L(u_j)$ and $v \in L(u_r) \cap L(u_s)$.

Note that this is equivalent to saying that $uv \notin E(X)$ if and only if $u, v \in L(u_i)$ for some $1 \leq i \leq 5$.

It turns out that the graph $X$ is the Petersen graph depicted in Fig. 1.

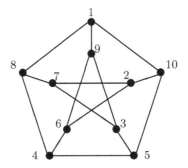

**Fig. 1.** The Petersen graph

Now, to every 3-path $\alpha\beta\gamma$ of $X$, let $\{\alpha,\,\beta,\,\gamma,\,x\}$ with $x \in L(u_6) = \{11,\,12,\,13,\,14\}$ be a list of colors to be assigned to some vertex in $V_2$ of $K_{6,\,125}$. Evidently, $X$ has thirty 3-paths (since each vertex $v$ in $X$ yields three 3-paths containing $v$ as the central vertex), and there are 4 choices for $x$ in $L(u_6)$, it follows that there are 120 such lists of 4 colors to be assigned to the vertices in $V_2$.

Note that there are precisely 5 independence sets of size 4 in $X$. In fact, these independence sets are induced by the lists $L(u_i)$, where $i = 1, 2, \ldots, 5$. Assign each of these lists of 4 colors to the remaining 5 vertices in $V_2$.

We shall show that $K_{6,\,125}$ is not list colorable with the above list assignment $L$. Suppose the contrary. Note that any $L$-coloring on $V_1$ must use up at least 3 colors from $\{1, 2, \ldots, 10\}$ on the vertices $u_1, u_2, \ldots, u_5$ (since each of these 10 colors appear exactly twice in the lists). Let $S$ denote this set of colors and let $X[S]$ denote the corresponding subgraph of $X$ induced by $S$.

Suppose $S = \{\alpha,\,\beta,\,\gamma\}$ and $u_6$ is colored with $x \in L(u_6)$. Then we may assume that $\alpha \in L(u_i) \cap L(u_j)$ and $\beta \in L(u_r) \cap L(u_s)$ and that $\gamma \in L(u_t)$ where $\{i, j, r, s, t\} = \{1, 2, \ldots, 5\}$. But this means that $\alpha\beta$ is an edge of $X[S]$. Since $\gamma$ must also appear exactly once in another color list (other than $L(u_t)$), it follows that $\gamma$ is adjacent to either $\alpha$ or $\beta$ and this gives rise to a 3-path in $X$. As such, the vertex $v \in V_2$ where $L(v) = \{\alpha,\beta,\,\gamma,\,x\}$ is not colorable from the list, a contradiction.

Suppose $S = \{\alpha,\,\beta,\,\gamma,\,\delta\}$. If $X[S]$ contains an edge, say $\alpha\beta$, then we may assume that $\alpha \in L(u_i) \cap L(u_j)$ and $\beta \in L(u_r) \cap L(u_s)$. Then at least one of $\gamma$ or $\delta$ must be in $L(u_i) \cup L(u_j) \cup L(u_r) \cup L(u_s)$ where $\{i, j, r, s\} \subseteq \{1, 2, \ldots, 5\}$, in which case, we have a 3-path in $X$ which yields a contradiction. Hence $X[S]$ is an independence set of size 4 in $X$. But then, the vertex $v \in V_2$ where $L(v) = \{\alpha,\,\beta,\,\gamma,\,\delta\}$ is not colorable from the list, a contradiction.

Suppose $|S| = 5$. Then $X[S]$ is a subgraph on 5 vertices. If $X[S]$ contains a 3-path or an independent set of size 4, then this yields a contradiction. Therefore $X[S]$ is an induced subgraph with 2 independent edges and an isolated vertex. But this means that the subgraph induced by $V(X) - S$ contains an independent set of size 4 (which is $L(u_i)$ for some $1 \leq i \leq 5$) which means that some vertex in $V_1$ ($u_i$ in this case) is not colored, a contradiction.

This finishes the proof.                                                 □

## 4   Tripartite Case

**Theorem 4.** *Let $L$ be a list assignment on $G = G(2, 2)$ such that $|L(v)| \geq 4$ for all vertices $v$ in $G$. Then $G$ is not $L$-colorable if and only if $L$ is equivalent to $L_{2,2}$.*

*Proof.* In this case, $G$ is the complete tripartite graph $K_{2,2,36}$. Suppose the partite sets of $G$ are $V_1$, $V_2$ and $V_3$ where $|V_1| = |V_2| = 2$ and $|V_3| = 36$.

We need only prove the necessity. Suppose $G$ is not $L$-colorable.

We first assert that (i) $L(u) \cap L(v) = \emptyset$ if $u$ and $v$ are distinct vertices in $V_i$ where $i = 1, 2$.

Suppose $c \in L(u) \cap L(v)$ where $u, v \in V_1$. Then an $L$-coloring of $G$ is possible: color $u$ and $v$ with $c$ and color the vertices in $V_2$ with $c_1$, $c_2$ ($\neq c$). Then each vertex $w$ in $V_3$ can be colored from the list because $|L(w) - \{c, c_1, c_2\}| \geq 1$. This contradiction proves the assertion.

Assertion (i) implies that (ii) any coloring of $V_1$ and $V_2$ must use up 4 colors with two from each $L(V_1)$ and $L(V_2)$.

Next, we assert that (iii) for any coloring of $V_1$ and $V_2$ which uses 4 colors $\alpha_1$, $\alpha_2$, $\alpha_3$, $\alpha_4$, there exists a vertex $w \in V_3$ such that $L(w) = \{\alpha_1, \alpha_2, \alpha_3, \alpha_4\}$.

If the assertion is not true, then there exists a set of 4 colors $\alpha_1$, $\alpha_2$, $\alpha_3$, $\alpha_4$ such that $|L(z) - \{\alpha_1, \alpha_2, \alpha_3, \alpha_4\}| \geq 1$ for any vertex $z \in V_3$ which means that each vertex in $V_3$ can be colored from the list, a contradiction.

Assertion (iii) implies that (iv) $L(V_3) = L(V_1) \cup L(V_2)$.

We assert that (v) $L(V_1) = L(V_2)$.

If the assertion is not true, then there exists a color $\alpha_i \in L(V_i)$ such that $\alpha_i \notin L(V_{3-i})$ for each $i = 1, 2$. Then the number of combinations of 4 colors, one from each $L(v)$ (where $v \in V_1 \cup V_2$), with $\alpha_1$ included (and $\alpha_2$ not included) is at least 24. Likewise the number of combination of 4 colors, one from each $L(v)$ (where $v \in V_1 \cup V_2$), with $\alpha_2$ included (and $\alpha_1$ not included) is at least 24. (Furthermore, we note that there are also combinations of 4 such colors which include both $\alpha_1$ and $\alpha_2$ and there are those that include neither). This means that we need at least 48 vertices in $V_3$ to exhaust all these combinations of 4 colors (by assertion (iii)), a contradiction.

Suppose $V_i = \{u_i, v_i\}$, $i = 1, 2$. We assert that (vi) $L(u_1) = A = L(u_2)$ and $L(v_1) = B = L(v_2)$ where $|A| = 4 = |B|$ and $A \cap B = \emptyset$.

If the assertion is not true, then we may assume that $L(u_1) = A = A_1 \cup A_2$, $L(v_1) = B = B_1 \cup B_2$, $L(u_2) = A_1 \cup B_1$ and $L(v_2) = A_2 \cup B_2$ where $A_i$ and $B_i$ are nonempty sets, $i = 1, 2$ and $A_1 \cap A_2 = \emptyset = B_1 \cap B_2$.

Consider the number of combinations of 4 colors, $\alpha_1$, $\alpha_2$, $\alpha_3$, $\alpha_4$, one from each $L(v)$ where $v \in V_1 \cup V_2$. There are three possible cases: (a) $\alpha_1 \in A$ and $\alpha_2$, $\alpha_3$, $\alpha_4 \in B$, (b) $\alpha_1 \in B$ and $\alpha_2$, $\alpha_3$, $\alpha_4 \in A$ and (c) $\alpha_1$, $\alpha_2 \in A$ and $\alpha_3$, $\alpha_4 \in B$. It is routine to check that the total sum of such combinations of colors in these three cases exceeds 36. But this is a contradiction because it means that we need more than 36 vertices in $V_3$ to exhaust all these combinations of 4 colors (by assertion (iii)).

It follows from assertions (i), (iii) and (vi) that $L$ is equivalent to $L_{2,2}$.     □

**Corollary 3.** *Let $G$ denote the complete tripartite graph $K_{2,2,m}$. Then the choice number of $G$ is given by*

$$ch(G) = \begin{cases} 5, & \text{if } m \geq 36, \\ 4, & \text{if } 5 \leq m < 36, \\ 3, & \text{if } m < 5. \end{cases}$$

*Proof.* Suppose $m \geq 36$. Then $ch(G) = 5$ by Theorem 1.

Suppose $m < 36$. Then, by Theorem 4, $G$ is 4-choosable so that $ch(G) \leq 4$.

If $m \geq 5$, it follows from the fact that $ch(K_{2,2,5}) = 4$ (by [1, Proposition 5]) that $ch(G) = 4$ (because $G$ contains $K_{2,2,5}$ as a subgraph in this case).

If $m \leq 4$, it follows from the fact that $ch(K_{2,2,4}) = 3$ (by [1, Theorem 1]) that $ch(G) = 3$ (because $G$ is a subgraph of $K_{2,2,4}$). $\qquad\square$

# References

1. Enomoto, E., Ohba, K., Ota, K., Sakamoto, J.: Choice Number of Some Complete Multi-Partite Graphs. Discrete Math. **244** (2002) 55–66
2. Erdös, P., Rubin, A.L., Taylor, H.: Choosability in Graphs. Congr. Numer. **26** (1980) 125–157
3. Gravier, S., Maffray, F.: Graphs Whose Choice Number is Equal to Their Chromatic Number. J. Graph Theory **27** (1998) 87–97
4. Hoffman, D.G., Johnson Jr., P.D.: On the Choice Number of $K_{m,n}$. Congr. Numer. **98** (1993) 105–111
5. Kierstead, H.A.: On the Choosability of Complete Multipartite Graphs with Part Size Three. Discrete Math. **211** (2000) 255–259
6. Mahadev, N.V.R., Roberts, F.S., Santhanakrishnan, P.: 3-Choosable Complete Bipartite Graphs. DIMACS, Tecnical Report (1991) 91–62
7. O'Donnell, P.: The Choice Number of $K_{6,q}$. Preprint Rutgers Univ. Math. Dept.
8. Shende, A.M., Tesman, B.: 3-Choosability of $K_{5,q}$. Congr. Numer. **111** (1995) 193–221
9. Tuza, Z.: Graph Colorings with Local Constraints—A Survey. Discuss. Math. Graph Theory **17** (1997) 161–228
10. Vizing, V.G.: Coloring the Vertices of a Graph in Prescribed Colors. Metody Diskret. Anal. v Teorii Kodov i Schem **29** (1976) 3–10 (In Russian)

# On Convex Quadrangulations of Point Sets on the Plane*

V.M. Heredia[1] and J. Urrutia[2]

[1] Facultad de Ciencias, Universidad Nacionál Autónoma de México, México D.F.
México
[2] Instituto de Matemáticas, Universidad Nacionál Autónoma de México, México D.F.
México

**Abstract.** Let $P_n$ be a set of $n$ points on the plane in general position, $n \geq 4$. A *convex quadrangulation* of $P_n$ is a partitioning of the convex hull $Conv(P_n)$ of $P_n$ into a set of quadrilaterals such that their vertices are elements of $P_n$, and no element of $P_n$ lies in the interior of any quadrilateral. It is straightforward to see that if $P$ admits a quadrilaterization, its convex hull must have an even number of vertices. In [6] it was proved that if the convex hull of $P_n$ has an even number of points, then by adding at most $\frac{3n}{2}$ Steiner points in the interior of its convex hull, we can always obtain a point set that admits a convex quadrangulation. The authors also show that $\frac{n}{4}$ Steiner points are sometimes necessary. In this paper we show how to improve the upper and lower bounds of [6] to $\frac{4n}{5} + 2$ and to $\frac{n}{3}$ respectively. In fact, in this paper we prove an upper bound of $n$, and with a long and unenlightening case analysis (over fifty cases!) we can improve the upper bound to $\frac{4n}{5} + 2$, for details see [9].

## 1  Introduction

Let $P_n$ be a set of $n$ points on the plane in general position. A triangulation $T$ of $P_n$ is a partitioning of the convex hull $Conv(P_n)$ into a set of triangles with disjoint interiors such that the vertices of the triangles in $T$ are elements of $P_n$, and no point in $P_n$ lies in the interior of a triangle of $T$. There is an extensive literature on the study of triangulations of point sets (see e.g. [3,8,11,12]); this is due to the fundamental nature of the subject, as well as to the many applications triangulations have in subjects such as mesh generation [4]. Triangulations are also highly relevant in numerous other application areas such as pattern recognition, computer graphics, solid modeling and geographic information systems.

A quadrangulation $Q$ of a point set $P_n$ is a partitioning of $Conv(P_n)$ into a set of quadrilaterals with disjoint interiors such that vertices of its elements are in $P_n$, and no point in $P_n$ lies in the interior of a quadrilateral of $Q$. In general, we do not require that quadrilaterals be convex. A quadrangulation $Q$ is called *convex* if all its elements are convex. The study of quadrangulations of point sets from

---

* Supported by CONACYT of Mexico, Proyecto SEP-2004-Co1-45876, and PAPIIT (UNAM), Proyecto IN110802.

J. Akiyama et al. (Eds.): CJCDGCGT 2005, LNCS 4381, pp. 38–46, 2007.

the point of view of the present paper was started by Bose and Toussaint [5]. A good reference on quadrangulations is Toussaint's paper [14], where applications of quadrangulations to a variety of problems such as mesh generation, finite element methods, scattered data, and interpolation are mentioned.

The problem of studying quadrangulations of bicolored point sets was started recently. A point set $P_n$ is called bicolored if each of its elements is colored red or blue. A quadrangulation $Q$ of a bicolored point set is a quadrangulation of $P_n$ in which each edge of $Q$ joins a red to a blue point. In a recent paper, Cortés, Marquez, Nakamoto and Valenzuela [7] showed that a necessary (but not sufficient) condition for a bicolored point set to admit a quadrangulation is that its convex hull have an even number of points, and that consecutive points in $Conv(P_n)$ have different colors. Their main results though, concern the the flip graph of the graph of quadrangulations of bicolored point sets. Bounds on the number of Steiner points needed to quadrangulate bicolored point sets are studied in [2] and [1]. They show that by adding at most $\lceil \frac{5m}{12} + \frac{7}{2} \rceil$ Steiner colored points in the interior of a bi-colored point set, we can always obtain a new bi-colored point set that admits a bi-colored quadrangulation, and that $\lfloor \frac{4m}{12} \rfloor$ Steiner points are sometimes sufficient.

It is easy to see that a point set $P_n$ admits a quadrangulation iff the convex hull of $P_n$ has an even number of vertices. Thus from now on we will assume that *the convex hull of all point sets considered here always has an even number of vertices.* In [5], Bose and Toussaint give an elegant method to quadrangulate a point set, although the set of quadrilaterals they obtain may contain *non-convex* quadrilaterals. First they define what they call a *spiral triangulation* (see Figure 1(a)) whose dual graph contains a path from which a quadrangulation is easily obtained, see Figure 1 (a) and (b).

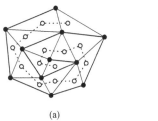
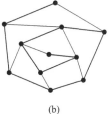

(a)                                    (b)

**Fig. 1.**

In this paper we study the problem of obtaining *convex quadrangulations* of point sets. It is straightforward to see that the condition that the convex hull of $P_n$ has an even number of vertices is not sufficient to guarantee the existence of a convex quadrangulation of a point set. For example, any set $P_5$ of five points such that four of them lie on $Conv(P_5)$ admits a quadrangulation, but not a convex quadrangulation. In [6] the problem of obtaining *convex quadrangulations* of point sets by adding Steiner points is studied. Following the terminology in [6], we say that a point set $P$ can be convex-quadrangulated with at most $k$ Steiner points if by adding at most $k$ Steiner points to $P$ (located in the interior of the convex hull of $P_n$), we obtain a point set that admits a convex quadrangulation. They show that any point set can be convex-quadrangulated with at most $3\lfloor\frac{n}{2}\rfloor$ Steiner points. They also show a point set in general position for which $\frac{n}{4}$ Steiner points are necessary.

In this paper we improve on the lower and upper bounds proved in [6]. We prove that $n-1$ Steiner points are always sufficient, and that $\frac{n}{3}$ are sometimes necessary to convex-quadrangulate any point set with $n$ elements. We then outline a method for improving the upper bound to $\frac{4n}{5}+2$. This requires an extensive and unenlightening case analysis that is skipped here. Full details appear in [9].

## 2   Convex Quadrangulations of Point Sets

In this section we prove the following result:

**Theorem 1.** *Any point set $P_n$ can be convex-quadrangulated with at most $n$ Steiner points placed in the interior of the convex hull of $P_n$.*

We now give the basic ideas of how to prove this result. The full details appear in [2]. To facilitate the presentation, we will allow the Steiner points to be placed on the boundary of $Conv(P_n)$. We then proceed to show how these points can be replaced by Steiner points in the interior of $Conv(P_n)$. We proceed as follows:

Choose the leftmost vertex on the convex hull of $P_n$, assuming without loss of generality that this point is unique, and let it be labeled $p$. Relabel the elements of $P_n - \{p\}$ by $\{p_0, \ldots, p_{n-2}\}$ in descending order according to the slope of the segments joining $p_i$ to $p$, $i = 1, \ldots, n-1$; see Figure 2(a)

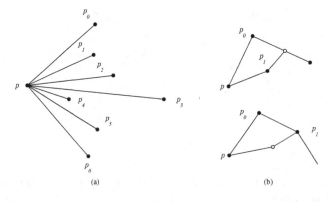

**Fig. 2.**

Consider $p_0$ and $p_1$. Two cases arise: $p_1$ lies in the interior of $Conv(P_n)$, or $p_1$ is a vertex of $Conv(P_n)$. We begin quadrangulating as shown in Figure 2(b).

Observe that in the first case, we insert a Steiner point slightly below the line segment joining $p$ to $p_1$, and in the second we place a Steiner point slightly above the line joining $p$ to $p_1$ and on the boundary of $Conv(P_n)$. We now proceed inductively, assuming that if $p_{2(i-1)+1} = p_{2i-1}$ is a vertex, there is a Steiner point slightly below the line joining $p$ to $p_{2i-1}$, or if $p_{2i-1}$ is an interior point to $Conv(P_n)$ then there is a Steiner point on the boundary of $Conv(P_n)$ slightly above the line segment joining $p$ to $p_{2i-1}$, $i \geq 0$.

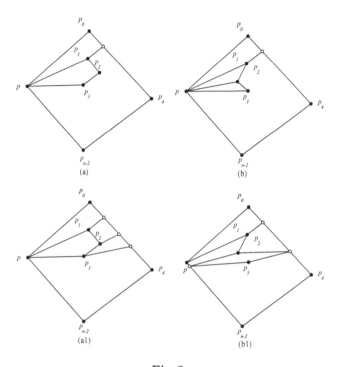

**Fig. 3.**

Assume without loss of generality that $p_{2i-1}$ is in the interior of $Conv(P_n)$. By the conditions stated in the previous paragraph, there is a Steiner point $s_j$ on the boundary of $Conv(P_n)$ slightly above the line joining $p$ to $p_{2i-1}$. Consider next $p_{2i}$ and $p_{2i+1}$, and assume that they are not vertices of $Conv(P_n)$. Two cases arise: $p_{2i+1}$ is below the line joining $p_{2i-1}$ to $p_{2i}$, or it is above it; see Figure 3(a) and 3(b), where Steiner points are shown as empty circles, and points in $P_n$ as solid small circles. The case shown in Figure 3(a) is solved as shown in Figure 3(a1).

The case shown in Figure 3(b) requires explanation. In this case, we place a Steiner point on the edge of $Conv(P_n)$ joining $p$ to $p_{n-2}$ close enough to $p$ and relabel it $p$. A second Steiner point is placed on the boundary of $P_n$ slightly above the line that passes through $p_{2i+1}$ and the Steiner point that was relabeled $p$. We then quadrangulate as shown in Figure 3(b1). From here on. the Steiner point relabeled $p$ substitutes for $p$ in the following iterations (until it is (possibly) replaced by another Steiner point). Observe that in both cases, two points in $P_n$

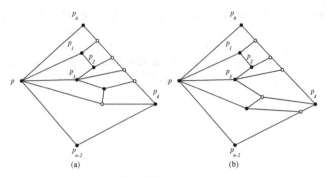

**Fig. 4.**

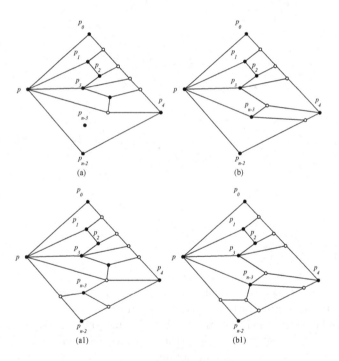

**Fig. 5.**

were processed and two Steiner points added. The case when one of $p_{2i}$ or $p_{2i+1}$ is a vertex can be solved in a similar way; see Figure 4.

Special care should be taken when the last points to be processed are $p_{(n-2)-1}$ and $p_{2n-2}$ or when all the points of $P_n$ except $p_{2n-2}$ have already been processed. In the latter case, it may be necessary to to introduce three Steiner points. Figure 5 shows how to handle these cases. It is now easy to see that we have used at most $n$ Steiner points.

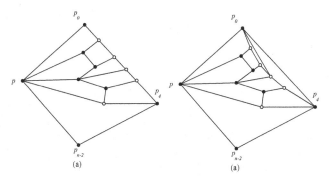

**Fig. 6.**

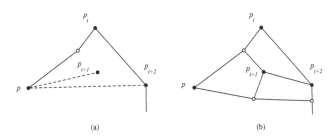

**Fig. 7.**

So far, we have proved that with the addition of at most $n$ Steiner points in the interior or the boundary of $Conv(P_n)$, we can convex-quadrangulate $P_n$. We now show how to modify the technique so that all the Steiner points used are located in the interior of $Conv(P_n)$. Let us assume that the edges of $Conv(P_n)$ are labeled in the clockwise direction along the boundary of $P_n$ by $e_0, \ldots, e_k$ such that $e_0$ is the edge joining $p$ to $p_0$. We observe that if an edge $e_i$ has an even number of Steiner points in it, it is straightforward to move these points to the interior of $Conv(P_n)$ and re-quadrangulate it. See Figure 6.

The problems arise when there is an edge that has an odd number of Steiner points in it, as in Figure 5(b). To solve this problem, we proceed as follows: Suppose that we have processed up to point $p_i$, and that the next pair of vertices involves a vertex of $Conv(P_n)$. Suppose that $p_{i+2}$ is a vertex of $Conv(P_n)$ (the case when $p_{i+1}$ is a vertex in $Conv(P_n)$ is handled in a similar way, and will be left to the reader). Several sub-cases arise:

1. $p_i$ and $p_{i+2}$ are vertices of $Conv(P_n)$. The normal procedure would place a single Steiner point on the edge $e_j$ joining $p_i$ to $p_{i+2}$. This case is solved as shown in Figure 7(a). In this case, we can place one Steiner point in the edge starting at $p_{i+2}$ and the other on the line joining this point to $p$, and quadrangulate as shown in Figure 7(b).

2. There are an even number of Steiner points on the edge $e_j$, ending (in clockwise order) at $p_{i+2}$; see Figure 8(b). Two sub-cases arise, namely the line determined by $p_i$ and $p_{i+1}$ does not intersect $e_j$, or it does. In the first case the problem is solved as shown in Figure 8(a); in the second, as shown in Figure 8(b).

To finish the proof, we observe that once we reach $p_{n-2}$, by a cardinality argument, the edge joining $p$ to $p_{n-2}$ must have an even number of Steiner points (introduced when the point $p$ was duplicated as in Figure 3(b1) ), which by our previous observations can be moved to the interior of $Conv(P_n)$. This completes the proof of Theorem 1. ☐

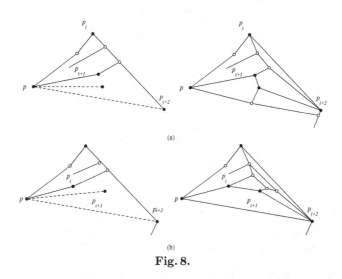

(a)

(b)

**Fig. 8.**

To conclude we mention that using the same technique, but taking groups of five elements of $P_n$ instead of two, we can always convex-quadrangulate $P_n$ using four Steiner points for each five elements of $P_n$. The analysis involves studying over 50 cases, and does not give any further insight into how to improve our upper bound to what we conjecture is the correct number of Steiner points, namely $\frac{n}{2} + c$, $c$ a constant. For this reason that result is not presented here. For complete details, the reader is again referred to [9]. Thus we have:

**Theorem 2.** *Any set $P_n$ of $n$ points can be convex-quadrangulated with at most $\lceil \frac{4n}{5} \rceil + 2$ Steiner points located in the interior of $Conv(P_n)$.*

## 2.1 Lower Bounds

In [6] it was proved that there are families of point sets (not in general position) with $n$ points for which $\lceil \frac{n-3}{2} \rceil - 1$ Steiner points are needed to convex-quadrangulate them. For points in general position, an example is also presented in which $\frac{n}{4}$ points are necessary. The $\frac{n}{4}$ lower bound can be improved as follows: Consider a convex polygon $Q$ with an even number of vertices, and for every other edge of $Q$, place a point in the interior of $Q$ at distance $\epsilon$ from the middle point of the edge. An example for an octagon is shown in Figure 9(a). It is easy now to see that in any convex-quadrangulation of the point set, a Steiner point must be placed in each of the shaded polygons shown in Figure 9(b), proving that there are point sets for which $\frac{n}{3}$ Steiner points are needed to convex-quadrangulate them.

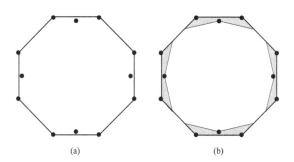

(a)                              (b)

**Fig. 9.**

## 3    Conclusions

We have proved that any point set $P_n$ whose convex hull contains an even number of points can be convex-quadrangulated by adding to it at most $n$ Steiner points placed in the interior of of the convex hull of $P_n$. Using the same technique involving a long, tedious and unenlightening process, our bound can be improved to $\frac{4n}{5} + 2$ Steiner points [9]. An example where $\frac{n}{3}$ Steiner points is also presented. We believe that our upper bound and lower bounds are not tight, and that the correct values for both of them are close to $\frac{n}{2}$.

## References

1. V.M. Alvarez-Amaya, *Gráficas Geométricas Sobre Conjuntos de Puntos Colore-ados*. B. Sc. Thesis, Facultad de Ciencias, Universidad Nacional Autónoma de México, July 2006.
2. V.M. Alvarez-Amaya, T. Sakai and J. Urrutia, *Bichromatic quadrangulations with Steiner points*, in preparation.
3. F. Aurenhammer, and R. Klein, *Voronoi Diagrams*, in *Handbook of Computational Geometry*, eds. J.R. Sack and J. Urrutia, North Holland, 2000, pp. 201-290.
4. M. Bern and D. Eppstein, *Mesh Generation and Optimal Triangulation*, in *Computing in Euclidean Geometry*, D.-Z. Du and F.K. Hwang, eds., World Scientific, 1992, pp 23–90.
5. P. Bose and G. Toussaint. *Characterizing and efficiently computing quadrangulations of planar point sets.*, Computer Aided Geometric Design, vol. 14, pp 763–785, 1997.

6. D. Bremner, F. Hurtado, S. Ramaswami, and V. Sacristan, *Small strictly convex quadrilateral meshes of point sets*, Algorithmica, Volume 38, Number 2, pp 317–339, November 2003.
7. C. Cortés, A. Márquez, A. Nakamoto and J. Valenzula. *Quadrangulations and 2-colorations*. 21th European Workshop on Computational Geometry, 2005.
8. S. Fortune, Voronoi diagrams and Delaunay triangulations, in Computing in Euclidean Geometry, 2nd edition, World Scientific, 1995.
9. M. Heredia, *Cuadrilaterizaciones convexas con pocos puntos Steiner*, B. Sc. Thesis, Facultad de Ciencias, Universidad Nacional Autónoma de México, July 2006.
10. M. Heredia and J. Urrutia, *Convex quadrangulations with Steiner points*, in preparation.
11. J.M. Keil, *Polygon decomposition*, in *Handbook of Computational Geometry*, eds. J.R. Sack and J. Urrutia, North Holland, 2000, pp 491–518.
12. F. Preparata, and I. Shamos, *Computational Geometry, an Introduction*. Springer Verlag, 1985.
13. S. Ramaswami, P. Ramos and G. Toussaint. *Converting triangulations to quadrangulations.*, Computational Geometry: Theory And Applications, vol. 9, pp 257–276, 1998.
14. G. Toussaint. *Quadrangulations of planar sets*. In Proceedings of the 4th International Workshop on Algorithms and Data Structures, pages 218–227. Springer-Verlag, 1995.

# Sufficient Conditions for the Existence of Perfect Heterochromatic Matchings in Colored Graphs

Lin Hu and Xueliang Li

Center for Combinatorics and LPMC Nankai University, Tianjin 300071, China
xjhulin@eyou.comm, lxl@nankai.edu.cn

**Abstract.** Let $G = (V, E)$ be an edge-colored graph. A matching of $G$ is called heterochromatic if its any two edges have different colors. Unlike uncolored matchings for which the maximum matching problem is solvable in polynomial time, the maximum heterochromatic matching problem is NP-complete. This means that to find both sufficient and necessary good conditions for the existence of perfect heterochromatic matchings should be not easy. In this paper, we obtain sufficient conditions of Hall-type and Tutte-type for the existence of perfect heterochromatic matchings in colored bipartite graphs and general colored graphs. We also obtain a sufficient and necessary condition of Berge-type to verify if a heterochromatic matching $M$ of $G$ is maximum.

## 1 Introduction

We use Bondy and Murty [2] for terminology and notations not defined here and consider simple graphs only.

Let $G = (V, E)$ be a graph. By an *edge-coloring* of $G$ we mean a surjective function $C \colon E \to N$, the set of nonnegative integers. If $G$ is assigned such a coloring, then $G$ is called an *edge-colored graph* or, simply *colored graph*. Denote the colored graph by $(G, C)$. We call $C(e)$ the *color* of the edge $e = uv \in E$, and $C(e)$ is also said to *present at* the edge $e$ and its two ends $u$ and $v$. Note that $C$ is not necessarily a proper edge-coloring, i.e., two adjacent edges may have the same color. For a vertex $v$ of $G$, the *color neighborhood* $CN(v)$ of $v$ is defined as the set $\{C(e) \colon e \text{ is incident with } v\}$. Then, $CN(S) = \bigcup_{v \in S} CN(v)$ for $S \subseteq V$. For a subgraph $H$ of $G$, let $C(H) = \{C(e) \colon e \in E(H)\}$. A subgraph $H$ of $G$ is called *monochromatic* if its any two edges have the same color, whereas $H$ is called *heterochromatic*, or *rainbow*, or *colorful* if its any two edges have different colors. There are many publications studying monochromatic or heterochromatic subgraphs, done mostly by the Hungarian school, see [1, 5] and the references in [3, 6]. Very often the subgraphs considered are paths, cycles, trees, etc. In this paper we study heterochromatic matchings of colored graphs.

A path $P$ is said to *start from* (or *end at*) a color $c \in C(G)$, if $c$ presents at the first (or last) edge of $P$. A connected graph is called a *star* if all but at most one (called the *center* of the star) of its vertices are leaves. We call a star $T$ the *tail* of a path $P$, if the center of $T$ is the last vertex of $P$. In the following we

J. Akiyama et al. (Eds.): CJCDGCGT 2005, LNCS 4381, pp. 47–58, 2007.

also regard a cycle as a path whose two ends coincide. A matching $M$ is called *heterochromatic* in a colored graph $G$, if its any two edges have different colors. The two ends of an edge in $M$ are said to be *matched under $M$*. $M$ *saturates* a vertex $u$, or $u$ is *$M$-saturated*, if there is an edge in $M$ incident with $u$; otherwise, $u$ is *$M$-unsaturated*. If every vertex of $G$ is $M$-saturated, the heterochromatic matching $M$ is called *perfect*. $M$ is a *maximum heterochromatic matching*, if $G$ has no heterochromatic matching $M'$ such that $|M'| > |M|$.

Unlike uncolored matchings for which the maximum matching problem is solvable in polynomial time (see [7]), the maximum heterochromatic matching problem is NP-complete (see [4]). This means that to find both sufficient and necessary good conditions for the existence of perfect heterochromatic matchings should be not easy. In the following we will present sufficient conditions of Hall-type and Tutte-type for the existence of perfect heterochromatic matchings in colored bipartite graphs and general colored graphs. We also obtain a sufficient and necessary condition of Berge-type to verify if a heterochromatic matching is maximum.

## 2   Preliminaries

In order to present our main results, we need some new notations and definitions.

For $S \subseteq V(G)$, denote $N_c(S)$ as one of the minimum set(s) $W$ satisfying $W \subseteq N(S)\backslash S$ and $[CN(S) \backslash C(G[S])] \subseteq CN(W)$. Note that there might be more than one such sets $W$, however, for any two such sets $W$ and $W'$ we have $CN(W) = CN(W')$ for the given $S$. In the following $A \vartriangle B$ denotes the symmetric difference of two sets $A$ and $B$, and $A|_B$ denotes $A \cap B$ .

**Definition 1.** *Let $M$ be a heterochromatic matching. $P = P_1 \cup P_2 \cup \cdots \cup P_k$ is called a heterochromatic $M$-alternating path system, if $P_1$ is an $M$-alternating path (in the common sense) such that any two edges in $M|_{P_1} \vartriangle P_1$ have different colors, and for every $i$ $(2 \leq i \leq k)$, $P_i$ is a set of paths such that*

**(1)** *every path in $P_i$ is an $M$-alternating path $p_{i_j}$ (in the common sense). Remind that a cycle is also regarded as a path (with the same start and end);*

**(2)** *the first edge of every path $p_{i_j}$ of $P_i$ is an edge $v_{i_\omega} u_{i_\omega}$ in $M$ such that*

$$C\left(v_{i_\omega} u_{i_\omega}\right) \in C\left(M|_{P_{i-1}} \vartriangle P_{i-1}\right) \cap C\left(M \backslash \bigcup_{j=1}^{i-1} P_j\right);$$

**(3)** *any two edges in $M |_{P_i} \vartriangle P_i$ have different colors;*

**(4)** *$C(M |_{P_s} \vartriangle P_s) \cap C(M |_{P_t} \vartriangle P_t) = \emptyset$ for any $s \neq t \in \{1, 2, \ldots, i\}$;*

**(5)** *$C(M) \cap C(M |_{P_k} \vartriangle P_k) \subseteq C\left(M \cap \bigcup_{j=1}^{k} P_j\right).$*

From Definition 1, one can see the following facts:

*Fact 1.* $P_1$ may be a signal edge $e$. If $e \in M$, then $P = P_1 = e$; if $e \notin M$ , then $M \vartriangle P_1 = e$. In particular, if $C(M |_{P_1} \vartriangle P_1) \cap C(M) \subseteq C(M \cap P_1)$, then $P = P_1$.

*Fact 2.* If, for $i \geq 2$, $\left|\left\{\omega \colon \omega \in C\left(M|_{P_{i-1}} \triangle P_{i-1}\right) \cap C\left(M \setminus \bigcup_{j=1}^{i-1} P_j\right)\right\}\right| = r$, then $P_i = p_{i_1} \cup p_{i_2} \cup \cdots \cup p_{i_r}$, where each $p_{i_\omega}$ is an $M$-alternating path (in the common sense) such that its first edge $v_{i_\omega} u_{i_\omega}$ is in $M$. We call each $p_{i_\omega}$ a *branch* of $P_i$. Obviously, $P_1$ has only one branch. If $p_{i_t} \cap p_{i_\omega} \neq \emptyset$ then $p_{i_t} \cap p_{i_\omega} = \{v\}$ for some vertex $v$, and $v$ is the start of $p_{i_t}$ and the end of $p_{i_\omega}$. So, $p_{i_t} \cup p_{i_\omega}$ is an $M$-alternating path. Hence, different branches must have different ends.

*Fact 3.* If $P_i \cap P_j \neq \emptyset$ for $i \neq j$, then $P_i \cap P_j = \{v\}$ for some vertex $v$, and $v$ is the start of a branch of $P_i$ and the end of a branch of $P_j$, and then $P_i \cup P_j$ is an $M$-alternating path system. If the start $u$ of $P_1$ is $M$-unsaturated and $P_1 \cap P_i = \{v\}$ for $i \geq 2$, then $u \neq v$.

Therefore, for $p_{i_t} \cap p_{j_s} \neq \emptyset$ such that $i$, $j$, $s$, $t$ are pairwise different, if we regard $p_{i_t}$ union $p_{j_s}$ as one path, then $p_{i_t} \cup p_{j_s}$ is again an $M$-alternating path, and so, the heterochromatic $M$-alternating path system $P$ can also be equivalently defined as follows:

**Definition 2.** $P = p_1 \cup p_2 \cup \cdots \cup p_l$ *is an $M$-alternating path system, if*

**(1)** *for all $i, j \in \{1, 2, \ldots, l\}$ and $i \neq j$, $p_i \cap p_j = \emptyset$;*
**(2)** *any two edges in $M|_P \triangle \bigcup_{i=1}^{l} p_i$ have different colors;*
**(3)** $C(M) \cap C\left(M|_P \triangle \bigcup_{i=1}^{l} p_i\right) \subseteq C\left(M \cap \bigcup_{i=1}^{l} p_i\right)$.

The following two definitions are important in the sequel.

**Definition 3.** *For an $M$-unsaturated vertex $u$, a heterochromatic $M$-augmenting path system starting from $u$ is a heterochromatic $M$-alternating path system $P = P_1 \cup P_2 \cup \cdots \cup P_k$ such that*

**(1)** *$u$ is the start of $P_1$, and the end of $P_1$ is $M$-unsaturated;*
**(2)** *for any $P_i$ ($i \geq 2$), its every branch $p_{i_j}$ is of even length, and the end of $p_{i_j}$ is $M$-unsaturated or the start of another branch.*

**Definition 4.** *A heterochromatic $M$-augmenting path system of $G$ is a heterochromatic $M$-alternating path system $P = p_1 \cup p_2 \cup \cdots \cup p_l$ such that $|M|_P \triangle P| > |M \cap P|$, where $p_i \cap p_j = \emptyset$ for all $i \neq j$ and $i, j \in \{1, 2, \ldots, l\}$.*

## 3   Main Results

Although we cannot find both sufficient and necessary conditions for the existence of perfect heterochromatic matchings, in this section we give a sufficient condition of Hall-type for colored bipartite graphs and of Tutte-type for general colored graphs for the existence of perfect heterochromatic matchings.

**Theorem 1. (Hall-type)** *Let $(B, C)$ be a colored bipartite graph with bipartition $(X, Y)$. Then, $B$ contains a heterochromatic matching that saturates every vertex in $X$, if*

$$|N_c(S)| \geq |S|, \tag{1}$$

*for all $S \subseteq X$.*

*Proof.* Suppose $(B, C)$ is a colored bipartite graph satisfying condition (1), then it is clear that $|N(S)| \geq |S|$ for all $S \subseteq X$ from the definition of $N_c(S)$. So, $B$ contains an uncolored matching that saturates every vertex in $X$ from [2, Theorem 5.2, p. 72].

By contradiction, suppose $B$ does not contain any heterochromatic matching that saturates every vertex in $X$. Let $X = \{u_1, u_2, \ldots, u_n\}$, $Y = \{v_1, v_2, \ldots, v_m\}$, and $M^*$ be a maximum heterochromatic matching of $B$ with $C(M^*) = \{1, 2, \ldots, r\}$. So, $M^*$ cannot saturate every vertex in $X$. Let $u$ be an $M^*$-unsaturated vertex in $X$. Without loss of generality, we may assume $u_i v_i \in M^*$ with $C(u_i v_i) = i$ for $i \in \{1, 2, \ldots, r\}$. Since $|N(S)| \geq |S|$ for any $S \subseteq X$, there is an $M^*$-alternating path (in the common sense) starting from $u$ whose end is an $M^*$-unsaturated vertex.

In the following we call a heterochromatic $M^*$-alternating path system $P$ a *possible $M^*$-augmenting path system* if any two edges of $M^*|_P \triangle P$ have different colors.

Next we will find a heterochromatic $M^*$-augmenting path system starting from $u$, which leads to a contradiction to the maximality of $M^*$. In the following, for a set $A$ of paths and a set of vertices $X$ we simply use $A \cap X$ to denote the set of vertices each of which is both in a path of $A$ and in $X$, i.e., $V(\bigcup_{p \in A} p) \cap X$. As usual, we use $l(p)$ to denote the length of a path $p$.

*Step 1.* Denote by $A_1$ the set of all possible $M^*$-augmenting paths starting from $u$ (see Definition 3) such that each such path is extended with a length as large as possible. Then, every $P_1 \in A_1$ has one of the following properties:

1) $l(P_1) \equiv 0 \pmod 2$, and any two edges in $M^*|_{P_1} \triangle P_1$ have different colors;
2) $l(P_1) \equiv 1 \pmod 2$, and the end of $P_1$ is $M^*$-saturated. Moreover, any two edges in $(M^*|_{P_1} \triangle P_1) \backslash e$ have different colors, and $C(e) \in C((M^*|_{P_1} \triangle P_1) \backslash e)$, where $e$ is the last edge of $P_1$;
3) $l(P_1) \equiv 1 \pmod 2$, and the end of $P_1$ is $M^*$-unsaturated. Moreover, any two edges in $M^*|_{P_1} \triangle P_1$ have different colors.

Define $\wp_1 = \{P_1 : P_1 \in A_1 \text{ has property 3)}\}$, then $\wp_1 \neq \emptyset$. Otherwise, every path in $A_1$ has either property 1) or 2), and so, every this kind of path is not augmenting. Let
$$S_1' = A_1 \cap X, \quad S_1'' = (M^* \cap A_1) \cap Y.$$
Clearly, $\left|S_1''\right| = \left|S_1' \backslash u\right|$ and $CN\left(S_1'\right) \subseteq CN\left(S_1''\right)$. So, $\left|N_c\left(S_1'\right)\right| \leq \left|S_1''\right| = \left|S_1' \backslash u\right| = \left|S_1'\right| - 1$, a contradiction to condition (1).

If some $P_1 \in \wp_1$ satisfies $C(M^*) \cap C(M^*|_{P_1} \triangle P_1) \subseteq C(M^* \cap P_1)$, then $P = P_1$ is a heterochromatic $M^*$-augmenting path starting from $u$. So, the matching $M' = (M^* \backslash P_1) \cup (M^*|_{P_1} \triangle P_1)$ is heterochromatic that saturates $u$, and hence, $|M'| > |M^*|$, which contradicts the maximality of $M^*$. So, for any $P_1 \in \wp_1$ we must have $C(M^*|_{P_1} \triangle P_1) \cap C(M^* \backslash P_1) \neq \emptyset$. Then, set $S_1^* = \wp_1 \cap X$, and go to the next step.

*Step 2.* For every $P_1 \in \wp_1$ and every $\omega \in C(M^*|_{P_1} \triangle P_1) \cap C(M^*\backslash P_1)$, let $P_2$ denote the set of all possible heterochromatic $M^*$-augmenting paths starting from $v_\omega$ with first edge $u_\omega v_\omega$, such that each such path is extended with a length as large as possible. Denote by $A_2$ the set of all these $P_2$'s corresponding to all $P_1$'s. Then, any branch $p_{2_\omega}$ of every $P_2 \in A_2$ has one of the following properties:

1) $l(p_{2_\omega}) \equiv 1 \pmod 2$, and any two edges in $M^*|_{p_{2_\omega}} \triangle p_{2_\omega}$ have different colors;
2) $l(p_{2_\omega}) \equiv 0 \pmod 2$, and any two edges in $(M^*|_{p_{2_\omega}} \triangle p_{2_\omega}) \backslash e$ have different colors, and $C(e) \in C(M^*|_{(p_{2_\omega} \cup P_1)} \triangle (P_1 \cup (p_{2_\omega}\backslash e)))$, where $e$ is the last edge of $p_{2_\omega}$;
3) $l(p_{2_\omega}) \equiv 0 \pmod 2$, and any two edges in $M^*|_{p_{2_\omega}} \triangle p_{2_\omega}$ have different colors.

Define

$$\wp_2 = \left\{ P_2 \colon P_2 = \bigcup_{\omega \in C(M^*|_{P_1} \triangle P_1) \cap C(M^*\backslash P_1)} p_{2_\omega} \right\},$$

where every $p_{2_\omega}$ is a path with property 3), and any two edges in $M^*|_{P_2} \triangle P_2$ have different colors. We claim $\wp_2 \neq \emptyset$. Otherwise, for every $P_1 \in \wp_1$, there is a corresponding set $P_2$ of paths in Step 2, and among the branches of $P_2$, there exists a branch with property 1) or 2), or there are at least two branches each of which has an even numbered edge with the same color $c$. So, these branches are not augmenting. For the latter case, we take a subbranch of each such branch to end at the color $c$. Particularly, if $p_1$, $p_2$ and $p_3$ are branches of a $P_2$ such that $p_1$ and $p_2$ have a common color $c_1$ and $p_2$ and $p_3$ have a common color $c_2$, and if the color of the first edge, respectively, in the three paths is $\omega_1$, $\omega_2$ and $\omega_3$, and the three colors appear in $P_1$ in the same ordering, then the subbranches of $p_1$ and $p_2$ are taken to end at the color $c_1$, and the subbranch of $p_3$ is taken to end at the color $c_2$. Let $A_2'$ denote the set of all these not augmenting branches and subbranches corresponding to the $P_2$'s. Denote $S_2' = A_2' \cap X$.

Note that if some $P_1 \in \wp_1$ has a tail, then the structure of the $P_2$ corresponding to $P_1$ in Step 2 is complicated, for which we analysis as follows:

1) $P_1 \in \wp_1$ has a tail $T$ such that $C(T) \cap C(M^*) = \emptyset$. If the $P_2$ corresponding to this $P_1$ has a branch $p_{2_\omega}$ with property 2), and $C(e) \in C(M^*|_{P_1} \triangle (P_1\backslash T))$ for the last edge $e$ of $p_{2_\omega}$, then we have $p_{2_\omega} \in A_2'$. Denote $\wp_1' = \{P_1 \colon P_1 \in \wp_1$ has a tail $T_{1,1}\}$, where $C(T_{1,1}) \cap C(M^*) = \emptyset$ and $|C(T_{1,1})| = |C(M^*|_{P_1} \triangle P_1) \cap C(M^*\backslash P_1)|$, and all the branches of the $P_2$ are in $A_2'$ and have property 1) or 2). Then, denote

$$T_1' = \{T_{1,1} \colon T_{1,1} \text{ is the tail of a } P_1 \in \wp_1'\},$$

$$X_1' = \{x \colon x \text{ is the center of a } T_{1,1} \in T_1'\}.$$

2) $P_1 \in \wp_1$ has a tail $T_{1,2}$ such that $C(T_{1,2}) \subseteq C(M^*\backslash P_1)$. Denote $\wp_1'' = \{P_1 \colon T_{1,2}$ is the tail of $P_1 \in \wp_1\}$, where for any $P_1$ with tail $T_{1,2}$, all branches

of $P_2$ corresponding to the $T_{1,2}$ are in $A_2'$. That is, all the branches of $P_2$ such that the color of the first edge is in $C(T_{1,2})$ belong to $A_2'$. Then, denote

$$T_1'' = \{T_{1,2} \colon T_{1,2} \text{ is the tail of a } P_1 \in \wp_1''\},$$

$$X_1'' = \{x \colon x \text{ is the center of a } T_{1,2} \in T_1''\}.$$

**3)** $P_1 \in \wp_1$ has a tail $T_{1,3}$ such that $C(T_{1,3}) \cap C(M^*) \neq \emptyset$ and $C(T_{1,3}) \backslash C(M^*) \neq \emptyset$, and

$$G[\{e \colon e \in E(T_{1,3}) \text{ and } C(e) \in C(M^*)\}] \subseteq T_1''$$

$$G[\{e \colon e \in E(T_{1,3}) \text{ and } C(e) \notin C(M^*)\}] \subseteq T_1'.$$

Then, denote

$$\wp_1''' = \{P_1 \colon P_1 \in \wp_1 \text{ has a tail } T_{1,3}\}$$

$$T_1''' = \{T_{1,3} \colon T_{1,3} \text{ is the tail of a } P_1 \in \wp_1'''\}$$

$$X_1''' = \{x \colon x \text{ is the center of a } T_{1,3} \in T_1'''\}.$$

Now we consider the colors of all paths of $A_2'$. If we do not consider the colors presenting in the path $P_1$, then we have the following two cases:

*Case 1.* There is a path $p \in A_2'$ with property 1) or 2).

*Case 2.* $A_2'$ has at least two branches in $P_2$ each of which has an even numbered edge with the same color $c$.

Then, $CN(A_2' \cap Y) \supseteq CN(A_2' \cap X)$, and the vertices of $(A_2' \cap X) \backslash \{u_\omega\}$ are perfectly matched with the vertices of $(A_2' \cap Y) \backslash \{v', v'', v_\omega\}$ under $M^*$, where $v'' \in Y$ is the end of a subbranch constructed in Case 2, $v' \in Y$ is the end of a path with property 2). If only Case 1 happens, then the vertices of $(A_2' \cap X) \backslash \{u_\omega\}$ are perfectly matched with the vertices of $(A_2' \cap Y) \backslash \{v', v_\omega\}$ under $M^*$. Since $|\{u_\omega\}| \geq 1$, we have $|N_c(S_2')| < |S_2'|$. Otherwise, the vertices of $(A_2' \cap X) \backslash \{u_\omega\}$ are perfectly matched with the vertices of $(A_2' \cap Y) \backslash \{v', v'', v_\omega\}$ under $M^*$. Among the branches of Case 2, consider those ending at a same color, and take the end $v$ of one of them and put it into $(A_2' \cap Y) \backslash \{v', v'', v_\omega\}$. Then, $CN((A_2' \cap X) \backslash \{u_\omega\}) \subseteq CN(((A_2' \cap Y) \backslash \{v', v'', v_\omega\}) \cup \{v\})$ and $|(A_2' \cap X) \backslash \{u_\omega\}| + 1 \geq |((A_2' \cap Y) \backslash \{v', v'', v_\omega\}) \cup \{v\}|$. Since $S_2' = A_2' \cap X$ and $|\{u_\omega\}| \geq 2$, we have $|N_c(S_2')| < |S_2'|$, which always holds provided we do not consider the colors presenting in $P_1$.

Therefore, if some $P_1$ has a tail $T \notin T_1' \cup T_1'' \cup T_1'''$ such that $C(T) \cap C(M^*) \neq \emptyset$ and $C(T) \backslash C(M^*) \neq \emptyset$, then let $A_2'' = A_2' \backslash \{p \colon p$ is a branch constructed in Step 2, corresponding to the edge $e \in E(T)$ such that $C(e) \in C(M^*)\}$ and let $S_2'' = A_2'' \cap X$. Since $|N_c(\{p\} \cap X)| \geq |\{p\} \cap X|$, we again have $|N_c(S_2'')| < |S_2''|$, provided we do not consider the colors presenting in $P_1$ of $A_2''$. Set

$$S_2 = S_2'' \cup ((S_1^* \backslash T_1) \cup X_1' \cup X_1'' \cup X_1'''),$$

where $S_1^* = \wp_1 \cap X$ and $T_1 = \{T \colon T$ is the tail of a $P_1 \in \wp_1\}$.

If $|N_c((S_1^*\setminus T_1)\cup X_1'\cup X_1''\cup X_1''')| \leq |(S_1^*\setminus T_1)\cup X_1'\cup X_1''\cup X_1'''|$, then $|N_c(S_2)| < |S_2|$. Otherwise, $|N_c((S_1^*\setminus T_1)\cup X_1'\cup X_1''\cup X_1''')| > |(S_1^*\setminus T_1)\cup X_1'\cup X_1''\cup X_1'''|$, and hence, their difference comes from the numbers of colors of $T_1'$, $T_1''$ and $T_1'''$. If $P_1 \in \wp_1'$, then $|N_c(P_1\cap X)| - |P_1\cap X| \leq |C(T_{1,1})| - 1$; if $P_1 \in \wp_1''$, then $|N_c(P_1\cap X)| - |P_1\cap X| \leq |C(T_{1,2})| - 1$; and if $P_1 \in \wp_1'''$, then $|N_c(P_1\cap X)| - |P_1\cap X| \leq |C(T_{1,3})| - 1$. Since $A_2''$ has $|C(T_{1,1})\cup C(T_{1,2})\cup C(T_{1,3})|$ branches such that their vertices belong to $S_2''$ whenever they also belong to $X$, we have $|N_c(S_2)| < |S_2|$, a contradiction to condition (3.1), and hence, $\wp_2 \neq \emptyset$.

Therefore, if there is a $P_2 \in \wp_2$ corresponding to $P_1 \in \wp_1$ such that $C(M^*)\cap C(M^*|_{P_2} \bigtriangleup P_2) \subseteq C(M^* \cap (P_2\cup P_1))$, then $P = P_1\cup P_2$ is a heterochromatic $M^*$-augmenting path starting from $u$, and then, the matching $M' = (M^*\setminus(P_2\cup P_1))\cup(M^*|_{(P_1\cup P_2)} \bigtriangleup (P_2\cup P_1))$ is heterochromatic that saturates $u$, and hence, $|M'| > |M^*|$, a contradiction to the maximality of $M^*$. So, for any $P_2 \in \wp_2$, we have $C(M^*|_{P_2} \bigtriangleup P_2)\cap C(M^*\setminus(P_2\cup P_1)) \neq \emptyset$. Then, set $S_2^* = (\wp_1\cup\wp_2)\cap X$, and proceed as above. Similarly, we can obtain $\wp_3, \wp_4, \ldots, \wp_{k-1}$, and arrive at the next step.

*Step $k$.* For every $P_{k-1} \in \wp_{k-1}$ and every $w \in C(M^*|_{P_{k-1}} \bigtriangleup P_{k-1})\cap C(M^*\setminus\bigcup_1^{k-1} P_i)$, let $P_k$ be the set of all possible $M^*$-augmenting paths starting from $v_w$ with first edge $u_w v_w$, such that each such path is extended with a length as large as possible. Denote by $A_k$ the set of all these $P_k$'s corresponding to all $P_{k-1}$'s. Then, any branch $p_{k_w}$ of every $P_k \in A_k$ has one of the following properties:

1) $l(p_{k_w}) \equiv 1 \pmod 2$, and any two edges in $M^*|_{p_{k_w}} \bigtriangleup p_{k_w}$ have different colors;
2) $l(p_{k_w}) \equiv 0 \pmod 2$, and any two edges in $(M^*|_{p_{k_w}} \bigtriangleup p_{k_w})\setminus e$ have different colors, where $e$ is the last edge $e$ of $p_{k_w}$ such that $C(e) \in C((M^*\cap \left(\bigcup_1^{k-1} P_i \cup p_{k_w}\right)) \bigtriangleup \left(\bigcup_1^{k-1} P_i \cup (p_{k_w}\setminus e)\right))$;
3) $l(p_{k_w}) \equiv 0 \pmod 2$, and any two edges in $M^*|_{p_{k_w}} \bigtriangleup p_{k_w}$ have different colors.

Define

$$\wp_k = \left\{ P_k : P_k = \bigcup_{\substack{w\in C\left(M^*|_{P_{k-1}} \bigtriangleup P_{k-1}\right)\cap \\ C(M^*\setminus(P_1\cup P_2\cup\cdots\cup P_{k-1}))}} p_{k_w} \right\},$$

where every path $p_{k_w}$ has property 3), and any two edges in $M^*|_{P_k} \bigtriangleup P_k$ have different colors. We claim $\wp_k \neq \emptyset$. Otherwise, for every $P_{k-1} \in \wp_{k-1}$, there is a set $P_k$ of paths in Step $k$, and among the branches of $P_k$, there exists a branch with property 1) or 2), or there are at least two branches each of which has an even numbered edge with the same color $c$. So, such branches are not augmenting. For the latter case, similar to Step 2 we construct their subbranches by cutting at the same color $c$. Let $A_k'$ denote the set of all these not augmenting branches and subbranches of all $P_k$'s, and let $S_k' = A_k'\cap X$.

Similar to Step 2, if a branch of $P_i$ $(1 \leq i \leq k-1)$ has a tail, then the structure of $P_k$ corresponding to $P_i$ in Step $k$ is complicated, for which we analysis as follows:

1) A branch of $P_i \in \wp_i$ $(1 \leq i \leq k-1)$ has a tail $T$ such that $C(T) \cap C(M^*) = \emptyset$. If a branch $p_{k_\omega}$ of $P_k$ corresponding to the $P_i$ is a path with property 2) and $C(e) \in C(M^*|_{P_i} \vartriangle (P_i \backslash T))$, where $e$ is the last edge of $p_{k_\omega}$, then $p_{k_\omega} \in A_k'$. Then, let $\wp_i' = \{P_i : P_i \in \wp_i \text{ has a tail } T_{i,1}\}$, where $C(T_{i,1}) \cap C(M^*) = \emptyset$, $|C(T_{i,1})| = |C(M^*|_{P_{k-1}} \vartriangle P_{k-1}) \cap C(M^* \backslash \bigcup_1^{k-1} P_i)|$, and all branches of $P_k$ corresponding to the $P_i$ are in $A_k'$ and have property 1) or 2). Define

$$T_i' = \{T_{i,1} : T_{i,1} \text{ is the tail of a } P_i \in \wp_i'\}$$

and

$$X_i' = \{x : x \text{ is center of a } T_{i,1} \in T_i'\}.$$

2) $T_{k-1,2}$ is the tail of a $P_{k-1} \in \wp_{k-1}$ and $C(T_{k-1,2}) \subseteq C(M^* \backslash \bigcup_1^{k-1} P_i)$. Define $\wp_{k-1}'' = \{P_{k-1} : P_{k-1} \in \wp_{k-1} \text{ has a tail } T_{k-1,2}\}$, where all branches of all $P_k$'s corresponding to all $P_{k-1}$'s with the tail $T_{k-1,2}$ belong to $A_k'$. Let

$$T_{k-1}'' = \{T_{k-1,2} : T_{k-1,2} \text{ is the tail of a } P_{k-1} \in \wp_{k-1}''\},$$

$$X_{k-1}'' = \{x : x \text{ is the center of a } T_{k-1,2} \in T_{k-1}''\}.$$

3) $T_{k-1,3}$ is the tail of a $P_{k-1} \in \wp_1$ such that $C(T_{k-1,3}) \cap C(M^*) \neq \emptyset$, $C(T_{1,3}) \backslash C(M^*) \neq \emptyset$, and

$$G[\{e : e \in E(T_{k-1,3}) \text{ and } C(e) \in C(M^*)\}] \subseteq T_{k-1}''$$

$$G[\{e : e \in E(T_{k-1,3}) \text{ and } C(e) \notin C(M^*)\}] \subseteq T_i'.$$

Then, define

$$\wp_{k-1}''' = \{P_{k-1} : P_{k-1} \in \wp_{k-1} \text{ has a tail } T_{k-1,3}\}$$

$$T_{k-1}''' = \{T_{k-1,3} : T_{k-1,3} \text{ is the tail of a } P_{k-1} \in \wp_{k-1}'''\}$$

$$X_{k-1}''' = \{x : x \text{ is the center of a } T_{k-1,3} \in T_{k-1}'''\}.$$

Now we consider the colors of all paths of $A_k'$. If we do not consider the colors presenting in $\bigcup_1^{k-1} P_t$, then we have the following two cases:

*Case 1.* $p \in A_k'$ is a path with property 1) or 2).

*Case 2.* $A_k'$ has at least two branches in $P_k$ each of which has an even numbered edge with the same color $c$.

Then, $CN(A_k' \cap Y) \supseteq CN(A_k' \cap X)$ and the vertices of $(A_k' \cap X) \backslash \{u_\omega\}$ are perfectly matched with the vertices of $(A_k' \cap Y) \backslash \{v', v'', v_\omega\}$ under $M^*$, where $v'' \in Y$ is the end of a subbranch constructed in Case 2, $v' \in Y$ is the end of a path with property 2). If only Case 1 happens, then the vertices of $(A_k' \cap X) \backslash \{u_\omega\}$

are perfectly matched with the vertices of $(A'_k \cap Y) \setminus \{v', v_\omega\}$ under $M^*$. Since $|\{u_\omega\}| \geq 1$, we have $|N_c(S'_k)| < |S'_k|$. Otherwise, the vertices of $(A'_k \cap X) \setminus \{u_\omega\}$ are perfectly matched with the vertices of $(A'_k \cap Y) \setminus \{v', v'', v_\omega\}$ under $M^*$. Among the branches of Case 2, consider those ending at a same color, and take the end $v$ of one of them and put it into $(A'_k \cap Y) \setminus \{v', v'', v_\omega\}$. Then, $CN((A'_k \cap X) \setminus \{u_\omega\}) \subseteq CN(((A'_k \cap Y) \setminus \{v', v'', v_\omega\}) \cup \{v\})$, and $|A'_k \cap X) \setminus \{u_\omega\}| + 1 \geq |((A'_k \cap Y) \setminus \{v', v'', v_\omega\}) \cup \{v\}|$. Since $|\{u_\omega\}| \geq 2$, we have $|N_c(S'_k)| < |S'_k|$, which always holds provided we do not consider the colors presenting in $\bigcup_1^{k-1} P_t$.

Therefore, if some $P_{k-1}$ has a tail $T \notin T'_{k-1} \cup T''_{k-1} \cup T'''_{k-1}$ such that $C(T) \cap C(M^*) \neq \emptyset$ and $C(T) \setminus C(M^*) \neq \emptyset$, then let $A''_k = A'_k \setminus \{p: p$ is a branch constructed in Step $k$, corresponding to the edge $e \in E(T)$ such that $C(e) \in C(M^*)\}$, and let $S''_k = A''_k \cap X$. Since $|N_c(\{p\} \cap X)| \geq |\{p\} \cap X|$, we again have $|N_c(S''_k)| < |S''_k|$, provided we do not consider the colors presenting in $\bigcup_1^{k-1} P_t$ of $A''_k$. Define

$$S_k = S''_k \cup \left( \left( S^*_{k-1} \setminus \bigcup_1^{k-1} T_t \right) \cup X'_i \cup X''_{k-1} \cup X'''_{k-1} \right),$$

where $S^*_{k-1} = (\wp_1 \cup \wp_2 \cup \cdots \cup \wp_{k-1}) \cap X$ and $T_t = \{T: T$ is the tail of a $P_k \in \wp_k\}$ for $t \in \{1, 2, \ldots, k-1\}$.

If $\left| N_c \left( \left( S^*_{k-1} \setminus \bigcup_1^{k-1} T_t \right) \cup X'_i \cup X''_{k-1} \cup X'''_{k-1} \right) \right| \leq \left| \left( S^*_{k-1} \setminus \bigcup_1^{k-1} T_t \right) \cup X'_i \cup X''_{k-1} \cup X'''_{k-1} \right|$, then $|N_c(S_k)| < |S_k|$. Otherwise, $\left| N_c \left( \left( S^*_{k-1} \setminus \bigcup_1^{k-1} T_t \right) \cup X'_i \cup X''_{k-1} \cup X'''_{k-1} \right) \right| > \left| \left( S^*_{k-1} \setminus \bigcup_1^{k-1} T_t \right) \cup X'_i \cup X''_{k-1} \cup X'''_{k-1} \right|$, and hence, their difference comes from the numbers of colors of $T'_i$, $T''_{k-1}$ and $T'''_{k-1}$. If $P_i \in \wp'_i$, then $|N_c(P_i \cap X)| - |P_i \cap X| \leq |C(T_{i,1})| - 1$; if $P_{k-1} \in \wp''_{k-1}$, then $|N_c(P_{k-1} \cap X)| - |P_{k-1} \cap X| \leq |C(T_{k-1,2})| - 1$; and if $P_{k-1} \in \wp'''_{k-1}$, then $|N_c(P_{k-1} \cap X)| - |P_{k-1} \cap X| \leq |C(T_{k-1,3})| - 1$. Since $A''_k$ has $|C(T_{i,1}) \cup C(T_{k-1,2}) \cup C(T_{k-1,3})|$ branches such that all their vertices belong to $S''_k$ whenever they also belong to $X$, we have $|N_c(S_k)| < |S_k|$, a contradiction to condition (3.1), and hence, $\wp_k \neq \emptyset$.

Therefore, there exists a $P_k \in \wp_k$ corresponding to $P_{k-1} \in \wp_{k-1}$, $P_{k-2} \in \wp_{k-2}, \ldots, P_2 \in \wp_2, P_1 \in \wp_1$ such that $C(M^*) \cap C(M^*|_{P_k} \triangle P_k) \subseteq C(M^* \cap (P_k \cup P_{k-1} \cup \cdots \cup P_1))$, and so, $P = P_1 \cup P_2 \cup \cdots \cup P_k$ is a heterochromatic $M^*$-augmenting path system starting from $u$, and then, the matching $M' = (M^* \setminus (P_k \cup P_{k-1} \cup \cdots \cup P_1)) \cup ((M^* \cap (P_k \cup P_{k-1} \cup \cdots \cup P_1)) \triangle (P_k \cup P_{k-1} \cup \cdots \cup P_1))$ is heterochromatic that saturates $u$, and hence, $|M'| > |M^*|$, a contradiction to the maximality of $M^*$. The proof is now complete. □

As one can see, Step $k$ is almost the same as Step 2, and so we should have omitted its details. But, for easy of understanding we prefer to repeat them. Also, note that our proof gives an algorithm to find a larger heterochromatic matching from a given one by extending from a series of constructible heterochromatic augmenting path systems, or find a subset $S$ of $X$ such that $|N_c(S)| < |S|$.

In the following we will give an analogue of Tutte's Theorem for general colored graphs. Before proceeding, we need more notations. Let $M$ be a heterochromatic matching of a general colored graph $G$. For $\omega \in C(M)$, define $E_\omega(M) = \{e \colon e \in E(M) \text{ and } e \text{ has the color } \omega\}$. We say that two edges $e_1$ and $e_2$ of $G$ are *connected with respect to the color* $\omega$ if $e_1$, $e_2 \in E_\omega(M)$ and there is an $M$-alternating path from $e_1$ to $e_2$, or there is an $e_3 \in E_\omega(M)$ such that there is an $M$-alternating path from $e_1$ to $e_3$ and another $M$-alternating path from $e_3$ to $e_2$. This sets up a *relation* between the edges of $G$, and it is easy to see that it is an equivalent relation. Let $E_\omega^s = \{e_1, e_2, \ldots, e_s\} \subseteq E_\omega(M)$ be the set of all connected edges with respect to the color $\omega$. Denote by $A$ the set of all $M$-alternating paths starting from an edge $e \in E_\omega^s$. Then, we say that $G[V(A)]$ is a *connected components with respect to the color* $\omega$.

**Theorem 2.** *A colored graph* $(G, C)$ *has a perfect heterochromatic matching, if*

**(1)** $o(G - S) \leq |S|$ *for all* $S \subset V$, *where* $o(G - S)$ *denotes the number of odd components in the remaining graph* $G - S$, *and*

**(2)** $|N_c(S)| \geq |S|$ *for all* $S \subseteq V$ *such that* $0 \leq |S| \leq \frac{|G|}{2}$ *and* $|N(S) \backslash S| \geq |S|$.

*Proof.* Suppose $(G, C)$ is a colored graph satisfying the conditions. Then, condition 1) guarantees that $G$ contains an uncolored perfect matching from [2, Theorem 5.4, p. 76]. In the following we will prove that $G$ also contains a perfect heterochromatic matching by deducing contradictions.

Suppose $G$ does not contain any perfect heterochromatic matching. Let $M^*$ be an uncolored perfect matching of $G$ such that there is a $\omega \in C(M^*)$ satisfying $|E_\omega(M^*)| = \max\limits_{M} \max\limits_{c \in C(M)} |E_c(M)|$, where $M$ runs over all the uncolored perfect matchings of $G$. Let $C_\omega^1, C_\omega^2, \ldots, C_\omega^r$ be the connected components with respect to the color $\omega$.

Next, we will find a set $S \subseteq V(G)$ such that for any $e \in E(M^*)$ we have $|V(e) \cap S| \leq 1$, discussed in the following cases:

*Case 1.* There is a $C_\omega^i$ $(1 \leq i \leq r)$ that has at least two edges in $E_\omega(M^*)$.

Then, there are two edges $e_1 = u_1 v_1$ and $e_2 = u_2 v_2 \in E_\omega(M^*)$ such that there is an $M^*$-alternating path $P$ between them. Assume that the path $P$ is oriented along with the direction of $\overrightarrow{u_1 v_1}$. Then, let $S_1 = \{v_1\} \cup \{u \colon u \in N(v_1)\}$, such that if $e \in E(M^*) \cap E(G[N(v_1)])$ then $|V(e) \cap S_1| = 1$. Find all $M^*$-alternating paths along with the direction of $\overrightarrow{u_1 v_1}$, and then proceed as the following steps:

*Step 1.* $\overrightarrow{u_1 v_1}$ is in an $M^*$-alternating cycle $L$.

Then, define $S_2 = S_1 \cup \{u \colon \text{there is a } v \text{ such that } uv \in E(M^* \cap L) \text{ and } u, v \notin N(u_1)\} \cup \{u \colon \text{there is a } v \text{ such that } uv \in E(M^* \cap L) \text{ and } u \notin N(u_1), v \in N(u_1)\}$, where $S_2$ is obtained by letting $L$ run over all the $M^*$-alternating cycles that contain the oriented edge $\overrightarrow{u_1 v_1}$.

*Step 2.* $\overrightarrow{u_1 v_1}$ is in an $M^*$-alternating path.

Then, check the even numbered vertices on this path. Similarly, by running over all the $M^*$-alternating paths that contain the oriented edge $\overrightarrow{u_1 v_1}$, we define $S_3' = \{u \colon uv \in E(M^*), v \text{ is an even numbered vertex on an } M^*\text{-alternating path, and } v \in N(u_1)\}$, and $S_3'' = \{u \colon u \text{ is an even numbered vertex on an } M^*\text{-alternating path, and there is a } v \notin S_2 \text{ such that } uv \in E(M^*)\}$.

Let $S_3 = S_3' \cup S_3''$ and $S = S_2 \cup S_3$. Then, from the definitions of $S_2$ and $S_3$ and the fact that $M^*$ is an uncolored perfect matching, we have $0 \le |S| \le \frac{|G|}{2}$ and $|N(S) \backslash S| = |S|$. Since $CN(S) \subseteq CN((N(S) \backslash S) \backslash \{u_1\})$ and $|(N(S) \backslash S) \backslash \{u_1\}| < |S|$, we have $|N_c(S)| < |S|$, a contradiction to condition 2).

*Case 2.* All $C_\omega^i$ ($1 \le i \le r$) contain only one edge with the color $\omega$.

Consider a fixed $C_\omega^i$, and let $e_i = u_i v_i$ is the only edge with the color $\omega$. Then, similar to Case 1, along with the direction of $\overrightarrow{u_i v_i}$, we can define sets $S_2$ and $S_3$. Let $S_4 = S_2 \cup S_3$. Then $|N_c(S_4)| = |S_4|$. For another $C_\omega^j$ ($i \ne j$), similar to $C_\omega^i$ we can define the last set $S_5$ such that $|N_c(S_5)| = |S_5|$. Let $S = S_4 \cup S_5$. Then $CN(S) \subseteq CN((N(S) \backslash S) \backslash \{u_i\})$. Since $|(N(S) \backslash S) \backslash \{u_i\}| < |S|$, we have $|N_c(S)| < |S|$, again a contradiction to condition 2). The proof is thus complete. $\square$

To conclude this paper, we give a sufficient and necessary condition to verify if a heterochromatic matching of a general colored graph is maximum, which is an analogue of the well-known Berge's Theorem [2].

**Theorem 3.** *A heterochromatic matching $M$ of a colored graph $(G, C)$ is maximum if and only if there is no heterochromatic $M$-augmenting path system in $G$.*

*Proof.* Let $M$ be a heterochromatic matching of $G$. Suppose there is a heterochromatic $M$-augmenting path system $P$ in $G$. Then, define $M' = (M \backslash P) \cup (M|_P \triangle P)$, and $M'$ is a heterochromatic matching of $G$ such that $|M'| > |M|$, and hence, $M$ is not a maximum heterochromatic matching of $G$.

Conversely, suppose $M$ is not a maximum heterochromatic matching of $G$. Let $M'$ be a maximum heterochromatic matching of $G$. Then, $|M'| > |M|$. Denote $H = G[M \triangle M']$, then every component $p$ of $H$ has one of the following properties:

**1)** the edges of $p$ appear alternately in $M$ and $M'$. Note that a cycle is also regarded as this kind of path;

**2)** the length of $p$ is odd and at least 3, and the edges of $p$ appear alternately in $M$ and $M'$;

**3)** the length of $p$ is 1, and the signal edge of $p$ is in $M$ or $M'$.

Since $\left| M' \right| > |M|$ and both $M'$ and $M$ are heterochromatic, $H$ contains more edges of $M'$ than of $M$, and then, $|M \triangle H| > \left| M' \triangle H \right|$ and $C(M \triangle H) \cap C(M) \subseteq C(M \cap H)$, and hence, we have $|M \triangle H| > |M \cap H|$. It is not difficult to see that $H = \bigcup\limits_{p \text{ is a component of } H} p$ is a heterochromatic $M$-augmenting path system of $G$. $\square$

Our conditions are only sufficient, but not necessary. The following are some simple examples:

*Example 1.* Let $B = (X, Y)$ with $X = \{v_1, v_2, v_3, v_4\}$ and $Y = \{v_5, v_6, v_7, v_8, v_9, v_{10}\}$ be a bipartite graph such that $E = \{v_1 v_5, v_1 v_6, v_2 v_6, v_2 v_7, v_2 v_9, v_3 v_5,$

$v_3v_8$, $v_4v_8$, $v_4v_{10}$} and $C(v_1v_5) = C(v_4v_8) = 1$, $C(v_1v_6) = C(v_2v_9) = 2$, $C(v_2v_6)$
$= C(v_3v_5) = C(v_4v_{10}) = 3$, $C(v_2v_7) = 4$, $C(v_3v_8) = 5$. Then, {$v_1v_6$, $v_2v_7$, $v_3v_8$,
$v_4v_{10}$} is a heterochromatic matching that saturates every vertex of $X$. But, if
we take $S = \{v_1, v_3, v_4\}$, then $N_c(S) = \{v_6, v_8\}$, and so, $|N_c(S)| < |S|$, which
tells us that the condition in our Theorem 1 is not necessary. This very example
is also valid for showing the un-necessity of the conditions in our Theorem 1.

*Example 2.* For non-bipartite graph, let $V = \{v_1, v_2, v_3, v_4, v_5, v_6\}$ and $E = \{v_1v_2, v_1v_4, v_1v_5, v_2v_3, v_2v_6, v_3v_4, v_3v_6, v_4v_5, v_5v_6\}$ such that $C(v_1v_2) = C(v_2v_6) = C(v_3v_4) = C(v_4v_5) = C(v_1v_4) = 1$, $C(v_2v_3) = C(v_3v_6) = 2$, $C(v_1v_5) = C(v_5v_6) = 3$. Take $S = \{v_1, v_4, v_5\}$. Then $N_c(S) = \{v_2, v_6\}$, and
so the conditions in our Theorem 2 are not necessary.

From these simple examples, we can see that our conditions of both theorems
need great improvement. Moreover, how to find both sufficient and necessary
conditions for the existence of perfect heterochromatic matchings in general colored graphs is a more interesting question.

## Acknowledgements

This work was supported by NSFC and the 973 project. The authors would like
to thank H. Chen and J. Tu for their helpful discussions.

## References

1. Alon, N., Jiang, T., Miller, Z., Pritikin, D.: Properly Colored Subgraphs and Rainbow Subgraphs in Edge-Colored Graphs with Local Constraints. Random Struct.
   Algorithoms **23** (4) (2003) 409–433
2. Bondy, J.A., Murty, U.S.R., Graph Theory with Applications, Macmillan, London,
   Elsevier, New York (1976)
3. Broersma, H.J., Li, X., Wöginger, G., Zhang, S.: Paths and Cycles in Colored
   Graphs, Australasian J. Combin. **31** (2005) 299–312
4. Garey, M.R., Johnson, D.S.: Computers and Intractabilty, Freeman, New York
   (1979)
5. Jamison, R., Jiang, T., Ling, A.: Constrained Ramsey Numbers of Graphs. J. Graph
   Theory **42** (1) (2003) 1–16
6. Jin, Z., Li, X.The Complexity for Partitioning Graphs by Monochromatic Trees,
   Cycles and Paths. Intern. J. Computer Math. **81** (11) (2004) 1357–1362
7. Lawler, E.L.: Combinatorial Optimization: Networks and Matroids. Holt, Rinehart
   and Winston, New York-Montreal Que.-London (1976)

# Impossibility of Transformation of Vertex Labeled Simple Graphs Preserving the Cut-Size Order

Hiro Ito

Department of Communications and Computer Engineering, School of Informatics, Kyoto University, Kyoto 606-8501, Japan
itohiro@i.kyoto-u.ac.jp

**Abstract.** Three partial orders, cut-size order, length order, and operation order, defined between labeled graphs with the same number of vertices are known to be equivalent. From the equivalence, $G$ precedes $G'$ in the order means that there is a sequence of cross-operations transforming $G$ into $G'$. However, even if both graphs are simple, non-simple graphs may appear on the way of the transformation. If both graphs have the same number of edges, we conjecture that there is a sequence in which only simple graphs appear. But for graphs having different number of edges, there is a counter example. That is, there is a pair of simple graphs $G$ and $G'$ such that $G$ precedes $G'$ in the order and $G$ can not be transformed into $G'$ by using only simple graphs. Thus we must introduce other operations than cross-operation for transforming them by using only simple graphs. Then we naturally reach to a question: is there a sufficient set of operations for this purpose? For this problem, this paper shows a negative result that there is no such finite set of operations.

## 1 Introduction

Let $N = \{x_0, x_1, \ldots, x_{n-1}\}$ be the set of vertices of a convex polygon $P$ in the plane, where the vertices are arranged in this order counter-clockwisely, and hence $(x_i, x_{i+1})$ is an edge of $P$ for $i = 0, 1, \ldots, n-1$ (We adopt the residue class on $n$ for treating integers in $N$, i.e., $i \pm j$ is $i' \in N$ such that $i' \equiv i \pm j$ (mod $n$)). An internal angle of $P$ may be $\pi$. We consider graphs whose node set corresponds to $N$, i.e., the node set is $\{0, 1, \ldots, n-1\}$ and each node $i$ is assigned to $x_i$, and each edge $e = (i, j)$ of the graph is represented by a line segment $x_i x_j$.

We adopt the cyclic order for treating integers (or numbered vertices) in $N$. Thus for $i, j \in N$,

$$[i, j] = \begin{cases} \{i, \ i+1, \ldots, j\}, & \text{if } i \leq j, \\ \{i, \ i+1, \ldots, \ n-1, 0, 1, \ldots, j\}, & \text{if } i > j; \end{cases}$$

for $i, j, k \in N$, $i \leq j \leq k$ means $j \in [i, k]$; for $i, j, k, h \in N$, $i \leq j \leq k \leq h$ means that $i, j, k, h$ appear in this order when we traverse the nodes of $[i, h]$

J. Akiyama et al. (Eds.): CJCDGCGT 2005, LNCS 4381, pp. 59–69, 2007.

from $i$ to $h$. For notational simplicity, $\{i\}$ may be written as $i$. For a graph $G$, $E(G)$ means the edge set of $G$.

In this paper all graphs are regarded as weighted graphs, i.e., we introduce a weight function $w_G\colon E(G) \to \boldsymbol{R}$ and a weighted graph $G$ always has a weight function $w_G$ in this paper. A simple graph is a graph which has a weight function $w_G(e) = 1$ for all $e \in E(G)$. The *size* of a graph $G$ is denoted by $\sum\{w_G(e) \mid e \in E(G)\}$. Note that it equals the number of edges for simple graphs.

Three relations, cut-size order, length order, and operation order, were introduced between vertex-labeled graphs in [3] and shown that they are equivalent [3,4].

From the equivalence of the three relations, $G$ precedes $G'$ in the order means that there is a sequence of cross-operations transforming $G$ into $G'$. However, even if both $G$ and $G'$ are simple graphs, non-simple graphs may appear on the way of the transformation. In fact, there are two simple graphs $G$ and $G'$ such that $G$ precedes $G'$ in the order and $G$ can not be transformed into $G'$ by using only simple graphs. Thus we must introduce other operations than cross-operation for transforming them by using only simple graphs. Thus we reach a question: Is there a sufficient set of operations for this purpose? To treat this problem rigorously, we must define what are operations. We define that an operation consists of finite number of adding an edge and removing an edge. Under this definition, this paper shows a negative result that there is no efficient finite set of operations.

## 2    Three Equivalent Relations

We introduce some terms as follows.

*Linear cuts.* For a graph $G$ and a pair of distinct nodes $i, j \in N$, a *linear cut* $C_G(i, j)$ is an edge set:

$$C_G(i, j) = \{(k, h) \in E(G) \mid k \in [i, j-1], \ h \in [j, i-1]\}.$$

Fig. 1 shows examples of linear cuts. The capacity of a linear cut $C_G(i, j)$ is defined as

$$c_G(i, j) = \sum_{e \in C_G(i, j)} w_G(e).$$

For two subsets $N'$ and $N''$ of nodes,

$$w_G(N', N'') = \sum_{i \in N', j \in N''} w_G(i, j).$$

The *degree* of a node $i \in N$ of a graph $G$ is defined as $c_G(i, i+1) = w_G(i, [i+1, i-1])$ and may be simply denoted by $d_G(i)$. As a generalization of degree, $d_G(N')$ denotes $w_G(N', N - N')$ for $N' \subset N$. From them, $c_G(i, j) = d_G([i, j-1])$, since they mean the same thing. We introduce a relation based on sizes of linear cuts as follows. For two weighted graphs $G$ and $G'$, $G \preceq_c G'$

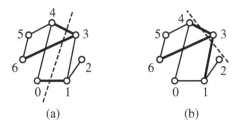

**Fig. 1.** Linear cuts: (a) $C_G(1, 4)$, (b) $C_G(3, 4)$

means that $c_G(i, j) \leq c_{G'}(i, j)$ for all $i, j \in N$, $i \neq j$. This relation is known to be a partial order. This result is easily obtained from the following property presented by Skiena [5].

**Theorem 1 (Skiena [5]).** *For two weighted graphs $G$ and $G'$, if $c_G(i, j) = c_{G'}(i, j)$ for all $i, j \in N$, $i \neq j$, then $G = G'$.*

*Sum of edge lengths.* For an edge $(i, j)$ of a weighted graph $G$ and a convex $n$-gon $P$, let $\mathrm{dist}(i, j)$ be the length of the line segment $x_i x_j$. We define a sum of weighted edge length of $G$ with respect to $P$ as

$$s_P(G) = \sum_{(i,j) \in E(G)} w(i, j) \cdot \mathrm{dist}(i, j).$$

We introduce a relation based on this measure as follows. For two weighted graphs $G$ and $G'$, $G \preceq_l G'$ means that $s_P(G) \leq s_P(G')$ for all convex $n$-gons $P$.

*Cross-operations.* We introduce an operation transforming a graph to another one. For a weighted graph $G$, two distinct $i, j \in N$ and a real value $\Delta$, $\mathrm{ADD}_G(i, j; \Delta)$ means adding $\Delta$ to $w(i, j)$ (if $(i, j) \notin E(G)$, adding an edge $(i, j)$ to $E(G)$ previously). The reverse operation of ADD can be defined, i.e., $\mathrm{REMOVE}_G(i, j; \Delta)$ means $\mathrm{ADD}_G(i, j; -\Delta)$. We extend these operations in the case $i = j$, i.e., both $\mathrm{ADD}_G(i, i; \Delta)$ and $\mathrm{REMOVE}_G(i, i; \Delta)$ mean doing nothing. For nodes $i, j, k, h \in N$ with $i \leq j \leq k \leq h$ and a positive $\Delta > 0$ (see, Fig. 2), a *cross-operation* $X_G(i, j, k, h; \Delta)$ is applying

$\mathrm{REMOVE}_G(i, j; \Delta), \mathrm{REMOVE}_G(k, h; \Delta), \mathrm{ADD}_G(i, k; \Delta),$ *and* $\mathrm{ADD}_G(j, h; \Delta)$.

If some of $\{i, j, k, h\}$ are equal, a cross-operation may increase edges. In fact, if $i = j < k < h < i$ or $i = j < k = h < i$ (or the cases symmetric with respect to one of them), then the total edge weights increases (see (a) and (b) of Fig. 3). If $j = k$ or $i = h$, the edge set is not changed (see (c) and (d) of Fig. 3). We introduce a relation based on cross-operations as follows. For two weighted graph $G$ and $G'$, $G \preceq_o G'$ means that $G'$ can be obtained from $G$ by applying finite number (including zero) of cross-operations.

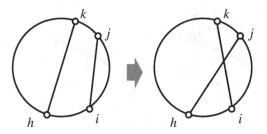

**Fig. 2.** Cross-operation $X(i, j, k, h; 1)$

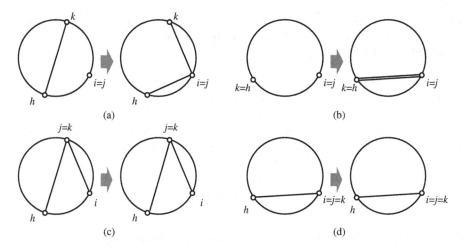

**Fig. 3.** These cross-operations $X(i, j, k, h; 1)$ when some of nodes are the same

We have the following theorem on these three relations.

**Theorem 2.** *Three relations* $\preceq_c$, $\preceq_l$, *and* $\preceq_o$ *are equivalent.*

The proof of this theorem was shown in [3] for graphs with the same number of edges, and in [4] for the general case.

From Theorem 2, these three partial orders can be denoted by $\preceq$ simply.

## 3　Operations Preserving Simpleness

Cross-operations sometimes violate the simpleness of graphs. Let $G$ and $G'$ be a pair of weighted graphs with $G \preceq G'$. This means there is a sequence of weighted graphs

$$G = G_0 \preceq G_1 \preceq \cdots \preceq G_k = G',$$

such that $G_i$ is obtained by applying a cross-operation on $G_{i-1}$ for $i = 1, 2, \ldots, k$. Even if $G$ and $G'$ are both simple, there may be a nonsimple graph on the sequence.

Cross-operations are very similar to 2-switches, presented by Hakimi [1,2] and developed by West [6]. Let $G$ be a graph and $(i, j)$ and $(k, h)$ be two distinct edges ($i$, $j$, $k$, $h$ are all distinct) of $G$. A *2-switch* of $G$ consists of removing $(i, j)$ and $(k, h)$ from $G$ and adding two edges $(i, k)$ and $(j, h)$. Note that if $(i, k)$ or $(j, h)$ are edges of the original $G$, the new graph includes parallel edges. The following result is known [6].

**Theorem 3 (West [6]).** *If $G$ and $G'$ are two simple graphs with a node set $N$, then $d_G(i) = d_{G'}(i)$ for all $i \in N$ if and only if there is a sequence of 2-switches such that it transforms $G$ to $G'$ and only simple graphs are appeared on the way of the transformation.*

Here we return to the subject on cross-operations. We have a conjecture that simple graphs are enough in this case too.

*Conjecture 1.* If $G$ and $G'$ are two simple graphs with a node set $N$ with $d_G(i) = d_{G'}(i)$ for all $i \in N$ then $G \preceq G'$ if and only if there is a sequence of cross-operations such that it transforms $G$ to $G'$ and only simple graphs are appeared on the way of the transformation.

The "if" part is directly obtained from Theorem 2, but "only if" part is open. The conjecture considers a pair of graphs with the same number of edges. However, Theorem 2 assures us cross-operations can transform any graph $G = (N, E)$ to any other graph $G' = (N, E')$ as long as $G \preceq G'$ even if $|E| < |E'|$.

Here we reached another question: Does Conjecture 1 hold even if the degree condition is deleted? To this question, we have a simple counter example shown in Fig. 4. If we don't mind breaking the simpleness, there is a sequence $X(0, 0, 1, 3; 0.5)$, $X(2, 2, 3, 1; 0.5)$, $X(1, 1, 2, 0; 0.5)$, $X(3, 3, 0, 2; 0.5)$ as shown in Fig. 5. However there is no efficient cross-operation that does not break the simpleness. Thus we should introduce other operations to complete transformations in general case. The example of Fig. 4 can be solved if we introduce a new operation, a *cross-square-operation*, consisting of REMOVE($i$, $k$), REMOVE($j$, $h$), ADD($i$, $j$), ADD($j$, $k$), ADD($k$, $h$), and ADD($h$, $i$), for $i \leq j \leq k \leq h$. Precisely, it may not be sufficient for general case. If we find another counter example, which can not be solved yet, then we should add another new operation.

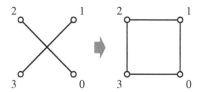

**Fig. 4.** Counter example

Then we reach another question: Is there a finite set of operations that transforms general pair of simple graphs without breaking simpleness of the graphs?

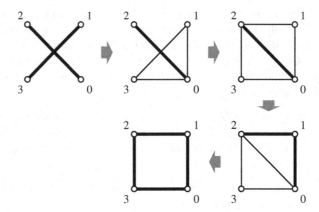

**Fig. 5.** Cross-operations: bold edges have weight 1 and thin edges have weight 0.5

For treating this question, we must define what is an "operation". We introduce the following natural definition.

**Definition 1.** *An* operation *is a set of finite number of* ADDs *and* REMOVEs, *which does not decrease the size of any linear-cut.*

For example, the cross-operation consists of two ADDs and two REMOVEs, and it does not decrease the size of any linear cut. The cross-square-operation consists of four ADDs and two REMOVEs, and it does not decrease the size of any linear cut also. Thus they satisfy the above definition.

Let $O$ be a finite set of operations and let $G$ and $G'$ be simple graphs with $G \preceq G'$. A *simple transformation sequence* from $G$ to $G'$ with $O$ is a sequence of simple graphs

$$G = G_0 \prec G_1 \prec \cdots \prec G_k = G'$$

such that $G_i$ ($i \in \{1, 2, \ldots, k\}$) is obtained from $G_{i-1}$ by applying an operation in $O$. The main theorem of this paper is the following.

**Theorem 4.** *There is no finite set of operations such that there is a simple transformation sequence from $G$ to $G'$ with the set, for any pair of simple graphs $G$ and $G'$ with $G \preceq G'$.*

## 4    Proof of Theorem 4

For proving Theorem 4, we define an *even-complete graph* $K_n^e$ and an *odd-complete graph* $K_n^o$ of even order $n$ as follows (see Fig. 6):

$$K_n^e = (N, E_n^e = \{(i, j) \mid i, j \in N, j - i \text{ is even}\}),$$

$$K_n^o = (N, E_n^o = \{(i, j) \mid i, j \in N, j - i \text{ is odd}\}).$$

By comparing the sizes of linear-cuts on them, we see that

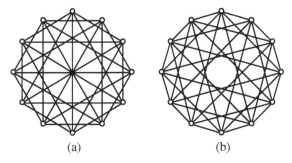

(a)                    (b)

**Fig. 6.** (a) $K^e_{12}$ and (b) $K^o_{12}$

$$c_{K^e_n}(i, j) < c_{K^o_n}(i, j) \quad \text{if } j - i \text{ is odd, and} \tag{1}$$

$$c_{K^e_n}(i, j) = c_{K^o_n}(i, j) \quad \text{if } j - i \text{ is even.} \tag{2}$$

Hence,

$$K^e_n \prec K^o_n.$$

We will show that $K^e_n$ can not be transformed into $K^o_n$ if $n$ is large enough as follows.

*Proof of Theorem 4:* Suppose otherwise, i.e., there is a finite set of operations $O$ satisfying the condition of the theorem. For an operation $o$ in $O$, the *order* of $o$ is the number of nodes such that one or more edges incident to the node are added or removed by $o$. For example, the order of cross-operation and the one of cross-square-operation are both four.

Let $k^*$ be the maximum order of the operations in $O$. Let $n \geq 2k^*$ be an even number. We will prove that any operation in $O$ can not transform $K^e_n$ to another graph $G$ such that $K^e_n \prec G \preceq K^o_n$. (This is enough to prove the theorem.)

Assume otherwise, i.e., there is an operation $o \in O$ that transform $K^e_n$ to another graph $G$ such that $K^e_n \prec G \preceq K^o_n$. We construct a $\{-1, 1\}$-weighted graph $G_o$ as follows. $G_o$ has a node set $N$, an edge set

$$E_o = \{(i, j) \mid \text{ADD}(i, j) \text{ or REMOVE}(i, j) \text{ is in } o\}$$

and a weight function $w_o$ such that $w_o(i, j) = 1$ if $\text{ADD}(i, j) \in o$ and $w_o(i, j) = -1$ if $\text{REMOVE}(i, j) \in o$.

We can easily see that the edge set of $G$ is

$$E = \{(i, j) \mid w_{E^e_n}(i, j) + w_o(i, j) = 1\}.$$

(Note that the weights of a simple graph are defined as 1 if there is an edge and 0 otherwise.) From $G \preceq K^o_n$, every linear-cut on $G_1$ has the size zero or one.

Let $i_0, i_1, \ldots, i_{k-1}$ ($i_0 < i_1 < \cdots < i_{k-1}$) be the nodes which have at least one edge incident to the nodes in $G_o$, where $k$ is the order of $o$. Since $k \leq n/2$, we can assume $i_0 - i_{k-1} \geq 2$ without loss of generality. (Note that the residue class is used for the difference.) Thus for any $j \in \{1, \ldots, k-1\}$ there is a linear-cut separating $i_0, \ldots, i_{j-1}$ and $i_j, \ldots, i_{k-1}$ such that the size on both $K_n^e$ and $K_n^o$ are the same, and hence, we see that the size of these linear-cuts on $G_o$ is zero, i.e.,

$$c_{G_o}(i_0, i_j) = 0, \quad \text{for every } j \in \{1, \ldots, k-1\}. \tag{3}$$

Thus

$$0 = c_{G_o}(i_0, i_1) = w_{G_o}(i_0, i_1) + w_{G_1}(i_0, [i_2, i_{k-1}]),$$

$$0 = c_{G_o}(i_0, i_2) = w_{G_o}(i_1, [i_2, i_{k-1}]) + w_{G_o}(i_0, [i_2, i_{k-1}])$$

(see, Fig. 7). Therefore,

$$w_{G_o}(i_0, i_1) = w_{G_o}(i_1, [i_2, i_{k-1}]). \tag{4}$$

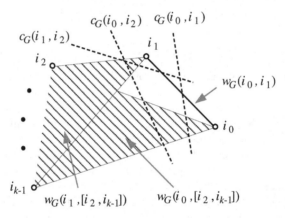

**Fig. 7.** $c_{G_o}(i_0, i_1)$, $c_{G_o}(i_0, i_2)$ and $c_{G_o}(i_1, i_2)$

Moreover,

$$0 \le c_{G_o}(i_1, i_2) = w_{G_o}(i_0, i_1) + w_{G_o}(i_1, [i_2, i_{k-1}]) \le 1 \tag{5}$$

(see, Fig. 7). From (4), (5), and that all edge weights are integers, we obtain

$$w_{G_o}(i_0, i_1) = w_{G_o}(i_1, [i_2, i_{k-1}]) = 0. \tag{6}$$

For an assumption for an induction, we assume that

$$w_{G_o}(i, j) = 0, \quad \text{for all } i, j \in \{i_0, \ldots, i_q\}. \tag{7}$$

By using this assumption, we will derive the result such that

$$w_{G_o}(i, j) = 0, \quad \text{for all } i, j \in \{i_0, \ldots, i_{q+1}\}$$

as follows:

$$a_h = w_{G_o}(i_h, i_{h+1}), \quad \text{for } h = 0, \ldots, q,$$

$$A_h = w_{G_o}(i_h, [i_{q+2}, i_{k_1-1}]), \quad \text{for } h = 0, \ldots, q+1,$$

(see, Fig. 8). From (7),

$$0 = c_{G_o}(i_0, i_1) = a_0 + A_0,$$

$$0 = c_{G_o}(i_0, i_2) = a_0 + a_1 + A_0 + A_1 = a_1 + A_1,$$

$$0 = c_{G_o}(i_0, i_3) = a_0 + a_1 + a_2 + A_0 + A_1 + A_2 = a_2 + A_2,$$

$$\vdots$$

Generally, we have

$$0 = a_h + A_h, \quad \text{for } h = 0, \ldots, q. \tag{8}$$

By adding them from $h = 1$ to $q$,

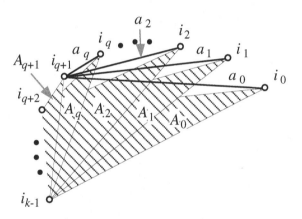

**Fig. 8.** $A_1, \ldots, A_{q+1}, a_1, \ldots, a_q$

$$0 = \sum_{h=0}^{q} (a_h + A_h) \tag{9}$$

Moreover,

$$0 = c_{G_o}(i_0, i_{q+2}) = \sum_{h=0}^{q+1} A_h. \tag{10}$$

By comparing (9) and (10), we have

$$\sum_{h=0}^{q} a_h = A_{q+1}. \tag{11}$$

The capacity of any linear-cut of $G_1$ is between zero and one, and hence

$$0 \leq c_{G_o}(i_{q+1}, i_{q+2}) = \sum_{h=0}^{q} a_h + A_{q+1} \leq 1. \tag{12}$$

From (11), (12), and that all edge weights are integers, we see

$$\sum_{h=0}^{q} a_h = A_{q+1} = 0. \tag{13}$$

From (13) and (8), we get

$$\sum_{h=0}^{q-1} a_h = A_q. \tag{14}$$

Moreover,

$$0 \leq c_{G_o}(i_q, i_{q+2}) = \sum_{h=0}^{q-1} a_h + A_q + A_{q+1} \leq 1.$$

Since equation 13,

$$0 \leq \sum_{h=0}^{q-1} a_h + A_q \leq 1. \tag{15}$$

From (14), (15), and that all edge weights are integers, we see

$$\sum_{h=0}^{q-1} a_h = A_q = 0. \tag{16}$$

By using similar arguments, we finally obtain the following equation:

$$A_q = A_{q-1} = \cdots = A_0 = 0. \tag{17}$$

From this and (8), we see

$$a_q = a_{q-1} = \cdots = a_0 = 0. \tag{18}$$

That is

$$w_{G_o}(i, j) = 0, \quad \text{for all } i, j \in \{i_0, \ldots, i_{q+1}\}.$$

By induction, we conclude that there is no edge between any pair of nodes in $\{i_0, i_1, \ldots, i_{k-1}\}$, i.e., $o$ consists of no ADD and REMOVE, contradiction. □

## 5   Concluding Remarks

This paper considers graph transformation preserving the partial order, which is equivalent to cut-size order, and simpleness of graphs simultaneously. For this problem, we proved that there is no efficient finite set of operations for the partial order's transformation in the general case. If the number of edges of the graphs are the same, we conjecture that a simple transformation sequence with cross-operations always exists (Conjecture 1). It remains for future research.

# References

1. Hakimi, S.L.: On Realizability of a Set of Integers as Degrees of the Vertices of a Linear Graph, I. J. Soc. Indust. Appl. Math. **10** (1962) 496–506
2. Hakimi, S.L.: On Realizability of a Set of Integers as Degrees of the Vertices of a Linear Graph II. J. Soc. Indust. Appl. Math. **11** (1963) 135–147
3. Ito, H.: Sum of Edge Lengths of a Multigraph Drawn on a Convex Polygon, Comput. Geom. **24** (2003) 41–47
4. Ito, H.: Three Equivalent Partial Orders on Graphs with Real Edge-Weights Drawn on a Convex Polygon. In: Discrete and Computational Geometry. Lecture Notes in Computer Science, Vol. 3742. Springer, Berlin, (2005) 123–130
5. Skiena, S.S.: Reconstructing Graphs from Cut-Set Sizes. Inform. Process. Lett. **32** (1989) 123–127
6. West, D.B.: Introduction to Graph Theory. Prentice Hall, Upper Saddle River, New Jersey (1996)

# A Neighborhood Condition for Graphs to Have [a, b]-Factors III

M. Kano[1] and Haruhide Matsuda[2]

[1] Department of Computer and Information Science, Ibaraki University
Hitachi, Ibaraki, 316-8511 Japan
kano@mx.ibaraki.ac.jp
[2] The Research Institute of Educational Development, Tokai University
2-28-4, Tomigaya, Shibuya-ku, Tokyo, 151-8611 Japan
hmatsuda@ried.tokai.ac.jp

**Abstract.** Let $a$, $b$, $k$, and $m$ be positive integers such that $1 \leq a < b$ and $2 \leq k \leq (b + 1 - m)/a$. Let $G = (V(G), E(G))$ be a graph of order $|G|$. Suppose that $|G| > (a+b)(k(a+b-1)-1)/b$ and $|N_G(x_1) \cup N_G(x_2) \cup \cdots \cup N_G(x_k)| \geq a|G|/(a+b)$ for every independent set $\{x_1, x_2, \ldots, x_k\} \subseteq V(G)$. Then for any subgraph $H$ of $G$ with $m$ edges and $\delta(G - E(H)) \geq a$, $G$ has an $[a, b]$-factor $F$ such that $E(H) \cap E(F) = \emptyset$. This result is best possible in some sense and it is an extension of the result of Matsuda (Discrete Mathematics **224** (2000) 289–292).

**Keywords:** Factor, neighborhood union, $[a, b]$-factor.

**AMS Subject Classifications:** 05C70.

## 1 Introduction

We consider finite undirected graphs without loops or multiple edges. Let $G$ be a graph with vertex set $V(G)$ and edge set $E(G)$. We denote by $|G|$ the order of $G$. For a vertex $v$ of $G$, let $\deg_G(v)$ and $N_G(v)$ denote the degree of $v$ in $G$ and the neighborhood of $v$ in $G$, respectively. Furthermore, $\delta(G)$ denotes the minimum degree of $G$, and $N_G(S) = \bigcup_{x \in S} N_G(x)$ for $S \subset V(G)$. We write $N_G[v]$ for $N_G(v) \cup \{v\}$. For two disjoint vertex subsets $A$ and $B$ of $G$, the number of edges of $G$ joining $A$ to $B$ is denoted by $e_G(A, B)$. For a subset $S \subset V(G)$, let $G - S$ denote the subgraph of $G$ induced by $V(G) - S$.

Let $a$ and $b$ be integers such that $1 \leq a \leq b$. An $[a, b]$-factor of $G$ is a spanning subgraph $F$ of $G$ such that

$$a \leq \deg_F(x) \leq b, \quad \text{for all } x \in V(G).$$

Note that if $a = b$, then an $[a, b]$-factor is a regular $a$-factor.

## 2 Background and Results

The following results on a $k$-factor are known.

J. Akiyama et al. (Eds.): CJCDGCGT 2005, LNCS 4381, pp. 70–78, 2007.

**Theorem 1. (Iida and Nishimura [1])** *Let $k \geq 2$ be an integer and let $G$ be a connected graph of order $|G|$ such that $|G| \geq 9k - 1 - 4\sqrt{2(k-1)^2 + 2}$, $k|G|$ is even, and $\delta(G) \geq k$. If $G$ satisfies $|N_G(x) \cup N_G(y)| \geq (|G| + k - 2)/2$ for all non-adjacent vertices $x$ and $y$ of $G$, then $G$ has a $k$-factor.*

**Theorem 2. (Niessen [4])** *Let $G$ be a connected graph of order $|G|$ and $\delta(G) \geq k \geq 2$, where $k$ is an integer with $k|G|$ is even and $|G| \geq 8k - 7$. If $|N_G(x) \cup N_G(y)| \geq |G|/2$ for all non-adjacent vertices $x$ and $y$ of $G$, then $G$ has a $k$-factor or $G$ belongs to some exceptional families.*

One of the authors showed a neighborhood condition for the existence of an $[a, b]$-factor.

**Theorem 3. (Matsuda [5])** *Let $a$ and $b$ be integers such that $1 \leq a < b$ and let $G$ be a graph of order $|G|$ with $|G| \geq 2(a + b)(a + b - 1)/b$ and $\delta(G) \geq a$. If*

$$|N_G(x) \cup N_G(y)| \geq \frac{a|G|}{a + b}$$

*for any two non-adjacent vertices $x$ and $y$ of $G$, then $G$ has an $[a, b]$-factor.*

The following theorem gurantees the existence of an $[a, b]$-factor which includes some specified edges.

**Theorem 4. (Matsuda [6])** *Let $a$, $b$, $m$, and $t$ be integers such that $1 \leq a < b$ and $2 \leq t \leq \lceil (b - m + 1)/a \rceil$. Suppose that $G$ is a graph of order $|G| > ((a + b)(t(a + b - 1) - 1) + 2m)/b$ and $\delta(G) \geq a$. Let $H$ be any subgraph of $G$ with $|E(H)| = m$. If*

$$|N_G(x_1) \cup N_G(x_2) \cup \cdots \cup N_G(x_t)| \geq \frac{a|G| + 2m}{a + b}$$

*for every independent set $\{x_1, x_2, \ldots, x_t\} \subseteq V(G)$, then $G$ has an $[a, b]$-factor including $H$.*

In this paper, we prove the following two theorems for the existence of an $[a, b]$-factor which excludes some specified edges.

**Theorem 5.** *Let $a$, $b$, $m$, and $k$ be positive integers such that $1 \leq a < b$ and $2 \leq k < (a + b + 1 - m)/a$. Let $G$ be a graph with $|G| > (a + b)((k + m)(a + b - 1) - 1)/b$. If*

$$|N_G(x_1) \cup N_G(x_2) \cup \cdots \cup N_G(x_k)| \geq \frac{a|G|}{a + b} \tag{1}$$

*for every independent set $\{x_1, x_2, \ldots, x_k\} \subseteq V(G)$, then for any subgraph $H$ of $G$ with $m$ edges and $\delta(G - E(H)) \geq a$, $G$ has an $[a, b]$-factor $F$ excluding $H$ (i.e., $E(H) \cap E(F) = \emptyset$).*

The condition (1) is best possible in the sense that we cannot replace $a|G|/(a+b)$ by $a|G|/(a + b) - 1$, which is shown in the following example: Let $t \geq 2m$ be a sufficiently large integer. Consider the join of two graphs $G = A + B$, where $A$ consists of $at - 2m$ isolated vertices and $m$ independent edges, and $B$ consists of $bt + 1$ isolated vertices. Then it follows that $|G| = |A| + |B| = (a + b)t + 1$ and

$$\frac{a|G|}{a+b} > |N_G(x_1) \cup N_G(x_2) \cup \cdots \cup N_G(x_k)| = at > \frac{a|G|}{a+b} - 1$$

for a subset $\{x_1, x_2, \ldots, x_k\} \subseteq B$ with $2 \leq k < (a + b + 1 - m)/a$. However, $G$ has no $[a, b]$-factor excluding the $m$ edges in $A$ because $b|A| < a|B|$.

The next theorem corresponds to the case $k = 1$ of Theorem 5.

**Theorem 6.** *Let $a$, $b$, and $m$ be integers such that $1 \leq a < b$ and $m \geq 1$. Suppose that $G$ is a graph with $\delta(G) \geq a|G|/(a + b)$ and $|G| > (a + b)((m + 1)(a + b + 1) - 5)/b$. Then for any subgraph $H$ of $G$ with $m$ edges, $G$ has an $[a, b]$-factor excluding $H$.*

## 3    Proofs of Theorems 5 and 6

For a vertex $v$ and a vertex subset $T$ of $G$, for convenience, we write $N_T(v)$ and $N_T[v]$ for $N_G(v) \cap T$ and $N_G[v] \cap T$, respectively. Our proofs of the theorems depend on the following criterion.

**Theorem 7. (Lam etc. [2])** *Let $1 \leq a < b$ be integers, and let $G$ be a graph and $H$ a subgraph of $G$. Then $G$ has an $[a, b]$-factor $F$ such that $E(H) \cap E(F) = \emptyset$ if and only if*

$$b|S| + \sum_{x \in T} \deg_{G-S}(x) - a|T| \geq \sum_{x \in T} \deg_H(x) - e_H(S, T)$$

*for all disjoint subsets $S$ and $T$ of $V(G)$.*

*Proof of Theorem 5.* Suppose that $G$ satisfies the assumption of the theorem, but has no desired $[a, b]$-factor for some subgraph $H$ with $m$ edges and $\delta(G - H) \geq a$. Then by Theorem 7, there exist two disjoint subsets $S$ and $T$ of $V(G)$ such that

$$b|S| + \sum_{x \in T} \left( \deg_{G-S}(x) - \deg_H(x) + e_H(x, S) - a \right) \leq -1. \tag{2}$$

We choose such subsets $S$ and $T$ so that $|T|$ is minimum.

**Claim 1.** $|S| \geq 1$.

If $S = \emptyset$, then by (2) we obtain

$$-1 \geq \sum_{x \in T} \left( \deg_G(x) - \deg_H(x) - a \right) \geq \sum_{x \in T} (\delta(G - E(H)) - a) \geq 0,$$

which is a contradiction.

**Claim 2.** $|T| \geq b+1$.

Suppose that $|T| \leq b$. Since $|S| + \deg_{G-S}(x) - \deg_H(x) \geq \deg_{G-H}(x) \geq \delta(G - E(H)) \geq a$ for all $x \in T$, it follows from (2) that

$$-1 \geq b|S| + \sum_{x \in T} \left( \deg_{G-S}(x) - \deg_H(x) + e_H(x, S) - a \right)$$

$$\geq \sum_{x \in T} \left( |S| + \deg_{G-S}(x) - \deg_H(x) + e_H(x, S) - a \right)$$

$$\geq 0.$$

This is a contradiction.

**Claim 3.** $\deg_{G-S}(x) - \deg_H(x) + e_H(x, S) \leq a - 1$, *for all* $x \in T$.

Suppose that there exists a vertex $u \in T$ such that $\deg_{G-S}(u) - \deg_H(u) + e_H(u, S) \geq a$. Then the subsets $S$ and $T - \{u\}$ satisfy (2), which contradicts the choice of $T$. Hence the claim holds.

By Claim 3, we obtain

$$|N_T[x]| \leq \deg_{G-S}(x) + 1 \leq \deg_H(x) - e_H(x, S) + a \qquad \text{for all } x \in T.$$

Now we obtain a set $\{x_1, x_2, \ldots, x_k\}$ of independent vertices of $G$ as follows: First define

$$h_1 = \min\left\{ \deg_{G-S}(x) - \deg_H(x) + e_H(x, S) \mid x \in T \right\},$$

and choose $x_1 \in T$ such that $\deg_{G-S}(x_1) - \deg_H(x_1) + e_H(x_1, S) = h_1$ and $\deg_H(x_1) - e_H(x_1, S)$ is minimum. Next, for $i = 2, \ldots, k$, where $k < (a + b + 1 - m)/a$, we define

$$h_i = \min\left\{ \deg_{G-S}(x) - \deg_H(x) + e_H(x, S) \mid x \in T - \bigcup_{j=1}^{i-1} N_T[x_j] \right\},$$

and choose $x_i \in T - \bigcup_{j=1}^{i-1} N_T[x_j]$ such that $\deg_{G-S}(x_i) - \deg_H(x_i) + e_H(x_i, S) = h_i$ and $\deg_H(x_i) + e_H(x_i, S)$ is minimum. Then we have $h_1 \leq h_2 \leq \cdots \leq h_k \leq a - 1$ by Claim 3 and we have $\sum_{i=1}^{k} \deg_H(x_i) \leq m$ since $|E(H)| = m$ and $\{x_1, x_2, \ldots, x_k\}$ is an independent set of $G$. Note that by Claim 3 and $k < (a + b + 1 - m)/a$, we have

$$\left| \bigcup_{j=1}^{k-1} N_T[x_j] \right| \leq \sum_{j=1}^{k-1} (\deg_{G-S}(x_j) + 1) \leq \sum_{j=1}^{k-1} (a + \deg_H(x_i))$$

$$\leq a(k-1) + |E(H)|$$

$$\leq a(k-1) + m$$

$$< b + 1 \leq |T|.$$

Hence we can take an independent set $\{x_1, x_2, \ldots, x_k\}$.

By the condition of Theorem 5, the following inequalities hold:

$$\frac{a|G|}{a+b} \le |N_G(x_1) \cup N_G(x_2) \cup \cdots \cup N_G(x_k)|$$

$$\le \sum_{i=1}^{k} \deg_{G-S}(x_i) + |S|$$

$$\le \sum_{i=1}^{k} (h_i + \deg_H(x_i) - e_H(x_i, S)) + |S|,$$

which implies

$$|S| \ge \frac{a|G|}{a+b} - \sum_{i=1}^{k} (h_i + \deg_H(x_i) - e_H(x_i, S)). \tag{3}$$

Since $|G| - |S| - |T| \ge 0$ and $a - h_k \ge 1$, we obtain $(|G| - |S| - |T|)(a - h_k) \ge 0$. This inequality together with (2) gives us the following:

$$(|G| - |S| - |T|)(a - h_k)$$

$$\ge b|S| + \sum_{x \in T} \left( \deg_{G-S}(x) - \deg_H(x) + e_H(x, S) - a \right) + 1$$

$$\ge b|S| + \sum_{i=1}^{k-1} h_i |N_T[x_i]| + h_k \left( |T| - \sum_{i=1}^{k-1} |N_T[x_i]| \right) - a|T| + 1$$

$$= b|S| + \sum_{i=1}^{k-1} (h_i - h_k)|N_T[x_i]| + (h_k - a)|T| + 1$$

$$\ge b|S| + \sum_{i=1}^{k-1} (h_i - h_k)(h_i + 1 + \deg_H(x_i)) + (h_k - a)|T| + 1$$

$$= b|S| + \sum_{i=1}^{k} (h_i - h_k)(h_i + 1 + \deg_H(x_i)) + (h_k - a)|T| + 1,$$

where $h_i - h_k \le 0$ and $h_i + 1 + \deg_H(x_i) \ge |N_T[x_i]|$. Then it follows from the above inequality that

$$0 \le (a - h_k)|G| - (a + b - h_k)|S| + \sum_{i=1}^{k-1} (h_k - h_i)(h_i + 1 + \deg_H(x_i)) - 1. \tag{4}$$

Substituting (3) into (4), we have

$$0 \leq (a - h_k)|G| - (a + b - h_k)\left(\frac{a|G|}{a + b} - \sum_{i=1}^{k}(h_i + \deg_H(x_i) - e_H(x_i, S))\right)$$

$$+ \sum_{i=1}^{k}(h_k - h_i)(h_i + 1 + \deg_H(x_i)) - 1$$

$$= -\frac{b|G|}{a + b}h_k$$

$$- \sum_{i=1}^{k}\left(h_i^2 - (a + b - 1 - \deg_H(x_i))h_i - h_t - (a + b)\deg_H(x_i)\right) - 1.$$

By the condition $2 < (a + b + 1 - m)/a$, we have $m < b - a + 1$ and hence $a + b - 1 - \deg_H(x_i) \geq 2(a - 1)$ for each $i = 1, 2, \ldots, k$. This together with the inequalities $h_1 \leq h_2 \leq \cdots \leq h_k \leq a - 1$ of Claim 3 yields the fact $h_i^2 - (a + b - 1 - \deg_H(x_i))h_i$ attains its minimum at $h_i = h_k$. Suppose that $h_k \geq 1$. By $|G| > (a + b)((k + m)(a + b - 1) - 1)/b$, we obtain

$$0 \leq -\frac{b|G|}{a + b}h_k$$

$$- \sum_{i=1}^{k}\left(h_i^2 - (a + b - 1 - \deg_H(x_i))h_i - h_k - (a + b)\deg_H(x_i)\right) - 1$$

$$\leq -\frac{b|G|}{a + b}h_k$$

$$- \sum_{i=1}^{k}\left(h_k^2 - (a + b - 1 - \deg_H(x_i))h_k - h_k - (a + b)\deg_H(x_i)\right) - 1$$

$$= -\frac{b|G|}{a + b}h_k - kh_k^2 + k(a + b)h_k + (a + b - h_k)\sum_{i=1}^{k}\deg_H(x_i) - 1$$

$$\leq -\frac{b|G|}{a + b}h_k - kh_k^2 + k(a + b)h_k + (a + b - h_k)m - 1$$

$$\leq -kh_k^2 + \left(k(a + b) - \frac{b|G|}{a + b} - m\right)h_k + (a + b)m - 1$$

$$< -kh_k^2 + (k - (a + b)m + 1)h_k + (a + b)m - 1$$

$$= -(h_k - 1)(kh_k + (a + b)m - 1) \leq 0.$$

This is a contradiction. Hence we consider the case $h_1 = h_2 = \cdots = h_k = 0$. By (3) and (4), $\sum_{i=1}^{k} (\deg_H(x_i) - e_H(x_i, S)) \geq 1$. By the choice of $\{x_1, x_2, \ldots, x_k\}$, one of (i) and (ii) holds for any $w \in T \backslash (\{x_1, x_2, \ldots, x_k\} \cup N_H(\{x_1, x_2, \ldots, x_k\})$:

(i) $\deg_{G-S}(w) - \deg_H(w) + e_H(w, S) \geq 1$, or
(ii) $\deg_{G-S}(w) - \deg_H(w) + e_H(w, S) = 0$ and $\deg_H(w) - e_H(w, S) \geq 1$.

Since $\{x_1, x_2, \ldots, x_k\} \cap V(H) \neq \emptyset$ and any vertices $v \in T \backslash (\{x_1, x_2, \ldots, x_k\} \cup V(H))$ satisfy (i), we have

$$\sum_{x \in T} \left(\deg_{G-S}(x) - \deg_H(x) + e_H(x, S)\right) \geq |T| - k - 2m + 1.$$

By this inequality, (3), $\sum_{i=1}^{k} \deg_H(x_i) \leq m$, $2 \leq k < (a+b+1-m)/a$, and $|G| > (a+b)((k+m)(a+b-1)-1)/b$, we obtain

$$-1 \geq b|S| + |T| - k - 2m + 1 - a|T|$$
$$= b|S| + (1-a)|T| - k - 2m + 1$$
$$\geq b|S| + (1-a)(|G| - |S|) - k - 2m + 1$$
$$= (a+b-1)|S| - (a-1)|G| - k - 2m + 1$$
$$\geq (a+b-1)\left(\frac{a|G|}{a+b} - m\right) - (a-1)|G| - k - 2m + 1$$
$$= \frac{b|G|}{a+b} - m(a+b+1) - k + 1$$
$$> (k+m)(a+b-1) - 1 - m(a+b+1) - k + 1$$
$$= k(a+b-2) - 2m$$
$$\geq k(a+b-2) - 2(a+b-ak)$$
$$= k(3a+b-2) - 2(a+b)$$
$$\geq 2(3a+b-2) - 2(a+b)$$
$$= 4(a-1) \geq 0.$$

Therefore Theorem 5 is proved.           □

*Proof of Theorem 6.* Suppose that $G$ satisfies the assumption of the theorem, but has no desired $[a, b]$-factor for some subgraph $H$ with $m$ edges. Note that $\delta(G - H) \geq a|G|/(a+b) - m \geq a$ hold by the conditions of Theorem 6. Then by Theorem 7, there exist two disjoint subsets $S$ and $T$ of $V(G)$ such that

$$b|S| + \sum_{x \in T} \left(\deg_{G-S}(x) - \deg_H(x) + e_H(x, S) - a\right) \leq -1. \tag{5}$$

We choose such subsets $S$ and $T$ so that $|T|$ is minimum.

By the argument of Claims 1, 2, and 3 in the proof of Theorem 5, we obtain $|S| \geq 1$, $|T| \geq b+1$, and $\deg_{G-S}(x) - \deg_H(x) + e_H(x, S) \leq a-1$ for all $x \in T$. We now define

$$u_1 = \min \left\{ \deg_{G-S}(x) - \deg_H(x) + e_H(x, S) \mid x \in T \right\},$$

and choose $x_1 \in T$ such that $\deg_{G-S}(x_1) - \deg_H(x_1) + e_H(x_1, S) = u_1$ and $\deg_H(x_1) - e_H(x_1, S)$ is minimum. For $i = 2, \ldots, |T|$, we define

$$u_i = \min \left\{ \deg_{G-S}(x) - \deg_H(x) + e_H(x, S) \mid x \in T \setminus \{u_1, \ldots, u_{i-1}\} \right\},$$

and choose $x_i \in T \setminus \{x_1, \ldots, x_{i-1}\}$ such that $\deg_{G-S}(x_i) - \deg_H(x_i) + e_H(x_i, S) = u_i$ and $\deg_H(x_i) + e_H(x_i, S)$ is minimum. Then we have $u_1 \leq u_2 \leq \cdots \leq u_{|T|} \leq a - 1$.

By the condition of Theorem 6, the following inequalities hold:

$$\frac{a|G|}{a+b} \leq \delta(G) \leq \deg_G(x_1) \leq \deg_{G-S}(x_1) + |S| \leq u_1 + \deg_H(x_1) - e_H(x_1, S) + |S|,$$

which implies

$$|S| \geq \frac{a|G|}{a+b} - (u_1 + \deg_H(x_1) - e_H(x_1, S)). \tag{6}$$

On the other hand, by (5) and $u_1 \leq u_2 \leq \cdots \leq u_{|T|}$, we have

$$0 \geq b|S| + \sum_{i=1}^{|T|} u_i - a|T|$$

$$\geq b|S| + (u_1 - a)|T| + 1$$

$$\geq b|S| + (u_1 - a)(|G| - |S|) + 1$$

$$= (a + b - u_1)|S| - (a - u_1)|G| + 1,$$

which implies

$$0 \geq (a + b - u_1)|S| - (a - u_1)|G| + 1. \tag{7}$$

By (6), (7), $u_1 \leq u_2 \leq \cdots \leq u_{|T|} \leq a-1$, and $|G| > (a+b)((m+1)(a+b+1)-5)/b$,

$$0 \geq (a + b - u_1) \left( \frac{a|G|}{a+b} - (u_1 + \deg_H(x_1) - e_H(x_1, S)) \right) - (a - u_1)|G| + 1$$

$$= \frac{bu_1}{a+b}|G| - (a + b - u_1)(u_1 + \deg_H(x_1) - e_H(x_1, S)) + 1$$

$$\geq \frac{bu_1}{a+b}|G| - (a + b - u_1)(u_1 + m) + 1$$

$$> u_1((m+1)(a+b+1) - 5) - (a + b - u_1)(u_1 + m) + 1$$

$$= u_1^2 + (m(a+b+2) - 4)u_1 - m(a+b) + 1$$

$$= (u_1 - 1)^2 + m(a+b)(u_1 - 1) + 2(m-1)u_1.$$

If $u_1 \geq 1$, then the above inequalities imply $0 > 0$, a contradiction. Hence we must consider the case $u_1 = 0$. By (6) and (7), $\deg_H(x_1) - e_H(x_1, S) \geq 1$. By the definition of $x_1, x_2, \ldots, x_{|T|}$, one of (i) and (ii) holds for any $w \in \{x_2, \ldots, x_{|T|}\}$:

**(i)** $\deg_{G-S}(w) - \deg_H(w) + e_H(w, S) \geq 1$, or

**(ii)** $\deg_{G-S}(w) - \deg_H(w) + e_H(w, S) = 0$ and $\deg_H(w) - e_H(w, S) \geq 1$.

Therefore we have

$$\sum_{x \in T} \left( \deg_{G-S}(x) - \deg_H(x) + e_H(x, S) \right) \geq |T| - 2m.$$

By this inequality, (5), and $|G| > (a+b)((m+1)(a+b+1) - 5)/b$, we obtain

$$-1 \geq b|S| + |T| - 2m - a|T| = b|S| + (1-a)|T| - 2m$$

$$\geq b|S| + (1-a)(|G| - |S|) - 2m$$

$$= (a+b-1)|S| - (a-1)|G| - 2m$$

$$\geq (a+b-1) \left( \frac{a|G|}{a+b} - m \right) - (a-1)|G| - 2m$$

$$= \frac{b|G|}{a+b} - m(a+b+1) > 0$$

$$> (m+1)(a+b+1) - 5 - m(a+b+1)$$

$$= a+b-4$$

$$\geq -1.$$

Finally the proof of Theorem 6 is complete.        $\square$

## Acknowledgements

The authors would like to thank the referees for their helpful comments and suggestions.

## References

1. Iida, T., Nishimura, T.: Neighborhood Conditions and $k$-Factors. Tokyo J. Math. **20** (1997) 411–418
2. Lam, P.B.C., Liu, G., Li, G., Shiu, W.C.: Orthogonal $(g, f)$-Factorizations in Networks. Networks **35** (2000) 274–278
3. Li, Y., Cai, M.: A Degree Condition for a Graph to Have $[a, b]$-Factors. J. Graph Theory **27** (1998) 1–6
4. Niessen, T.: Neighborhood Unions and Regular Factors. J. Graph Theory **19** (1995) 45–64
5. Matsuda, H.: A Neighborhood Condition for Graphs to Have $[a, b]$-Factors. Discrete Mathematics **224** (2000) 289–292
6. Matsuda, H.: A Neighborhood Condition for Graphs to Have $[a, b]$-Factors II. Graphs and Combinatorics **18** (2003) 763–768

# General Balanced Subdivision of Two Sets of Points in the Plane

M. Kano[1] and Miyuki Uno[2]

[1] Department of Computer and Information Sciences,
Ibaraki University, Hitachi, Ibaraki 316-8511, Japan
kano@mx.ibaraki.ac.jp
http://gorogoro.cis.ibaraki.ac.jp/
[2] Department of Computer and Information Sciences,
Ibaraki University, Hitachi, Ibaraki 316-8511, Japan
umyu@mug.biglobe.ne.jp

**Abstract.** Let $R$ and $B$ be two disjoint sets of red points and blue points, respectively, in the plane such that no three points of $R \cup B$ are co-linear. Suppose $ag \leq |R| \leq (a+1)g$, $bg \leq |B| \leq (b+1)g$. Then without loss of generality, we can express $|R| = a(g_1 + g_2) + (a + 1)g_3$, $|B| = bg_1 + (b + 1)(g_2 + g_3)$, where $g = g_1 + g_2 + g_3$, $g_1 \geq 0$, $g_2 \geq 0$, $g_3 \geq 0$ and $g_1 + g_2 + g_3 \geq 1$. We show that the plane can be subdivided into $g$ disjoint convex polygons $X_1 \cup \cdots \cup X_{g_1} \cup Y_1 \cup \cdots \cup Y_{g_2} \cup Z_1 \cup \cdots \cup Z_{g_3}$ such that every $X_i$ contains $a$ red points and $b$ blue points, every $Y_i$ contains $a$ red points and $b+1$ blue points and every $Z_i$ contains $a+1$ red points and $b + 1$ blue points.

## 1 Introduction

We consider two disjoint sets $R$ and $B$ of red points and blue points in the plane, respectively. We always assume that $R \cup B$ is in general position, that is, no three points of $R \cup B$ lie on the same line. We want to subdivide the plane into some disjoint convex polygons so that each polygon contains prescribed numbers of red points and blue points. We begin with some known results on this problem.

The following Theorem 1, which was conjectured in [5] and proved for $a = 1, 2$ in [5] and [6], has been established in full generality by Bespamyatnikh, Kirkpatrick and Snoeyink [2], Sakai [9] and by Ito, Uehara and Yokoyama [3], independently. Note that this theorem with $g = 2$ is equivalent to the famous Ham-sandwich Theorem for the plane [4].

**Theorem 1 (The Equitable Subdivision Theorem [2], [3], [9]).** *Let $a \geq 1$, $b \geq 1$ and $g \geq 2$ be integers. If $|R| = ag$ and $|B| = bg$, then there exists a subdivision $X_1 \cup X_2 \cup \cdots \cup X_g$ of the plane into $g$ disjoint convex polygons such that every $X_i$ contains exactly $a$ red points and $b$ blue points.*

The next theorem shows that if $a = 1$ in the above Theorem 1, we can obtain more general subdivision.

J. Akiyama et al. (Eds.): CJCDGCGT 2005, LNCS 4381, pp. 79–87, 2007.
© Springer-Verlag Berlin Heidelberg 2007

**Theorem 2 (Kaneko and Kano [6]).** *Let $b \geq 1$ be an integer. Suppose that $R$ is a disjoint union of $R_1$ and $R_2$. If $|R_1| = g_1$, $|R_2| = g_2$ and $|B| = (b - 1)g_1 + bg_2$, then we can subdivide the plane into $g_1 + g_2$ disjoint convex polygons $X_1 \cup \cdots \cup X_{g_1} \cup Y_1 \cup \cdots \cup Y_{g_2}$ so that every $X_i$ contains exactly one red point of $R_1$ and $b - 1$ blue points, and every $Y_j$ contains exactly one red point of $R_2$ and $b$ blue points.*

The next theorem shows another result on balanced subdivisions.

**Theorem 3 (Kaneko, Kano and Suzuki [8]).** *Let $a \geq 1$, $g \geq 0$ and $h \geq 0$ be integers such that $g + h \geq 1$. If $|R| = ag + (a + 1)h$ and $|B| = (a + 1)g + ah$, then there exists a subdivision $X_1 \cup \cdots \cup X_g \cup Y_1 \cup \cdots \cup Y_h$ of the plane into $g + h$ disjoint convex polygons such that every $X_i$ contains exactly $a$ red points and $a + 1$ blue points and every $Y_j$ contains exactly $a + 1$ red points and $a$ blue points.*

In this paper, we consider balanced subdivision problem in the case that $ag \leq |R| \leq a(g + 1)$ and $bg \leq |B| \leq b(g + 1)$. As we shall show in Lemma 1, in this case we can express $|R| = a(g_1 + g_2) + (a + 1)g_3$ and $|B| = bg_1 + (b + 1)(g_2 + g_3)$, or $|R| = ag_1 + (a + 1)(g_2 + g_3)$ and $|B| = b(g_1 + g_2) + (b + 1)g_3$ for some integers $g_1, g_2, g_3 \geq 0$ with $g_1 + g_2 + g_3 \geq 1$. By symmetry, we may assume that

$$|R| = a(g_1 + g_2) + (a + 1)g_3 \quad and \quad |B| = bg_1 + (b + 1)(g_2 + g_3) \qquad (1)$$

holds. The following theorem is our main result.

**Theorem 4.** *Let $a \geq 1$, $b \geq 1$, $g_1 \geq 0$, $g_2 \geq 0$ and $g_3 \geq 0$ be integers such that $g_1 + g_2 + g_3 \geq 1$. If $|R| = a(g_1 + g_2) + (a + 1)g_3$ and $|B| = bg_1 + (b + 1)(g_2 + g_3)$, then there exists a subdivision $X_1 \cup \cdots \cup X_{g_1} \cup Y_1 \cup \cdots \cup Y_{g_2} \cup Z_1 \cup \cdots \cup Z_{g_3}$ of the plane into $g_1 + g_2 + g_3$ disjoint convex polygons such that every $X_i$ contains exactly $a$ red points and $b$ blue points, every $Y_i$ contains exactly $a$ red points and $b + 1$ blue points, and every $Z_i$ contains exactly $a + 1$ red points and $b + 1$ blue points, if $g_1 \geq 1$, $g_2 \geq 1$ and $g_3 \geq 1$, respectively (see Figure 1).*

We call the subdivision of the plane given in the above theorem a *general balanced subdivision*. Moreover, we notice that our proof of the above Theorem 4 gives $O(n^4)$ time algorithm for finding a balanced subdivision of the plane, where $n$ is the total number of red and blue points.

Before giving proofs, we remark that it seems to be impossible to derive our Theorem 4 from Theorem 1. Namely, someone may consider in the following way. Add some new imaginary red points and blue points so that in the resulting plane, there are exactly $a(g + 1)$ red points and $(b + 1)g$ blue points. Then we apply Theorem 1 to obtain an equitable subdivision of the plane, and remove the imaginary points. However, it seems to be impossible to guarantee that we can add new points so that each polygon of an equitable subdivision contains at most one imaginary red point and at most one blue point. Namely, some polygon may contain more than one imaginary red point or more than one imaginary blue point. Therefore, it seems to be impossible to derive Theorem 4 from Theorem 1.

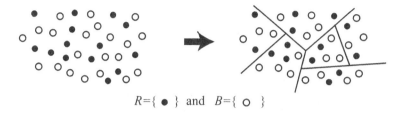

$R=\{\,\bullet\,\}$ and $B=\{\,\circ\,\}$

**Fig. 1.** $g_1 = 3$, $g_2 = 1$, $g_3 = 2$, $|R| = 2(g_1 + g_2) + 3g_3$ and $|B| = 3g_1 + 4(g_2 + g_3)$; and a balanced subdivision of the plane

## 2    Proof of Theorem 4

We begin with some definitions and notation. We deal only with *directed lines* in order to define the right side of a line and the left side of it. Thus a *line* means a directed line. A line $l$ dissects the plane into three pieces: $l$ and the two open half-planes $right(l)$ and $left(l)$ that are bounded to the left and to the right of $l$, respectively (see Figure 2). For a line $l$, we define $l^*$ as the line lying on $l$ and having the opposite direction of $l$ (see Figure 2). Furthermore, we always assume that a line does not pass through any point in $R \cup B$.

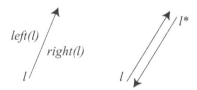

**Fig. 2.** $right(l)$, $left(l)$ and the line $l^*$

**Lemma 1.** *Let $a \geq 1$, $b \geq 1$ and $g \geq 1$ be integers. If $ag \leq |R| \leq (a+1)g$ and $bg \leq |B| \leq (b+1)g$, then we can uniquely express either $|R| = a(g_1+g_2)+(a+1)g_3$ and $|B| = bg_1 + (b+1)(g_2 + g_3)$, or $|R| = ag_1 + (a+1)(g_2 + g_3)$ and $|B| = b(g_1 + g_2) + (b+1)g_3$, where $g_1 \geq 0$, $g_2 \geq 0$, $g_3 \geq 0$ and $g = g_1 + g_2 + g_3$.*

*Proof.* We can uniquely express $|R| = ax + (a+1)(g-x)$ and $|B| = by + (b+1)(g-y)$ for some integers $0 \leq x \leq g$ and $0 \leq y \leq g$. If $x \geq y$, then by letting $g_1 = y$, $g_2 = x - y$ and $g_3 = g - x$, we can express $|R| = a(g_1 + g_2) + (a+1)g_3$ and $|B| = bg_1 + (b+1)(g_2 + g_3)$. Otherwise, by letting $g_1 = x$, $g_2 = y - x$ and $g_3 = g - y$, we have $|R| = ag_1 + (a+1)(g_2 + g_3)$ and $|B| = b(g_1 + g_2) + (b+1)g_3$. The uniqueness of the expression can be easily proved. □

The following theorem, called the 3-cutting Theorem, plays an important role. This theorem was proved by Bespamyatnikh, Kirkpatrick and Snoeyink [2] under the assumption that

$$\frac{m_1}{n_1} = \frac{m_2}{n_2} = \frac{m_3}{n_3}.$$

However this condition can be removed without changing the arguments in the proof given in [2]. This relaxation is necessary to prove our Theorem 4. Note that similar results, which seem to be essentially equivalent to the original 3-cutting Theorem, were obtained in [3] and [9], respectively.

**Theorem 5 (The 3-cutting Theorem [2]).** *Let $m_1, m_2, m_3, n_1, n_2, n_3$ be positive integers such that $|R| = m_1 + m_2 + m_3$ and $|B| = n_1 + n_2 + n_3$. Suppose that one of the following statements (i) or (ii) is true:*

*(i) For every integer $i \in \{1, 2, 3\}$ and for every line $l$ such that $|right(l) \cap R| = m_i$, we have $|right(l) \cap B| < n_i$ (Figure 3 (a)).*

*(ii) For every integer $i \in \{1, 2, 3\}$ and for every line $l$ such that $|right(l) \cap R| = m_i$, we have $|right(l) \cap B| > n_i$.*

*Then there exists three rays emanating from a certain same point such that the three open polygons $W_i$ ($1 \leq i \leq 3$) defined by these three rays are convex, and each $W_i$ ($1 \leq i \leq 3$) contains exactly $m_i$ red points and $h_i$ blue points (Figure 3 (b)). Moreover, one of the three rays can be chosen to be a vertically downward ray.*

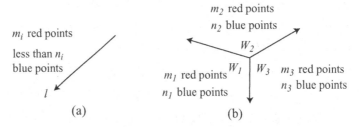

**Fig. 3.** A subdivision $W_1 \cup W_2 \cup W_3$ of the 3-cutting Theorem

Two different proofs of the next lemma can be found in [5] and [2].

**Lemma 2.** *If there exist two lines $l_1$ and $l_2$ such that $|right(l_1) \cap R| = |right(l_2) \cap R|$ and $|right(l_1) \cap B| < |right(l_2) \cap B|$, then for every integer $n$, $|right(l_1) \cap B| \leq n \leq |right(l_2) \cap B|$, there exists a line $l_3$ such that $|right(l_3) \cap R| = |right(l_1) \cap R|$ and $|right(l_3) \cap B| = n$.*

**Proof of Theorem 4.**    Let $g = g_1 + g_2 + g_3$. We shall prove the theorem by induction on $g$. It is trivial that the theorem holds for $g = 1$, and so we may assume $g \geq 2$. Moreover, by Theorem 1, if $g_1 = g_2 = 0$, $g_2 = g_3 = 0$ or $g_1 = g_3 = 0$ then the theorem is true. So we may assume that

$$\text{at least two of } g_1, g_2 \text{ and } g_3 \text{ are greater than or equal to } 1. \qquad (2)$$

Assume that there exist three integers $r \geq 0$, $s \geq 0$, $t \geq 0$ and two lines $l_1$ and $l_2$ such that $1 \leq r+s+t \leq g-1$, $|right(l_1) \cap R| = |right(l_2) \cap R| = a(r+s)+(a+1)t$,

$|right(l_1) \cap B| \leq br + (b+1)(s+t)$ and $|right(l_2) \cap B| \geq br + (b+1)(s+t)$. Then by Lemma 2, there exists a line $l_3$ that satisfies

$$|right(l_3) \cap R| = a(r+s) + (a+1)t \quad \text{and} \quad |right(l_3) \cap B| = br + (b+1)(s+t). \quad (3)$$

By applying the inductive hypotheses to $right(l_3)$ and $left(l_3)$ respectively, we can obtain the desired balanced subdivision of the plane. Therefore we may assume that the next claim holds.

**Claim 1.** *Let $(r, s, t)$ be a triple of integers such that $0 \leq r \leq g_1$, $0 \leq s \leq g_2$, $0 \leq t \leq g_3$ and $1 \leq r + s + t \leq g - 1$. Then for every line $l$ with $|right(l) \cap R| = a(r+s) + (a+1)t$, we always have either*

$$|right(l) \cap B| < br + (b+1)(s+t) \quad or \quad |right(l) \cap B| > br + (b+1)(s+t),$$

*in particular, $|right(l) \cap B| \neq br + (b+1)(s+t)$.*

By Claim 1, we can define the sign of every triple $(i, j, k)$ with $0 \leq i \leq g_1$, $0 \leq j \leq g_2$, $0 \leq k \leq g_3$ and $1 \leq i + j + k \leq g - 1$ as follows: For every line $l$ with $|right(l) \cap R| = a(i + j) + (a + 1)k$,

if $|right(l) \cap B| > bi + (b+1)(j + k)$, then $sign(i, j, k) = +$; and
if $|right(l) \cap B| < bi + (b+1)(j + k)$, then $sign(i, j, k) = -$.

The next claim gives an easy but useful property of $sign(i, j, k)$.

**Claim 2.** *Let $i$, $j$ and $k$ be integers such that $0 \leq i \leq g_1$, $0 \leq j \leq g_2$, $0 \leq k \leq g_3$ and $1 \leq i + j + k \leq g - 1$. Then*

$$sign(g_1 - i, g_2 - j, g_3 - k) = -sign(i, j, k). \quad (4)$$

We may assume that $sign(i, j, k) = +$ since otherwise we can similarly prove the claim. Let $l$ be a line such that $|right(l) \cap R| = a(i + j) + (a + 1)k$. Then $|right(l) \cap B| > bi + (b+1)(j + k)$ by $sign(i, j, k) = +$. This implies that

$$|right(l^*) \cap R| = a(g_1 - i + g_2 - j) + (a + 1)(g_3 - k), \quad \text{and}$$
$$|right(l^*) \cap B| = |left(l) \cap B| = |B| - |right(l) \cap B|$$
$$< b(g_1 - i) + (b + 1)(g_2 - j + g_3 - k).$$

Hence $sign(g_1 - i, g_2 - j, g_3 - k) = - = -sign(i, j, k)$.

**Claim 3.** *We may assume that the following four statements hold. (i) If $g_1 \geq 1$ $g_2 \geq 1$ and $g_3 \geq 1$, then $sign(1, 0, 0) = sign(0, 1, 0) = sign(0, 0, 1)$. (ii) If $g_1 = 0$, then $g_2 \geq 1$, $g_3 \geq 1$ and $sign(0, 1, 0) = sign(0, 0, 1)$. (iii) If $g_2 = 0$, then $g_1 \geq 1$, $g_3 \geq 1$ and $sign(1, 0, 0) = sign(0, 0, 1)$. (iv) If $g_3 = 0$, then $g_1 \geq 1$, $g_2 \geq 1$ and $sign(1, 0, 0) = sign(0, 1, 0)$.*

*Proof.* We first consider the case where $g_1 \geq 1$, $g_2 \geq 1$, $g_3 \geq 1$ and $sign(1, 0, 0) = -$. Let $l_1$ and $l_2$ be two parallel lines such that $|right(l_1) \cap R| = a$, $|right(l_2) \cap R| =$

$a + 1$ and that $l_1$ and $l_2$ pass very close to the $a$-th red point and $(a+1)$-th red point, respectively (see Figure 4). Then $|right(l_1) \cap B| < b$ by $sign(1,0,0) = -$.

The existence of $l_1$ implies $sign(0,1,0) = -$ by Claim 1. Assume $sign(0,0,1) = +$. Then $|right(l_2) \cap B| > b + 1$, and thus we can find a line $l_3$ between $l_1$ and $l_2$ such that $right(l_3)$ contains exactly $a$ red points and $b + 1$ blue points. This contradicts Claim 1 with $(0,1,0)$. Hence $sign(0,0,1) = -$.

We next consider the case where $g_1 \geq 1$, $g_2 \geq 1$, $g_3 \geq 1$ and $sign(1,0,0) = +$. Let $l_1$ and $l_2$ be the two parallel lines given above. Then $|right(l_1) \cap B| > b$ by $sign(1,0,0) = +$. Since $|right(l_1) \cap B| \neq b + 1$ by Claim 1 with $(0,1,0)$, we have $|right(l_1) \cap B| \geq b + 2$. Hence $sign(0,1,0) = +$. Furthermore, since $|right(l_2) \cap B| \geq |right(l_1) \cap B| \geq b + 2$, we have $sign(0,0,1) = +$.

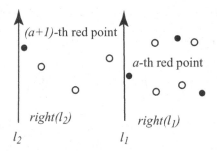

**Fig. 4.** Lines $l_1$ and $l_2$

By the above two results, we can say that for every $\alpha \in \{+, -\}$, $sign(0,1,0) = \alpha$ implies $sign(1,0,0) = \alpha$ and $sign(0,0,1) = \alpha$. Consequently, the statement (i) is proved.

Suppose $g_1 = 0$. Then $g_2 \geq 1$ and $g_3 \geq 1$ by (2). Let $l_1$ and $l_2$ be the same lines as given above. Assume $sign(0,1,0) = -$. Then $|right(l_1) \cap B| < b + 1$ by $sign(0,1,0) = -$. If $sign(0,0,1) = +$, then $|right(l_2) \cap B| > b + 1$, and so we can find a line $l_4$ between $l_1$ and $l_2$ such that $|right(l_4) \cap R| = a$ and $|right(l_4) \cap B| = b + 1$, which contradicts Claim 1 with $(0,1,0)$. Hence $sign(0,0,1) = -$. Next, assume $sign(0,1,0) = +$. Then $|right(l_1) \cap B| > b + 1$, and so $|right(l_2) \cap B| > b + 1$. Hence $sign(0,0,1) = +$. Therefore (ii) is proved.

Suppose that $g_2 = 0$. Then $g_1 \geq 1$ and $g_3 \geq 1$ by (2). Let $l_1$ and $l_2$ be the same lines as given above. Assume $sign(1,0,0) = -$. Then $|right(l_1) \cap B| < b$. If $sign(0,0,1) = +$, then $|right(l_2) \cap B| > b + 1$, and so we can find a line $l_5$ between $l_1$ and $l_2$ such that $|right(l_5) \cap R| = a$ and $|right(l_5) \cap B| = b$, which contradicts Claim 1 with $(1,0,0)$. Hence $sign(0,0,1) = -$. Assume $sign(1,0,0) = +$. Then $|right(l_1) \cap B| > b$, and thus $|right(l_2) \cap B| > b + 1$ by Claim 1 with $(0,0,1)$. Hence $sign(0,0,1) = +$. Consequently (iii) is true.

We finally consider the case where $g_3 = 0$. Then $g_1 \geq 1$ and $g_2 \geq 1$. Let $l_1$ and $l_2$ be the same line as given above. If $sign(1,0,0) = -$, then $sign(0,1,0) = -$ as $|right(l_1) \cap B| < b$. If $sign(1,0,0) = +$, then $|right(l_1) \cap B| > b$, and so $|right(l_1) \cap B| > b + 1$ by Claim 1 with $(0,1,0)$, which implies $sign(0,1,0) = +$. Consequently, (iv) holds, and hence Claim 3 is proved.

Because of symmetry, hereafter we assume

$$sign(1,0,0) = sign(0,1,0) = sign(0,0,1) = -, \tag{5}$$

when we can consider these signs. We say that three triples $(r_1, s_1, t_1)$, $(r_2, s_2, t_2)$, $(r_3, s_3, t_3)$ satisfy the condition of the 3-cutting Theorem if

$$g_1 = r_1 + r_2 + r_3, \quad g_2 = s_1 + s_2 + s_3, \quad g_3 = t_1 + t_2 + t_3$$
$$sign(r_1, s_1, t_1) = sign(r_2, s_2, t_2) = sign(r_3, s_3, t_3), \quad \text{and}$$
$$0 \leq r_i, s_i, t_i \quad \text{and} \quad 1 \leq r_i + s_i + t_i \quad \text{for every} \quad i \in \{1,2,3\}.$$

If a set of three triples $\{(r_i, s_i, t_i) \mid 1 \leq i \leq 3\}$ satisfies the above conditions, then

$$m_i = a(r_i + s_i) + (a+1)t_i \quad \text{and} \quad n_i = br_i + (b+1)(s_i + t_i), \quad (1 \leq i \leq 3)$$

satisfy the condition of the 3-cutting Theorem, and so we can subdivide the plane into three convex polygons, each of which contains exactly $m_i$ red points and $n_i$ blue points, and thus we can obtain the desired subdivision by applying the inductive hypotheses to each convex polygon.

**Claim 4.** *If $g_1 = 0$, then there exist three triples $(r_i, s_i, t_i)$ $(1 \leq i \leq 3)$ that satisfy the condition of the 3-cutting Theorem, and thus the theorem 4 holds.*

Since $g_1 = 0$, we have $g_2 \geq 1$ and $g_3 \geq 1$ by (2). Then $sign(0, g_2 - 1, g_3) = -sign(0,1,0)$ by Claim 2. Choose the lowest element $(0, s, t)$ in lexicographical order such that $sign(0, s, t) \neq sign(0, 1, 0)$, that is, $sign(0, s', t') = sign(0, 1, 0)$ if $s' < s$ or $s' = s$ and $t' < t$. If $s = 0$, then $t \geq 2$ by (5), and so we obtain the following three triples that satisfy the conditions of the 3-cutting Theorem.

$$(0, g_2, g_3 - t), \quad (0, 0, t - 1), \quad (0, 0, 1),$$

where $sign(0, g_2, g_3 - t) = -sign(0, 0, t) = sign(0, 1, 0) = sign(0, 0, 1) = -$ by Claims 2 and 3, and $sign(0, 0, t - 1) = sign(0, 1, 0) = -$ by the choice of $(0, s, t)$ and $s = 0$. If $s \geq 1$, then $s \leq g_2 - 1$ by the above fact that $sign(0, g_2 - 1, g_3) = -sign(0, 1, 0)$, and so $1 \leq g_2 - s$. Furthermore, if $s = 1$, then $1 \leq t$ by $sign(0, s, t) \neq sign(0, 1, 0)$ and (5). Hence by claim 2, we can obtain the following desired three triples:

$$(0, g_2 - s, g_3 - t), \quad (0, s - 1, t), \quad (0, 1, 0),$$

where $sign(0, g_2 - s, g_3 - t) = -sign(0, s, t) = sign(0, 1, 0) = -$. Therefore Claim 4 is proved.

**Claim 5.** *If $g_1 \geq 1$, then we may assume that $sign(0, j, k) = sign(1, 0, 0)$ for all $0 \leq j \leq g_2$ and $0 \leq k \leq g_3$ with $1 \leq j + k$ since otherwise the theorem 4 holds.*

Assume that $sign(0, j, k) \neq sign(1, 0, 0)$ for some $0 \leq j \leq g_2$ and $0 \leq k \leq g_3$ with $1 \leq j + k$. Choose the lowest element $(0, s, t)$ in lexicographical order such that $sign(0, s, t) \neq sign(1, 0, 0)$. By (5), it follows that $2 \leq s + t$. By Claim 2, we have $sign(g_1, g_2 - s, g_3 - t) = -sign(0, s, t) = sign(1, 0, 0) = sign(0, 1, 0)$. If $s \geq 2$, then by (5) we obtain the following three triples that satisfy the conditions of the 3-cutting Theorem.

$$(g_1, g_2 - s, g_3 - t), \quad (0, s - 1, t), \quad (0, 1, 0).$$

If $s = 1$, then $t \geq 1$ and so we have the following three triples:

$$(g_1, g_2 - 1, g_3 - t), \quad (0, 1, t - 1), \quad (0, 0, 1).$$

If $s = 0$, then $t \geq 2$, and we obtain the desired three triples as follows:

$$(g_1, g_2, g_3 - t), \quad (0, 0, t - 1), \quad (0, 0, 1).$$

In each case, the theorem is proved by induction, and hence we may assume that Claim 5 holds.

**Claim 6.** *If $g_1 \geq 1$, then there exist three triples $(r_i, s_i, t_i)$ $(1 \leq i \leq 3)$ that satisfy the condition of the 3-cutting Theorem, and hence the theorem 4 holds.*

By (2), we have $g_2 \geq 1$ or $g_3 \geq 1$. Here we assume $g_2 \geq 1$ since we can similarly prove the claim in the case of $g_3 \geq 1$. Since $sign(g_1, g_2 - 1, g_3) = -sign(0, 1, 0) = -sign(1, 0, 0)$, there exists $(i, j, k)$ such that $sign(i, j, k) \neq sign(1, 0, 0)$, $0 \leq i \leq g_1$, $0 \leq j \leq g_2$, $0 \leq k \leq g_3$ and $1 \leq i + j + k \leq g - 1$. Choose the lowest element $(r, s, t)$ in lexicographical order such that $sign(r, s, t) \neq sign(1, 0, 0)$, in particular, $sign(r', s', t') = sign(1, 0, 0)$ if $r' < r$. By Claim 5, we have $r \geq 1$, and if $r = 1$, then $s + t \geq 1$ by $sign(1, s, t) \neq sign(1, 0, 0)$. Hence we obtain the following desired three triples that satisfy the conditions of the 3-cutting Theorem.

$$(g_1 - r, g_2 - s, g_3 - t), \quad (r - 1, s, t), \quad (1, 0, 0),$$

where $sign(g_1 - r, g_2 - s, g_3 - t) = -sign(r, s, t) = sign(1, 0, 0)$ and $sign(r - 1, s, t) = sign(1, 0, 0)$.

By Claims 4 and 6, the proof is complete. □

We now analyze the time complexity of an algorithm for finding a balanced subdivision of the plane, which is directly obtained from our proof. We first consider an algorithm for finding a line $l_3$ that satisfies (3). Since there are $O(n^2)$ lines passing through two points of $R \cup B$, we can find a line $l_3$ satisfying (3), if any, in $O(n^3)$ time. Notice that for a line $l$ passing through two points $x$ and $y$ of $R \cup B$, we consider the four cases; (i) $x, y \in right(l)$, (ii) $x \in right(l)$ and $y \notin right(l)$, (iii) $x \notin right(l)$ and $y \in right(l)$; and (iv) $x, y \notin right(l)$. If there exists no such line $l$, then we can define $sign(i, j, k)$, and there exist three rays given in the 3-cutting Theorem. We can find such three rays in $O(n^{\frac{4}{3}}(\log n)^2)$ by [2]. Therefore in any case, if we denote by $f(n)$ the time complexity of finding a balanced subdivision of the plane with $n = |R| + |B|$ points, then we have

$$f(n) \leq O(n^3) + f(n_1) + f(n_2) + f(n_3),$$

where $n_1 + n_2 + n_3 = n$, $n_1 \geq a + b$, $n_2 \geq a + b$, $n_3 \geq 0$, and $n_3 = 0$ if and only if there exists a line $l_3$. Consequently, we obtain $f(n) \leq O(n^4)$.

# References

1. Bárány, I., Matoušek, J.: Simultaneous partitions of measures by $k$-fans. Discrete Comput. Geom. **25** (2001)317–334.
2. Bespamyatnikh, S., Kirkpatrick, D., Snoeyink, J.: Generalizing ham sandwich cuts to equitable subdivisions. Discrete Comput. Geom. **24** (2000) 605–622.
3. Ito, H., Uehara, H., Yokoyama, M.: 2-dimensional ham-sandwich theorem for partitioning into three convex pieces. Discrete Comput. Geom. LNCS **1763** (2000) 129-157.
4. *Handbook of Discrete and Computational Geometry*, edited by J. Goodman and J. O'Rourke, CRC Press, (2004) Chapter 14 Topological Methods written by Rade T. Živaljević , 305–329.
5. Kaneko, A., Kano, M.: Balanced partitions of two sets of points in the plane. Computational Geometry: Theory and Applications, **13** (1999), 253–261.
6. Kaneko, A., Kano, M.: Semi-balanced partitions of two sets of points in the plane and embeddings of rooted forests. Internat. J. Comput. Geom. Appl. in print.
7. Kaneko, A., Kano, M.: Discrete Geometry on Red and Blue Points in the Plane – A Survey –. Discrete and Computational Geometry, Algorithms Combin., **25**, Springer (2003) 551-570.
8. Kaneko, A., Kano, M., Suzuki, H.: Path Coverings of Two Sets of Points in the Plane. Towards a theory of geometric graphs, ed by J. Pach Contemporary Mathematics series of AMS, **342** (2004) 99-111.
9. Sakai, T.: Balanced Convex Partitions of Measures in $R^2$. Graphs and Combinatorics, **18** (2002), 169–192.

# Coverage Problem of Wireless Sensor Networks

G.L. Lan, Z.M. Ma, and S.S. Sun

[1] Institute of Applied Mathematics, Academy of Mathematics and Systems Science,
Chinese Academy of Sciences, Beijing 100080, China.
{langl,sunsuy}@amss.ac.cn, mazm@amt.ac.cn
[2] Graduate School of the Chinese Academy of Sciences

**Abstract.** In this paper we study the limiting achievable coverage problems of sensor networks. For the sensor networks with uniform distributions we obtain a complete characterization of the coverage probability. For the sensor networks with non-uniform distributions, we derive two different necessary and sufficient conditions respectively in the situations that the density function achieves its minimum value on a set with positive Lebesgue measure or at finitely many points. We propose also an economical scheme for the coverage of sensor networks with empirical distributions.

**Keywords:** Coverage, sensor networks, large-scale, uniform distribution.

## 1 Introduction

Recently, there has been a growing interest in studying large-scale wireless sensor networks. Such a network consists of a large number of sensors which are densely deployed in a certain area. For some reasons (reducing radio interference, limited battery capacity, etc.), these sensors are small in size and they communicate with each other in short distance. There are many fundamental problems that arise in the research of wireless sensor networks. Among them one important issue is that of limiting achievable coverage. A point in an area can be detected by a sensor provided the point is within the distant $r$ of the sensor, where $r$ is the sensing radius of the sensor. The area is said to be covered if every point in the area can be detected by a sensor. In the literature there have been several discussions concerning the minimum sensing radius, depending on the numbers of (active) sensors per unit area, which guarantees that the area is covered in a limiting performance. In [12], the authors considered the problem of covering a square of area $A$ with randomly located circles whose centers are generated by a two-dimensional Poisson point process of density $D$ points per unit area. Suppose that each Poisson point represents a sensor with sensing radius $R$ which may depend on $D$ and $A$. They proved that, for any $\varepsilon > 0$, if $R = \sqrt{(1 + \varepsilon) \ln A / \pi D}$, then $\lim_{A \to \infty} \Pr[\text{square covered}] = 1$. On the other hand, if $R = \sqrt{(1 - \varepsilon) \ln A / \pi D}$, then $\lim_{A \to \infty} \Pr[\text{square covered}] = 0$. Therefore the authors observed that, to guarantee that the area is covered, a node must have $\pi[(1 + \varepsilon) \ln A / \pi D]D$ or a little more than $\ln A$ nearest neighbors (Poisson point that lie at a distant of $R$ or less from it) on the average. In the paper [14],

J. Akiyama et al. (Eds.): CJCDGCGT 2005, LNCS 4381, pp. 88–100, 2007.
© Springer-Verlag Berlin Heidelberg 2007

the authors studied the coverage of a grid-based unreliable sensor network. They derived necessary and sufficient conditions for the random grid network to cover a unit square area. Their result shows that the random grid network asymptotically cover a unit square area if and only if $p_n r_n^2$ is of the order $(\ln n)/n$, where $r_n$ is the sensing radius and $p_n$ is the probability that a sensor is "active" (not failed). In connection with the above two results we mention that Hall [3] has considered the coverage problem of the following model: Circles of radius $r$ are placed in a unit-area disc $\mathcal{D}$ at a Poisson intensity of $\lambda$. Let $V(\lambda; r)$ denote the vacancy within $\mathcal{D}$, i.e., $V(\lambda; r)$ is the region of $\mathcal{D}$ not covered by the circles. It has been shown ([3, Theorem 3.11]) that

$$\frac{1}{20} \min \left\{ 1, \left( 1 + \pi r^2 \lambda^2 \right) e^{-\pi r^2 \lambda} \right\} < \Pr \left\{ |V(\lambda; r)| > 0 \right\}$$

$$< \min \left\{ 1, 3 \left( 1 + \pi r^2 \lambda^2 \right) e^{-\pi r^2 \lambda} \right\},$$

where $|V(\lambda; r)|$ is the area of $V(\lambda; r)$. Note that by Hall's result, if we set $\lambda = n$ and $r_n = \sqrt{(\ln n + \ln \ln n + a_n)/\pi n}$, then $\lim_{n \to \infty} \Pr[\text{square covered}] = 1$ for $a_n \to +\infty$, and $\lim_{n \to \infty} \Pr[\text{square covered}] < 19/20$ for $a_n \to -\infty$. However, it was not clear whether the limit $\lim_{n \to \infty} \Pr[\text{square covered}] = 0$ for $a_n \to -\infty$. In the literature there are also various other discussions about the coverage problems of wireless sensor networks, see [4–11, 13, 15, 16] and references therein.

In this paper, we employ some results in Aldous [1] to study the coverage problems. Among other results, we show that in the above mentioned Hall's model, if we set $\lambda = n$ and $r_n = \sqrt{(\ln n + \ln \ln n + a_n)/\pi n}$ with $a_n = o(\ln n)$, then

$$\Pr[\text{square covered}] = \exp[-\exp(-a_n)] + o(1),$$

provided the limit of $\{a_n\}$ exists in $[-\infty, +\infty]$ (see Theorem 1 and Remark 1 below). In particular, if $a_n \to -\infty$, then $\lim_{n \to \infty} \Pr[\text{square covered}] = 0$. Thus we clarify the above mentioned question. Moreover, the above equality implies that if $a_n \to a$ for $a \in (-\infty, +\infty)$, then $\lim_{n \to \infty} \Pr[\text{square covered}] = \exp(-e^{-a}) \in (0, 1)$.

The structure of this paper is as follows. In section 2 we investigate the coverage of sensor networks with uniform distributions, which is asymptotically the same as the sensor networks with sensors located according to a Poisson distribution. In section 3 we investigate the coverage of sensor networks with non-uniform distributions. Our research shows that in non-uniform case the minimum sensing radius in the coverage problem relies mainly on the behavior of the distribution density function around its minimum point. We give first a sufficient condition in Theorem 2. We then derive two different necessary and sufficient conditions in Theorem 3 and Theorem 4, respectively in the situations that the density function taking its minimum value on a set with positive Lebesgue measure or at finitely many points. Finally in Section 4 we propose an economical scheme for the coverage of sensor networks with empirical distributions.

## 2   Coverage of Sensor Networks with Uniform Distributions

Let $\mathcal{A} = [0, 1]^2$ be a unit square. Suppose $\{X_1, \ldots, X_n\}$ is a set of $n$ independent random points with the uniform distribution on $\mathcal{A}$ and $r_n$ is a positive number. We consider $\{r_n; X_1, \ldots, X_n\}$ as a sensor network on the region $\mathcal{A}$ with sensing radius $r_n$. Let $\mathcal{M}(n, r_n) = \bigcup_{k=1}^{n} B(X_k, r_n)$ be the union of $n$ circles with centers $X_1, \ldots, X_n$ and common radius $r_n$. Then the region is covered by the sensor networks if and only if $\mathcal{A} \subset \mathcal{M}(n, r_n)$. The task in this section is to find the smallest $r_n$ that guarantees

$$\lim_{n \to \infty} \Pr[\mathcal{A} \subset \mathcal{M}(n, r_n)] = 1. \tag{1}$$

We have the following complete result.

**Theorem 1.** *Let $\{X_1, \ldots, X_n\}$ be $n$ independent random points with the uniform distribution on $\mathcal{A}$. Suppose that $r_n = \sqrt{(\ln n + \ln \ln n + a_n)/\pi n}$, where $\{a_n\}$ is a sequence of real numbers such that $a_n = o(\ln n)$ and has a limit in $[-\infty, +\infty]$. Let $\mathcal{M}(n, r_n) = \bigcup_{k=1}^{n} B(X_k, r_n)$ be as above. Then we have*

$$\Pr[\mathcal{A} \subset \mathcal{M}(n, r_n)] = \exp[-\exp(-a_n)] + o(1). \tag{2}$$

*In particular, $\lim_{n \to \infty} \Pr[\mathcal{A} \subset \mathcal{M}(n, r_n)] = 1$ if and only if $a_n \to +\infty$, and $\lim_{n \to \infty} \Pr[\mathcal{A} \subset \mathcal{M}(n, r_n)] = 0$ if and only if $a_n \to -\infty$.*

*Proof.* Define a random variable $L_n$ as the radius of the largest circle that lies inside $\mathcal{A}$ containing no points of $\{X_1, \ldots, X_n\}$. Then $\mathcal{A}$ is covered by $\mathcal{M}(n, r_n)$ if and only if $L_n < r_n$. Applying the approximation method used by Aldous in the study of stochastic geometry, we can get ([1, p. 150, H1d])

$$\Pr[L_n < r_n] \approx \exp\left[-n^2 \pi r_n^2 \exp(-n \pi r_n^2)(1 - 2r_n)^2\right], \tag{3}$$

where the relation "$\approx$" is understood as

$$\Pr[L_n < r_n] = \exp\left[-n^2 \pi r_n^2 \exp(-n \pi r_n^2)(1 - 2r_n)^2\right] + O(e^{-n \pi r_n^2}).$$

By the assumption of $r_n$, we have $n \pi r_n^2 \to \infty$. Therefore,

$$\Pr[\mathcal{A} \subset \mathcal{M}(n, r_n)] = \Pr[L_n < r_n] = \exp[-I(n)] + o(1), \tag{4}$$

where

$$I(n) = n^2 \pi r_n^2 (1 - 2r_n)^2 \exp(-n \pi r_n^2).$$

Replacing $n \pi r_n^2$ by $\ln n + \ln \ln n + a_n$ in the expression of $I(n)$, we get

$$(1 - 2r_n)^{-2} I(n) = n(\ln n + \ln \ln n + a_n) \exp(-\ln n - \ln \ln n - a_n)$$

$$= n(\ln n + \ln \ln n + a_n)(n \ln n)^{-1} \exp(-a_n)$$

$$= \exp(-a_n)\left[1 + \ln \ln n (\ln n)^{-1} + a_n (\ln n)^{-1}\right].$$

By our assumption $a_n = o(\ln n)$, which implies $a_n(\ln n)^{-1} = o(1)$ and $r_n = o(1)$. Therefore, from the above formula we get $I(n) = \exp(-a_n) + o(1)$. Thus by (4)

$$\lim_{n \to \infty} \Pr[\mathcal{A} \subset \mathcal{M}(n, r_n)] = \lim_{n \to \infty} \exp[-I(n)] = \lim_{n \to \infty} \exp[-\exp(-a_n)],$$

verifying (2).                                                                                                   $\square$

*Remark 1.* The conclusion of Theorem 1 remains true in the model that the sensors put in the unit square according to a Poisson point process with intensity $n$ points per unit area. Because the difference of these two schemes is negligible in the limiting procedure. In fact, Aldous's result [1, H1d] (see (3) in this paper) was obtained via the approximation of large number of uniformly distributed points by Poisson point processes with large intensity.

## 3   Coverage of Sensor Networks with Non-uniform Distributions

In this section we investigate the coverage of sensor networks which are deployed non-uniformly according to a general distribution density function $g(x)$ on the unit square $\mathcal{A}$. We assume that $g$ is a bounded smooth function. Our results show that the minimum sensing radius in the coverage problem relies mainly on the minimum value of $g$ and the behavior of $g$ around its minimum value. We derive first a sufficient condition as follows.

**Theorem 2.** *Suppose* $\{X_1, \ldots, X_n\}$ *is a set of $n$ independent random points in the unit square $\mathcal{A}$ according to a smooth distribution density function $g(x)$, where $g(x)$ satisfies $0 < \beta \le g(x) \le \gamma < \infty$. Let $r_n = \sqrt{(\ln n + \ln \ln n + a_n)/\beta \pi n}$, where $\{a_n\}$ is a sequence in $\mathbb{R}$ tending to $+\infty$. Then we have*

$$\lim_{n \to \infty} \Pr[\mathcal{A} \subset \mathcal{M}(n, r_n)] = 1. \tag{5}$$

(Here and henceforth $\mathcal{M}(n, r_n)$ is the same as defined in the beginning of Section 2.)

*Proof.* Analogously, we define a random variable $L_n$ as the radius of the largest circle that lies inside $\mathcal{A}$ containing no points. Following Aldous [1, H9c, p. 160] and its argument, we can get

$$\Pr[L_n < r_n] = \exp\left(-\int_{\mathcal{A}} J_n(x)dx\right) + O\left(e^{-n\pi r_n^2}\right), \tag{6}$$

where

$$J_n(x) = n^2 \pi r_n^2 g(x)^2 \exp\left(-ng(x)\pi r_n^2\right).$$

Since $n\beta\pi r_n^2 = \ln n + \ln \ln n + a_n$, we have

$$J_n(x) = ng(x)^2\beta^{-1}(\ln n + \ln \ln n + a_n)$$

$$\cdot \exp\left[-g(x)\beta^{-1}(\ln n + \ln \ln n + a_n)\right]$$

$$\le n\gamma^2\beta^{-1}(\ln n + \ln \ln n + a_n)\exp\left[-(\ln n + \ln \ln n + a_n)\right]$$

$$= n\gamma^2\beta^{-1}(\ln n + \ln\ln n + a_n)(n\ln n)^{-1}\exp(-a_n)$$

$$= \gamma^2\beta^{-1}\left[\exp(-a_n) + \ln\ln n(\ln n)^{-1}\exp(-a_n) + a_n(\ln n)^{-1}\exp(-a_n)\right].$$

This yields that $J_n(x)$ tend to 0 as $a_n \to \infty$. By the dominated convergence theorem,

$$\lim_{n\to\infty}\int_A J_n(x)dx = \int_A \lim_{n\to\infty} J_n(x)dx = 0.$$

Thus,

$$\lim_{n\to\infty}\Pr[\mathcal{A} \subset \mathcal{M}(n, r_n)] = \lim_{n\to\infty}\Pr[L_n < r_n] = \lim_{n\to\infty}\exp\left(-\int_A J_n(x)dx\right) = 1.\square$$

In the remainder of this section, we explore the necessary and sufficient condition that guarantees (5) holds. It turns out that the behavior of $g$ around its minimum point will affect the coverage rate of $r_n$. See Theorem 3 and Theorem 4 for details.

**Theorem 3.** *Let* $\{X_1, \ldots, X_n\}$ *be* $n$ *independent random points in the unit square* $\mathcal{A}$ *according to a smooth distribution density function* $g(x)$. *Suppose that* $g(x)$ *satisfies* $0 < \beta \le g(x) \le \gamma < \infty$ *and* $\lambda(\{x\colon g(x) = \beta\}) > 0$, *where* $\lambda$ *is the Lebesgue measure. Let* $r_n = \sqrt{(\ln n + \ln\ln n + a_n)/\beta\pi n}$, *where* $\{a_n\}$ *is a sequence in* $\mathbb{R}$ *such that* $a_n = o(\ln n)$. *Then,*

$$\lim_{n\to\infty}\Pr[\mathcal{A} \subset \mathcal{M}(n, r_n)] = 1 \tag{7}$$

*if and only if* $a_n$ *tends to* $+\infty$ *as* $n \to \infty$. *Moreover,*

$$\lim_{n\to\infty}\Pr[\mathcal{A} \subset \mathcal{M}(n, r_n)] = 0 \tag{8}$$

*if and only if* $a_n$ *tends to* $-\infty$ *as* $n \to \infty$.

*Proof.* By the approximation formula (6),

$$\Pr[\mathcal{A} \subset \mathcal{M}(n, r_n)] = \exp\left[-I(n)\right] + O\left(e^{-n\pi r_n^2}\right), \tag{9}$$

where

$$I(n) = \int_A n^2\pi r_n^2 g(x)^2\exp\left(-ng(x)\pi r_n^2\right)dx.$$

Letting $J_n(x) = n^2\pi r_n^2 g(x)^2\exp\left(-ng(x)\pi r_n^2\right)$ and $S = \{x\colon g(x) = \beta\}$, we have

$$I(n) = I_1(n) + I_2(n) \tag{10}$$

with

$$I_1(n) := \int_S J_n(x)dx = \lambda(S)n^2\pi r_n^2\beta^2\exp\left(-n\beta\pi r_n^2\right) \tag{11}$$

and

$$I_2(n) := \int_{A\setminus S} J_n(x)dx \le \lambda(A\setminus S)n^2\pi r_n^2\gamma^2\exp\left(-n\beta\pi r_n^2\right). \tag{12}$$

Since $\lambda(S) > 0$, we have $I_2(n) \leq \alpha I_1(n)$ for some $\alpha \in (0, \infty)$. This implies that

$$I_1(n) \leq I(n) \leq (1+\alpha)I_1(n). \tag{13}$$

Since $n\beta\pi r_n^2 = \ln n + \ln\ln n + a_n$, hence we obtain

$$I_1(n) = \lambda(S)\beta\exp(-a_n) + o(1). \tag{14}$$

By (9), (13) and (14), we have

$$\lim_{n\to\infty} \Pr[\mathcal{A} \subset \mathcal{M}(n, r_n)] = 1 \Leftrightarrow \lim_{n\to\infty} I(n) = 0 \Leftrightarrow \lim_{n\to\infty} I_1(n) = 0 \Leftrightarrow a_n \to +\infty.$$

Similarly, we can check that

$$\lim_{n\to\infty} \Pr[\mathcal{A} \subset \mathcal{M}(n, r_n)] = 0 \Leftrightarrow \lim_{n\to\infty} I_1(n) = +\infty \Leftrightarrow a_n \to -\infty. \qquad \square$$

In the following we assume that the distribution density function $g(x)$ has finitely many minimum points at which some regular conditions hold. For a real function $g(x) = g(x_1, x_2)$ on $\mathbb{R}^2$ which is secondly differentiable at $x^*$, we define $B(x^*)$ as the $2 \times 2$ matrix with entries $b_{i,j} = \frac{\partial^2 g}{\partial x_i \partial x_j}(x^*)$, $i, j = 1, 2$.

**Theorem 4.** *Let $\{X_1, \ldots, X_n\}$ be $n$ independent random points according to a smooth distribution density function $g(x)$ on $\mathcal{A}$. Suppose that $g(x)$ has finitely many minimum points $\{x_i^*\} \subset \mathcal{A}$, $0 < \beta = g(x_i^*) \leq g(x) \leq \gamma < \infty$, and $B(x_i^*)$ are well defined and strictly positive definite. Let $r_n = \sqrt{(\ln n + a_n)/\beta\pi n}$, where $\{a_n\}$ is a sequence in $\mathbb{R}$ such that $a_n = o(\ln n)$. Then*

$$\lim_{n\to\infty} \Pr[\mathcal{A} \subset \mathcal{M}(n, r_n)] = 1 \tag{15}$$

*if and only if $a_n$ tends to $+\infty$ as $n \to \infty$. Moreover,*

$$\lim_{n\to\infty} \Pr[\mathcal{A} \subset \mathcal{M}(n, r_n)] = 0 \tag{16}$$

*if and only if $a_n$ tends to $-\infty$ as $n \to \infty$.*

Before proving Theorem 4, let us prepare several lemmas.

**Lemma 1.** *Suppose $B$ is a $d$-dimensional matrix which is symmetric and strictly positive definite. Then*

$$\int_{\mathbb{R}} \exp\left(-\frac{1}{2}x^T B^{-1} x\right) dx = (2\pi)^{d/2} |B|^{1/2}, \tag{17}$$

*where $|B|$ is the determinant of the matrix $B$.*

*Proof.* To prove (17), it suffices to note that

$$(2\pi)^{-d/2} |B|^{-1/2} \exp\left(-\frac{1}{2}x^T B^{-1} x\right)$$

is the density function of a normal distribution. $\qquad \square$

**Lemma 2.** *Suppose B is a $d \times d$ strictly positive definite matrix. Let*

$$D_n = \int_{\mathbb{R}^d} \exp\left(-\frac{n}{2} x^T B\, x\right) dx$$

*and*

$$D_n(r) = \int_{|x| \le r} \exp\left(-\frac{n}{2} x^T B\, x\right) dx.$$

*Then for $r > 0$,*

$$\lim_{n \to \infty} D_n(r)/D_n = 1.$$

*Proof.* By Lemma 1, it can be calculated that

$$D_n = (2\pi)^{d/2} n^{-d/2} |B|^{-1/2}. \tag{18}$$

Since $B$ is a strictly positive definite matrix, there exists $\delta > 0$ such that $x^T B\, x \ge \delta |x|^2$, where $|\cdot|$ is the Euclidean norm in $\mathbb{R}^d$. Thus,

$$D_n - D_n(r) = \int_{|x|>r} \exp\left(-\frac{1}{2} n x^T B\, x\right) dx$$

$$\le \int_{|x|>r} \exp\left(-\frac{n}{4} \delta r^2 - \frac{n}{4} x^T B\, x\right) dx$$

$$\le \exp\left(-n\delta r^2/4\right) \int_{\mathbb{R}^d} \exp\left(-n/4 \cdot x^T B\, x\right) dx$$

$$= \exp\left(-n\delta r^2/4\right) (2\pi)^{d/2} (n/2)^{-d/2} |B|^{-1/2}$$

$$= \exp\left(-n\delta r^2/4\right) \cdot 2^{d/2} D_n.$$

Therefore,

$$D_n(r)/D_n = 1 - \exp\left(-n\delta r^2/4\right) \cdot 2^{d/2} = 1 - o(1). \qquad \square$$

In what follows for two sequences $\{x_n\}$ and $\{y_n\}$ in $\mathbb{R}^+$, we write $x_n \sim y_n$ to mean that $a \le \liminf_{n\to\infty} x_n/y_n \le \limsup_{n\to\infty} x_n/y_n \le b$ for some $a, b \in (0, +\infty)$. It is clear that "$\sim$" is a equivalence relation.

**Lemma 3.** *Suppose that a real function $g(x)$ on $\mathbb{R}^2$ has a unique minimum points $x = 0$ and $B(0)$ is well defined and strictly positive definite. Define*

$$E_n(\Omega) = \int_{\Omega} \exp[-ng(x)] dx$$

*for $\Omega \subseteq \mathbb{R}^2$. Let $F_n(\varepsilon) = E_n(\{x \colon |x| \le \varepsilon\})$ and $G_n(\alpha) = E_n\left(\{x \colon |x| \le n^{-\alpha}\}\right)$ for $\varepsilon, \alpha > 0$. Then*

**(a)** *for any $\alpha \in (0, 1/2]$ and $\varepsilon > 0$, $F_n(\varepsilon) \sim G_n(\alpha)$;*
**(b)** *for any bounded set $\Omega$ containing a neighborhood of $x = 0$, $E_n(\Omega) \sim F_n(\varepsilon)$.*

*Proof.* We prove only the assertion (a). The proof of the assertion (b) is similar. Since $B(0)$ is a strictly positive definite matrix, there exist $\delta_1, \delta_2 > 0$ such that

$$\delta_1 |x|^2 \leq 1/2 \cdot x^T B(0)x \leq \delta_2 |x|^2.$$

By the Taylor's formula, $g(x) = g(0) + 1/2 \cdot x^T B(0)x + o(|x|^2)$. Moreover, we can choose $\delta_*$ and $\delta^*$ such that for $|x| \leq \varepsilon$,

$$\delta_* |x|^2 \leq 1/2 \cdot x^T B(0)x + o(|x|^2) \leq \delta^* |x|^2.$$

Therefore,

$$F_n(\varepsilon) = \int_{|x| \leq \varepsilon} \exp[-ng(x)] dx$$

$$= \int_{|x| \leq \varepsilon} \exp\left[-n\left(g(0) + 1/2 \cdot x^T B(0)x + o\left(|x|^2\right)\right)\right] dx$$

$$\leq \int_{|x| \leq \varepsilon} \exp\left[-n\left(\delta_* |x|^2 + g(0)\right)\right] dx$$

$$= \exp[-ng(0)] \int_{|x| \leq \varepsilon} \exp\left(-n\delta_* |x|^2\right) dx.$$

Now we calculate the integral in the above right hand side . Let $x = (\rho cos\theta, \rho sin\theta)$, where $\theta \in [0, 2\pi]$ and $\rho \in [0, \varepsilon]$. Then the Jacobi determinant $|J| = \rho$. Thus,

$$\int_{|x| \leq \varepsilon} \exp\left(-n\delta_* |x|^2\right) dx$$

$$= \int_0^\varepsilon \int_0^{2\pi} \exp\left(-n\delta_* \rho^2\right) \rho d\rho d\theta$$

$$= \frac{\pi}{n\delta_*} \left[1 - \exp\left(-n\delta_* \varepsilon^2\right)\right].$$

Therefore,

$$F_n(\varepsilon) \leq \frac{\pi}{n\delta_*}[1 - \exp\left(-n\delta_* \varepsilon^2\right)] \exp[-ng(0)]. \tag{19}$$

Analogously, for $G_n(\alpha)$ we have

$$G_n(\alpha) = \int_{|x| \leq n^{-\alpha}} \exp[-ng(x)] dx$$

$$= \int_{|x| \leq n^{-\alpha}} \exp\left[-n\left(g(0) + \frac{1}{2}x^T B(0)x + o\left(|x|^2\right)\right)\right] dx$$

$$\geq \int_{|x| \leq n^{-\alpha}} \exp\left[-n\left(\delta^* |x|^2 + g(0)\right)\right] dx$$

$$= \exp[-ng(0)] \int_{|x| \leq n^{-\alpha}} \exp\left(-n\delta^* |x|^2\right) dx.$$

A similar calculation leads to

$$G_n(\alpha) \geq \frac{\pi}{n\delta*} \left[1 - \exp(-n^{1-2\alpha}\delta*)\right] \exp[-ng(0)]. \tag{20}$$

Comparing (19) and (20), we obtains

$$\frac{F_n(\varepsilon)}{G_n(\alpha)} \leq \frac{\delta^*}{\delta_*} \cdot \frac{1 - \exp\left(-n\delta_*\varepsilon^2\right)}{1 - \exp\left(-n^{1-2\alpha}\delta*\right)}.$$

Since $\alpha \leq 1/2$, letting $n \to \infty$, we obtain

$$1 \leq \liminf_{n\to\infty} F_n(\varepsilon)/G_n(\alpha) \leq \limsup_{n\to\infty} F_n(\varepsilon)/G_n(\alpha) \leq \frac{\delta^*}{\delta_*} \cdot \frac{1}{1 - e^{-\delta*}} < \infty. \quad \square$$

*Proof of Theorem 4.* Without loss of generality, we assume that $g(x)$ has a unique minimum point $x^*$. By the approximation formula (6),

$$\Pr[\mathcal{A} \subset \mathcal{M}(n, r_n)] = \exp\left[-I(n)\right] + O\left(e^{-n\pi r_n^2}\right), \tag{21}$$

where

$$I(n) = \int_{\mathcal{A}} n^2 \pi r_n^2 g(x)^2 \exp\left[-ng(x)\pi r_n^2\right] dx.$$

Since $0 < \beta \leq g(x) \leq \gamma < +\infty$, we have

$$I(n) \sim n^2 \pi r_n^2 \int_{\mathcal{A}} \exp\left[-ng(x)\pi r_n^2\right] dx. \tag{22}$$

Note that $c_n := n\pi r_n^2 \to \infty$. By Lemma 3 we get,

$$I(n) \sim nc_n \int_{|x-x^*|\leq c_n^{-1/2}} \exp[-c_n g(x)] dx. \tag{23}$$

By the Taylor's formula, $g(x) = g(x^*) + 1/2 \cdot (x - x^*)^T B(x^*)(x - x^*) + o(|x|^2)$. Then the right hand side of (23) can be written as

$$nc_n \exp[-c_n g(x^*)] \int_{|x|\leq c_n^{-1/2}} \exp\left[-\frac{c_n}{2}x^T B(x^*)x - c_n o(|x|^2)\right] dx. \tag{24}$$

Note that $g(x^*) = \beta$. Applying Lemma 1-3, one obtains

$$I(n) \sim nc_n \exp(-c_n\beta) \int_{|x|\leq c_n^{-1/2}} \exp\left[-\frac{c_n}{2}x^T B(x^*)x\right] dx$$

$$\sim nc_n \exp(-c_n\beta) \int_{|x|\leq\varepsilon} \exp\left[-\frac{c_n}{2}x^T B(x^*)x\right] dx$$

$$\sim nc_n \exp(-c_n\beta) \int_{\mathbb{R}^2} \exp\left[-\frac{c_n}{2}x^T B(x^*)x\right] dx$$

$$= nc_n \exp(-c_n\beta) \cdot 2\pi c_n^{-1}|B(x^*)|^{-1/2}$$

$$\sim n \exp(-c_n\beta).$$

Since $c_n\beta = \pi\beta n r_n^2 = \ln n + a_n$, we have $n\exp(-c_n\beta) = \exp(-a_n)$. Hence,

$$I(n) \sim \exp(-a_n). \tag{25}$$

Therefore,

$$\lim_{n\to\infty} \Pr[\mathcal{A} \subset \mathcal{M}(n, r_n)] = 1 \Leftrightarrow \exp[-I(n)] \to 1 \Leftrightarrow I(n) \to 0 \Leftrightarrow a_n \to +\infty.$$

Thus the first assertion is verified. The second assertion can be checked similarly.

$\square$

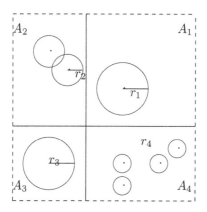

**Fig. 1.** Sensors with different sensing radii in different regions

## 4    Coverage of Sensor Networks with Empirical Distributions

From the discussion of the above section, we see that if sensors are deployed non-uniformly, then the minimum value of the density function $g(x)$ plays an important role in determining the smallest sensing radius for the coverage problem. In this section we propose a more economical scheme for non-uniformly distributed sensor networks. That is, sensors in different subregions could have different sensing radii, which mainly rely on the "local minimum value" of $g(x)$ in each subregion. Obviously, the scheme that sensors deployed with different sensing radii in non-uniform case saves more energy than that sensors distributed with common radii. The remainder of this section illustrate this scheme in details. Note that in practice, the density function $g(x)$ can always be approximated by empirical functions. Thus we assume that $g$ is an empirical distribution functions, that is, $g(x) = \sum_{i=1}^{K} c_i I_{A_i}$, where $\{A_1, \ldots, A_K\}$ is a partition of $\mathcal{A}$, i.e. $\bigcup_{i=1}^{K} A_i = \mathcal{A}$ and $A_i \cap A_j = \phi$ for $i \neq j$. In this situation, we can choose different sensing radius $r_i(n)$ for different $A_i$. We deploy sensors by the following procedure. (See Fig. 1.)

**(a)** Divide $n$ sensors into $K$ groups. The number $n_i$ of sensors in each group is proportional to $c_i S_i$, where $S_i$ is the area of the region $A_i$.

**(b)** Deploy the sensors of each group into the corresponding area $A_i$ independently and according to the uniform distribution on $A_i$.

With the above scheme we have the following result.

**Proposition 1.** *Let $\mathcal{A} = [0, 1]^2$ be divided as $\mathcal{A} = \bigcup_{i=1}^{K} A_i$ with $A_i \bigcap A_j = \phi$ for $i \neq j$, $i, j = 1, \cdots, K$, where each $A_i$ is a rectangle with area $\mu(A_i)$. For $n$ large enough, we put $m_i = \lceil nc_i\mu(A_i)\rceil$ sensors in $A_i$, which are located as $m_i$ independent random points with uniform distribution on $A_i$. Suppose that each sensor in $A_i$ has a sensing radius $r_i(n) = \sqrt{(\ln n + \ln\ln n + a_i(n))/c_i\pi n}$. Suppose further that $a_i(n) = o(\ln n)$ and $\lim_{n\to\infty} a_i(n)$ exists in $[-\infty, +\infty]$. Then we have*

$$\Pr[\mathcal{A} \text{ is covered}] = \exp\left[-\sum_{i=1}^{K} \exp(-a_i(n))\right] + o(1). \tag{26}$$

*Proof.* In the limiting procedure we may ignore the boundary effects. Thus,

$$P[\mathcal{A} \text{ is covered}] = \prod_{i=1}^{K} P[A_i \text{ is covered}] + o(1).$$

It is enough to check that

$$P[A_1 \text{ is covered}] = \exp\left[-\exp(-a_1(n))\right] + o(1). \tag{27}$$

For convenience we assume that $c_1\mu(A_1)n = m_1$ is an integer. Scaling the area and the sensing radius simultaneously, we can apply Theorem 1 to obtain

$$P[A_1 \text{ is covered}] = \exp[-I(m_1)] + o(1),$$

where

$$I(m_1) = \frac{1}{\mu(A_1)} m_1^2 \pi r_1(n)^2 \exp\left[-m_1\pi r_1(n)^2 \frac{1}{\mu(A_1)}\right]. \tag{28}$$

Under our assumption on $r_1(n)$, we have

$$\pi r_1(n)^2 = \frac{1}{c_1 n}\left(\ln n + \ln\ln n + a_1(n)\right)$$

$$= \frac{\mu(A_1)}{c_1 n\mu(A_1)}\left[\ln(c_1\mu(A_1)n) + \ln\ln(c_1\mu(A_1)n) + a_1(n) + O(1)\right]$$

$$= \frac{\mu(A_1)}{m_1}[\ln m_1 + \ln\ln m_1 + a_1(n) + O(1)]. \tag{29}$$

Combining (28) and (29) we obtain

$$I(m_1) = \exp[-a_1(n)]\left[1 + \ln\ln m_1(\ln m_1)^{-1} + a_1(n)(\ln m_1)^{-1}\right]. \tag{30}$$

Thus $I(m_1) = \exp[-a_1(n)] + o(1)$. Hence (27) follows. □

# 5    Conclusions

In this paper we studied the limiting achievable coverage problems of sensor networks. Based on our results we can estimate the coverage probability which might be useful for the design and analysis of sensor networks. In particular, the results provide us certain functional relations between the sensing radius and the number of sensors which ensures a preassigned region to be covered with high probability. we found that if sensors are deployed non-uniformly, then the minimum value of the density function plays an important role in determining the smallest sensing radius for the coverage problem. Based on this observation an economical scheme for non-uniformly distributed sensor networks was proposed in the paper.

# Acknowledgements

We thank Yibing Cai for his talk on wireless sensor networks at our seminar. We thank also Qingyang Guan, Xue Cheng, Shaojun Guo, Yongsheng Song for helpful discussions. Zhi-Ming Ma is grateful to the organizers of the Conference CJCDGCGT2005 for inviting him to give a talk. The work is partly supported by NSFC and 973 Project. We thank the two anonymous referees for their valuable comments.

# References

1. Aldous, D.: Probability Approximations via the Poisson Clumping Heuristic, Spring-Verlag, New York, (1989)
2. Gupta, P., Kumar, P.R.: Critical Power for Asymptotic Connectivity in Wireless Networks. In: McEneaney, W.M., Yin, G.G., Zhang, Q.(eds.): Stochastic Analysis, Control, Optimization and Applications. Systems Control Found. Appl., Birkhäuser Boston, Boston (1998) 547–566
3. Hall, P.: Introduction to the Theory of Coverage Processes. John Wiley & Sons, New York (1988)
4. Huang, C.-F., Tseng, Y.-C.: The Coverage Problem in a Wireless Sensor Network. In: ACM Intl. Workshop on Wireless Sensor Networks and Applications (2003) 115–121
5. Kar, K., Banerjee, S.: Node Placement for Connected Coverage in Sensor Networks. In: Proceedings of WiOpt : Modeling and Optimization in Mobile. Sophia Antipolis, France (2003)
6. Koskinen, H.: On the Coverage of a Random Sensor Network in a Bounded Domain, In: Proceedings of 16th ITC Specialist Seminar (2004) 11–18
7. Liu, B., Brass, P., Dousse, O., Nain, P., Towsley, D.: Mobility Improves Coverage of Sensor Networks. Submitted to the ACM MobiHoc 2005 Symposium (2004)
8. Liu, B., Towsley, D.: A Study of the Coverage of Large-Scale Sensor Networks. In: IEEE International Conference on Mobile Ad-hoc and Sensor Systems (2004)
9. Li, X.Y., Wan, P.J., Frieder, O.: Coverage in Wireless Ad Hoc Sensor Networks. IEEE Trans. on Computers, Vol. 52. (2003) 753–763

10. Meguerdichian, S., Koushanfar, F., Potkonjak, M., Srivastava, M.B.: Coverage Problems in Wireless Ad-hoc Sensor Networks. In: Proc. IEEE Infocom (2001) 1380-C1387
11. Miorandi, D., Altman, E.: Coverage and Connectivity of Ad-Hoc Networks in Presence of Channel Randomness. In: Proc. of IEEE INFOCOM 2005. Miami, USA (2005)
12. Philips, T.K., Panwar, S.S., Tantawi, A.N.: Connectivity Properties of a Packet Radio Network Model. IEEE Transactions on Information Theory **35** (5) (1989) 1044–1047
13. Poduri, S., Sukhatme., G.S.: Constrained Coverage for Mobile Sensor Networks. IEEE International Conference on Robotics and Automation. LA, USA (2004) 165–172
14. Shakkottai, S., Srikant, R., Shroff, N.: Unreliable Sensor grids: Coverage, Connectivity and Diameter. In: Proc. IEEE Infocom (2003)
15. Tian, D., Georganas, N.D.:A coverage-preserving Node Scheduling Scheme for Large Wireless Sensor Networks. In: First ACM International Workshop on Wireless Sensor Networks and Applications. (2002) 32–41
16. Zhang, H.h., Hou, J.C.: Maintaining Sensing Coverage and Connectivity in Large Sensor Networks. Ad Hoc & Sensor Wireless Networks, Vol. 1. March 3, (2005) 89C-124

# Some Topics on Edge-Coloring*

Guizhen Liu[1] and Changqing Xu[2]

[1] Department of Applied Mathematics, Shandong University,
Weihai 264209, P.R. China
gzliu@sdu.edu.cn
[2] Department of Applied Mathematics, Hebei University of Technology, Tianjin
300130, P.R. China
chqxu@hebut.edu.cn

**Abstract.** In this paper some new results on edge coloring of graphs are introduced. This paper deals mainly with edge cover coloring, $g$-edge cover coloring, $(g, f)$-coloring and equitable edge coloring. Some new problems and conjectures are presented.

## 1   Introduction

Throughout this paper, let $G(V, E)$ be a graph, which allows parallel (multiple) edges but no loops and has a finite vertex set $V(G)$ and a finite nonempty edge set $E(G)$. If $G$ has no multiple edges, $G$ is called a simple graph. For a vertex $v$ of $G$ the degree of $v$ in $G$ is denoted by $d_G(v)$. Let $\Delta(G)$ and $\delta(G)$ respectively denote the maximum degree and the minimum degree of $G$. Given two vertices $u, v \in V(G)$, let $E(uv)$ be the set of edges joining $u$ and $v$ in $G$. The multiplicity $\mu(uv)$ of edge $uv$ is the size of $E(uv)$. Set $\mu(v) = \max\{\mu(uv) \colon u \in V\}$, $\mu(G) = \max\{\mu(v) \colon v \in V\}$, which are respectively called the multiplicity of vertex $v$ and the multiplicity of graph $G$. An edge-coloring of $G$ is an assignment of colors to the edges of $G$. Associate positive integer $1, 2, \ldots$ with colors, and we call $C$ a $k$-edge-coloring of $G$ if $C \colon E \to \{1, 2, \ldots, k\}$. Let $i_C(v)$ denote the number of edges of $G$ incident with vertex $v$ that receive color $i$ by the coloring $C$. For simplification, we write $i(v) = i_C(v)$ if there is no obscurity.

The edge-coloring problem is one of the fundamental problems on graphs. The ordinary edge-coloring is to color the edges of $G$ such that no two adjacent edges are assigned the same color. The minimum number of colors needed to color edges of $G$ is called the chromatic index $\chi'(G)$ of $G$. In 1964, Vizing gave the famous Vizing's theorem [16]. In 1974, Gupta gave the concept of edge cover coloring, which is parallel to the ordinary edge-coloring, and in some sense is a generalization of the ordinary edge-coloring. The edge cover coloring of $G$ is to color the edges of $G$ so that each color appears at each vertex $v$ at least once. The maximum number of colors needed to an edge coloring of $G$ is called the edge cover chromatic index of $G$ and is denoted by $\chi'_c(G)$. Corresponding to the ordinary edge-coloring problems, there are many interesting problems about

---

* This research was supported by NNSF(10471078)and RSDP(20040422004) of China.

J. Akiyama et al. (Eds.): CJCDGCGT 2005, LNCS 4381, pp. 101–108, 2007.

edge cover coloring to consider. In this paper, we introduce some new results of edge coloring which are related to the edge cover coloring.

## 2    Edge Cover Coloring

For edge cover chromatic index, Gupta gave the following result in [2].

**Theorem 1. (Gupta's Theorem [2])** *For any graph $G$,*

$$\min\{d(v) - \mu(v) \colon v \in V\} \leq \chi_c'(G) \leq \delta.$$

Clearly, Gupta's theorem parallels to Vizing's theorem. In [22] we improved the lower bound of $\chi_c'(G)$ when $2 \leq \delta(G) \leq 5$.

**Theorem 2.** [22] *Let $G$ be a graph and $2 \leq \delta(G) \leq 5$. Then $\chi_c'(G) \geq \delta(G) - 1$.*

A graph $G$ is said to be of *Type CI* if $\chi_c'(G) = \delta$ and of *Type CII* if $\chi_c'(G) < \delta$. For two vertices $u, v \in V$, an edge $e = uv \notin E(G)$ is said to be *critical* if $\chi_c'(G + e) > \chi_c'(G)$. A graph $G$ is said to be *edge cover critical* if it is of Type $CII$ and every edge with vertices $V(G)$ not belonging to $E(G)$ is critical.

Miao studied the classification of simple graphs based on the edge cover chromatic index in [8].

**Theorem 3.** [8] *Let $G$ be a simple graph with $n$ vertices, $m$ edges and with minimum degree $\delta$, then $G$ is of Type CII if $m < \delta\lceil n/2 \rceil$.*

**Theorem 4.** [8] *Let $G$ be a simple graph of odd order with minimum degree $\delta$. If $\Sigma_{v \in V}(d(v) - \delta) < \delta$, then $G$ is of Type CII.*

**Theorem 5.** [8] *If $G$ is a regular simple graph with an odd number of vertices, then $G$ is of Type CII.*

**Theorem 6.** [8] *Let $G$ be a simple graph. If the subgraph of $G$ induced by the vertices of degree $\delta(G)$ is a forest, then $G$ is of Type CI.*

A deep noregular graph is a graph such that the degrees of any two vertices adjacent to a same vertex are different.

**Theorem 7.** [8] *All deep noregular simple graphs are of Type CI.*

Miao also discussed the classification of planar graphs, considered the edge cover chromatic indices of Halin graphs, Composition graphs and Cartesian product graphs in [8].

Song and Liu discussed some properties of edge cover critical simple graphs [14]. They gave the following adjacent theorem which is in some sense parallel to the famous Vizing's adjacency lemma.

**Theorem 8.** [14] *Let $G$ be an edge cover critical simple graph. Then for any $u, v \in V$, $e = uv \notin E(G)$, there exists $w \in \{u, v\}$ such that $d(w) \leq 2\delta - 2$ and $w$ is adjacent with at least $d(w) - \delta + 1$ vertices of degree $\delta$.*

We extended this result to multigraphs.

**Theorem 9.** **[23]** *Let $G$ be an edge cover critical graph. Then for any $u$, $v \in V$, $e = uv \notin E(G)$, there exists $w \in \{u, v\}$ such that $d(w) \leq 2\chi_c'(G)$. And let $D(w) = \{x \colon x \text{ and } w \text{ is adjacent in } G \text{ and } d(x) - \mu(xw) \leq \chi_c'(G)\}$ and let $d^*(w)$ denote the number of edges of $G$ joining $w$ to vertices of $D(w)$, then*

$$d^*(w) \geq d(w) - \chi_c'(G).$$

Miao and Liu considered the fractional edge cover coloring and gave an exact formula of the fractional edge cover chromatic index in [9].

## 3    *g*-Edge Cover-Coloring

Song and Liu generalized the edge cover coloring to $g$-edge cover coloring in [13]. An edge coloring $C$ of $G$ is called a $g$-edge cover coloring, if each color appears at least $g(v)$ times at each vertex $v \in V(G)$, where $g(v)$ is a positive integer-valued function. The maximum number of colors need in a $g$-edge cover-coloring of a graph $G$ is called the $g$-edge cover chromatic index of $G$ and is denoted by $\chi_{gc}'(G)$.

Song and Liu determined the $g$-edge cover chromatic index of some kinds of graphs and gave a lower bound of the $g$-edge cover chromatic index.

Set $\delta_g(G) = \min\{\lfloor d_G(v)/g(v)\rfloor \colon v \in V(G)\}$. Let $E(i)$ be the set of edges receive color $i$ in an edge-coloring $C$ of $G$.

**Theorem 10.** **[13]** *Let $G$ be a bipartite graph. Then $\chi_{gc}'(G) = \delta_g(G)$. Furthermore if $\chi_{gc}'(G) = k \geq 2$, there exists a g-edge cover-coloring $C$ of $G$ for which $\|E(i) - E(j)\| \leq 1$ for $i$ and $j \in \{1, 2, \ldots, k\}$ and $|i_C(v) - j_C(v)| \leq 1$ for all $v \in V$, $i$ and $j \in \{1, 2, \ldots, k\}$.*

**Theorem 11.** **[13]** *Let $G$ be a graph. If $g(v)$ is positive and even for all $v \in V$, then*

$$\chi_{gc}'(G) = \delta_g(G).$$

**Theorem 12.** **[13]** *Let $G$ be a graph. Let $1 \leq g(v) \leq d(v)$ for all $v \in V$. Then*

$$\chi_{gc}'(G) \geq \min_{v \in V}\left\{\lfloor (d(v) - \mu(v))/g(v)\rfloor\right\}.$$

Set $\delta_g' = \min_{v \in V}\{\lfloor (d(v) - 1)/g(v)\rfloor\}$. We obtained the following result in [19].

**Theorem 13.** **[19]** *Let $G$ be a graph. Let $1 \leq g(v) \leq d(v)$ for all $v \in V$ and $1 \leq \delta_g' \leq 4$. Then $\chi_{gc}'(G) \geq \delta_g'$.*

## 4   Super $g$-Edge Cover-Coloring

A super $g$-edge cover-coloring of $G$ is a $g$-edge cover-coloring with the property that parallel edges receive distinct colors. Let $\chi''_{gc}(G)$ denote the maximum positive integer $k$ for which a super $g$-edge cover-coloring of $G$ exists. $\chi''_{gc}(G)$ is called the super $g$-edge cover chromatic index of $G$.

A super $g$-edge cover-coloring is a generalization of a $g$-edge cover-coloring. Liu, Xu and Xin discussed the properties of the super $g$-edge cover-coloring. A super $g$-edge cover-coloring is called a super edge cover-coloring when $g(v) = 1$ for all $v \in V(G)$.

**Theorem 14.** [19] *Let $G$ be a graph. Then there exists a super edge cover-coloring if and only if $\chi'_c(G) \geq \mu(G)$. Furthermore, if there exists a super edge cover coloring, then $\chi''_c(G) = \chi'_c(G)$.*

**Theorem 15.** [18] *Let $G$ be a graph and $\mu(G) = 2$. If $1 \leq g(v) \leq \lfloor (d(v)-1)/2 \rfloor$ for each $v \in V(G)$, then there exists a super $g$-edge cover coloring.*

**Theorem 16.** [18] *Let $G$ be a graph. If $1 \leq g(v) \leq \lfloor (d(v) - \mu(v))/\mu \rfloor$ for each $v \in V(G)$, then*

$$\chi''_{gc}(G) \geq \min\{\lfloor (d(v) - \mu(v))/g(v) \rfloor\}.$$

**Theorem 17.** [19] *Let $G(V, E)$ be a bipartite graph. Let $g$ be an integer-valued function defined on $V(G)$ such that $1 \leq g(v) \leq d(v)/\mu$ for all $v \in V$. Then $\chi''_{gc}(G) = \delta_g$.*

*Furthermore, if $\delta_g = k \geq 2$, then there exists a super $g$-edge cover-coloring $C$ of $G$ for which $\|E(i)| - |E(j)\| \leq 1$ for $i$ and $j \in \{1, 2, \ldots, k\}$ and $|i_C(v) - j_C(v)| \leq 1$ for all $v \in V$, $i$ and $j \in \{1, 2, \ldots, k\}$.*

**Corollary 1.** [19] *Let $G(V, E)$ be a bipartite graph and $g$ be a positive integer-valued function defined on $V(G)$. Then $G$ exists a super $g$-edge cover coloring if and only if $\mu(G)g(v) \leq d(v)$ for all $v \in V$.*

For $g$-edge cover-coloring, it is proved that if $g(v)$ is even and positive for all $v \in V$, then $\chi'_{gc}(G) = \delta_g$ ([13]). But for super $g$-edge cover-coloring, if $g(v)$ is even and positive for all $v \in V$, and $g(v) \leq d(v)/\mu$, the conclusion $\chi''_{gc}(G) = \delta_g$ does not always holds.

**Theorem 18.** [19] *If $\chi''_{gc}(G)$ exists and $d(v)(\bmod\ g(v)) \geq \mu(v)$ for each $v \in V$, then*

$$\chi''_{gc}(G) = \chi'_{gc}(G).$$

## 5   Equitable Edge-Coloring in Multigraphs

An edge-coloring $C$ of $G$ with $k$ colors is equitable if $|i(v) - j(v)| \leq 1$ ($1 \leq i < j \leq k$) for any $v \in V(G)$; and nearly equitable if $|i(v) - j(v)| \leq 2$ ($1 \leq i < j \leq k$) for any $v \in V(G)$.

Hilton and de Werra studied the equitable edge-coloring problem of simple graphs in [4]. The following theorem is one of the main results in [4].

**Theorem 19.** [4] *Let $G$ be a simple graph and let $k \geq 2$. If $k \nmid d(v)$ for all $v \in V(G)$, then $G$ has an equitable edge-coloring with $k$ colors.*

If $k = \Delta(G) + 1$, then Theorem 19 reduces to Vizing's theorem of simple graphs [16]. If $k = \delta(G) - 1$, then Theorem 19 implies Gupta's theorem of simple graphs [2]. We discussed equitable edge-coloring of multigraphs.

**Theorem 20.** [24] *Let $G$ be a graph and $k \geq 2$. If $\mu(v) \leq d(v)(\mathrm{mod}\, k) \leq k - \mu(v)$ for all $v \in V(G)$, then $G$ has an equitable edge-coloring with $k$ colors.*

Theorem 20 implies Theorem 19. Furthermore, let $k = \Delta + \mu$, then $r(v) = d(v)$. So $r(v) \geq \mu(v)$ and $k - r(v) \geq \mu(v)$ for all $v \in V$. By Theorem 20, $G$ has an equitable edge-coloring with $k$ colors. That is, Theorem 20 can deduce to Vizing's theorem in multigraphs [16].

# 6    $(g, f)$-Coloring

Let $G$ be a graph. Let $g$ and $f$ be two integer-valued functions defined on $V(G)$ such that $0 \leq g(v) \leq f(v)$ for each vertex $v$ of $V(G)$. A $(g, f)$-coloring is to color all the edges of $G$ such that each color appears at each vertex $v$ at least $g(v)$ and at most $f(v)$ times. The minimum number of colors used by a $(g, f)$-coloring of $G$ is denoted by $\chi'_{gf}(G)$ which is called the $(g, f)$-chromatic index of $G$. The maximum number of colors used by a $(g, f)$-coloring of $G$ is denoted by $\overline{\chi}'_{gf}(G)$, which is called the upper $(g, f)$-chromatic index of $G$. The $(g, f)$-coloring is a generalization of the edge-coloring and the edge cover coloring, which arises in many applications, such as the network design, the file transfer problem on computer networks and so on [7, 10, 12]. The $(g, f)$-chromatic index was first posed in [26]. Liu gave the concept of the upper $(g, f)$-chromatic index.

An $f$-coloring of a graph $G$ is a $(g, f)$-coloring for the case $g(v) = 0$ for each vertex $v \in V(G)$. The minimum number of colors needed to an $f$-coloring of $G$ is called the $f$-chromatic index of $G$ and is denoted by $\chi'_f(G)$. The $f$-coloring of a graph $G$ was considered in [3].

A $(g, f)$-factor of $G$ is a spanning subgraph $H$ of $G$ satisfying $g(v) \leq d_H(v) \leq f(v)$ for each $v \in V(G)$. If a graph $G$ itself is a $(g, f)$-factor, then $G$ is called a $(g, f)$-graph. Set $\Delta_f(G) = \max\{\lceil d_G(v)/f(v) \rceil : v \in V(G)\}$.

**Theorem 21.** [17] *Let $G$ be a bipartite graph. Then $G$ has a $(g, f)$-coloring if and only if $G$ is an $(mg, mf)$-graph for some positive integer $m$.*

**Theorem 22.** [20] *Let $G$ be a bipartite graph. Then $G$ has a $(g, f)$-coloring if and only if $\Delta_f(G) \leq \delta_g(G)$. And if $G$ has a $(g, f)$-coloring, then*

$$\chi'_{gf}(G) = \Delta_f(G), \quad \overline{\chi}'_{gf}(G) = \delta_g(G).$$

**Lemma 1.**  [20] *Let $G$ be a graph and let $g$ and $f$ be respectively nonnegative and positive integer-valued functions defined on $V(G)$ such that $g(v) \leq f(v)$ for all $v \in V(G)$. If $G$ has a $(g, f)$-coloring, then*

$$\chi'_f(G) \leq \chi'_{gf}(G) \leq \overline{\chi}'_{gf}(G) \leq \chi'_{gc}(G).$$

**Theorem 23.**  [20] *Let $G$ be a graph with $\chi'_f(G) \leq \chi'_{gc}(G)$. If there is an equitable coloring of $G$ with $\chi'_f(G)$ colors, then $G$ has a $(g, f)$-coloring and $\chi'_{gf}(G) = \chi'_f(G)$.*

**Theorem 24.**  [20] *Let $G$ be a graph with $\chi'_f(G) \leq \chi'_{gc}(G)$. If there is an equitable coloring of $G$ with $\chi'_{gc}(G)$ colors, then $G$ has a $(g, f)$-coloring and $\overline{\chi}'_{gf}(G) = \chi'_{gc}(G)$.*

**Theorem 25.**  [20] *Let $G$ be a graph with $\chi'_f(G) \leq \chi'_{gc}(G)$. Let $k$ be an integer such that $\chi'_f(G) \leq k \leq \chi'_{gc}(G)$. If one of the following conditions holds,*

**(1)** $d(v) \geq k(g(v) + 1) - 1$, *for all $v \in V(G)$,*
**(2)** $d(v) \leq k(f(v) - 1) + 1$, *for all $v \in V(G)$,*

*then $G$ has a $(g, f)$-coloring.*

**Theorem 26.**  [19] *Let $G$ be a graph. If $g(v)$ and $f(v)$ are positive and even for all $v \in V(G)$, and $\delta_g(G) \geq \Delta_f(G)$, then $\chi'_{gf}(G) = \Delta_f(G)$ and $\overline{\chi}'_{gf}(G) = \delta_g(G)$.*

**Theorem 27.**  [19] *Let $G$ be a graph and $m$ be a positive integer.*

**(1)** *If $g(v)$ is positive and even for all $v \in V(G)$ and $G$ is an $(mg, mf - m + 1)$-graph, then $\overline{\chi}'_{gf}(G) = \delta_g(G)$.*
**(2)** *If $f(v)$ is positive and even for all $v \in V(G)$ and $G$ is an $(mg + m - 1, mf)$-graph, then $\chi'_{gf}(G) = \Delta_f(G)$.*

**Theorem 28.**  [19] *Let $G$ be a simple graph. If $\chi'_f(G) = \max_{v \in V}\{\lceil(d(v) + 1)/f(v)\rceil\}$, $\chi'_{gc}(G) = \min_{v \in V}\{\lfloor(d(v) - 1)/g(v)\rfloor\}$ and $\chi'_{gc}(G) \geq \chi'_f(G)$, then $G$ has a $(g, f)$-coloring and $\chi'_{gf}(G) = \chi'_f(G)$, $\overline{\chi}'_{gf}(G) = \chi'_{gc}(G)$.*

# 7    Further Research Problems

Finally, we present some problems for further research as follows.

*Problem 1.* If $\chi''_{gc}(G)$ exists, does $\chi'_{gc}(G) = \chi''_{gc}(G)$ hold?

*Problem 2.* Is $\chi'_{gc}(G) \geq \chi'_f(G)$ a necessary and sufficient condition for a graph $G$ to have a $(g, f)$-coloring?

*Problem 3.* If $G$ has a $(g, f)$-coloring, do $\chi'_{gf}(G) = \chi'_f(G)$ and $\overline{\chi}'_{gf}(G) = \chi'_{gc}(G)$ hold?

*Problem 4.* What is the sufficient condition for equitable edge-coloring of multigraph for a general $k$?

# References

1. Balister, P.N., Riordan, O.M., Schelp, R.H.: Vertex Distinguishing Edge Colorings of Graphs. J. Graph Theory **42** (2003) 95–109
2. Gupta, R.P.: On Decompositions of a Multigraph into Spanning Subgraphs. Bull. Amer. Math. Soc. **80** (1974) 500–502
3. Hakimi, S.L., Kariv, O.: A Generalization of Edge-Coloring in Graphs, J. Graph Theory **10** (1986) 139–154
4. Hilton, A.J., de. Werra, D.: A Sufficient Condition for Equitable Edge-Coloring of Simple Graphs, *Discrete Math.*, **128** (1994) 179–201
5. Li, J., Zhang, Z., Chen, X., Sun, Y.: A Note on Adjacent Strong Edge Coloring of K(n, m). Acta Math. Appl. (English Series) **22** (2) (2006) 283–286
6. Liu, G. On $(g, f)$-Factors and Factorizations. Acta Math. Sinica **37** (1994) 230–237
7. Liu, G., Deng, X.: A Polynomial Algorithm for Finding $(g, f)$-Colorings Orthogonal to Stars in Bipartite Graphs. Science in China, Ser A **35** (3) (2005) 334–344
8. Miao, L.: The Edge Cover Coloring of Graphs. Ph.D. Thesis of Shandong University (2000)
9. Miao, L., Liu, G.: A Edge Cover Coloring and Fractional Edge Cover Coloring. J. of Systems Science and Complexing **15** (2) (2002) 187–193
10. Nakano, S., Nishizeki, T., Saito, N.: On the $f$-Coloring of Multigraphs. IEEE Trans. Circuit and Syst. **35** (3) (1988) 345–353
11. Nakano, S., Nishizeki, T., Saito, N.: On the $fg$-Coloring of Graphs. Combinatorica **10** (1) (1990) 67–80
12. Nakano, S., Nishizeki, T.: Scheduling File Transfers under Port and Channel Constraints. Internat. J. Found. Comput. Sci. **4** (2) (1993) 425–431
13. Song, H., Liu, G.: On $f$-Edge Cover-Coloring in Multigraphs. Acta Math. Sin. **48** (5) (2005) 910–919
14. Song, H., Liu, G.: Some Properties of Edge Cover Critical Graphs. Advances in Math. **33** (2004) 96–102
15. Song, H., Liu, G.: Applications of an Equitable Edge-Coloring Theorem. The International Conference on Mathematical Programming (2002) 19–22
16. Vizing, V.G.: On an Estimate of the Chromatics Class of a $p$-Graph (Russian). Diskret. Analiz. **3** (1964) 25–30
17. de. Werra, D.: Equitable Colorations of Graphs. Rev. Fran?aise Informat. Recherche Oprationnelle **R-3** (1971) 3–8
18. Xin, Y.: On Edge-Colouring Problems of Graphs. Ph.M. Thesis of Shandong University (2005)
19. Xu, C.: Some Topics on Edge Coloring Problems. Post-doctor Report of Shandong University (2005)
20. Xu, C., Liu, G.: The Existence of $(g, f)$-Coloring. Submitted to the Electronic J. Comb.
21. Xu, C., Liu, G.: On Super $f$-Edge Cover-Coloring in Multigraphs. Submitted to Discrete Math.
22. Xu, C., Liu, G.: A Note on Edge Covered Chromatic Index of Multigraphs. Submitted to Appl. Math. Letters
23. Xu, C., Liu, G.: Edge Covered Critical Multigraphs. Submitted to Discrete Math.

24. Xu, C., Liu, G.: On Equitable Edge-Coloring in Multigraphs. Submitted to Science in China, Ser A
25. Zhang, Z., Liu, L., Wang, J.: Adjacent Strong Edge Coloring of Graphs. Applied Math. Letters **15** (2002) 623–626
26. Zhou, X., Fuse, K., Nishizeki, T.: A Linear Algorithm for Finding $[g, f]$-Colorings of Partial $k$-Trees. Algorithmica **27** (2000) 227–243

# Hamiltonicity of Complements of Total Graphs

Guoyan Ma and Baoyindureng Wu

College of Mathematics and System Science, Xinjiang University,
Urumqi, Xinjiang, 830046, P.R. China
baoyin@xju.edu.cn

**Abstract.** For a graph $G$, the total graph $T(G)$ of $G$ is the graph with vertex set $V(G) \cup E(G)$ in which the vertices $x$ and $y$ are joined by an edge if $x$ and $y$ are adjacent or incident in $G$. In this paper, we show that the complement of total graph $T(G)$ of a simple graph $G$ is hamiltonian if and only if $G$ is not isomorphic to any graph in $\{K_{1,r} \mid r \geq 1\} \cup \{K_{1,s} + K_1 \mid s \geq 1\} \cup \{K_{1,t} + e \mid t \geq 2\} \cup \{K_2 + 2K_1, K_3 + K_1, K_3 + 2K_1, K_4\}$.

**Keywords:** total graph; complement; Hamilton cycle.

## 1 Introduction

All graphs considered here are finite, undirected and simple. We refer to [2] for unexplained terminology and notations. Let $G = (V(G), E(G))$ be a graph. $|V(G)|$ and $|E(G)|$ are called the order and the size of $G$, respectively. For a vertex $v$ of $G$, if there is no confusion, the degree $d_G(v)$ is simply denoted by $d(v)$. $I_G(v)$ denotes the set of all edges incident with $v$. The neighborhood $N_G(v)$ of $v$ is the set of all vertices of $G$ adjacent to $v$. The symbols $\Delta(G)$, $\delta(G)$, $\kappa(G)$ and $\lambda(G)$ denote the maximum degree, the minimum degree, the connectivity and the edge connectivity, respectively. Let $S$ be a nonempty subset of $V(G)$. The subgraph of $G$ induced by $S$, denoted by $G[S]$, is the subgraph of $G$ with vertex set $S$, in which two vertices are adjacent if and only if they are adjacent in $G$. If $G[S]$ has no edge, $S$ is called an independent set; if $G[S]$ is a complete graph, then $S$ is called a clique. The maximum cardinality of an independent set is called the independence number of $G$, denoted by $\alpha(G)$; the maximum cardinality of a clique is called the clique number of $G$, denoted $\omega(G)$.

As usual, $P_n$, $C_n$ and $K_n$ are respectively, the path, cycle, and complete graph of order $n$. For two positive integers $r$ and $s$, $K_{r,s}$ is the complete bipartite graph with two partite sets containing $r$ and $s$ vertices. In particular, $K_{1,s}$ is called a star. For $s \geq 2$, $K_{1,s} + e$ is the graph obtained from $K_{1,s}$ by adding a new edge which joins two vertices of degrees one, while $K'_{1,s}$ denotes the graph by adding a new vertex and joining it to exactly one vertex of degree one of $K_{1,s}$. For a star $K_{1,s}$ with $s \geq 3$, when we add two more edges which join the vertices of degrees 1, there are only two possibilities. $(K_{1,s} + 2e)_1$ denotes the resulting graph if the two edges are adjacent, and $(K_{1,s} + 2e)_2$, otherwise. $K_n^-$ denotes the graph resulting from $K_n$ by deleting an edge. We say two graphs $G$ and $H$ are disjoint if they have no vertex in common, and denotes their union by $G + H$;

J. Akiyama et al. (Eds.): CJCDGCGT 2005, LNCS 4381, pp. 109–119, 2007.

it is called the disjoint union of $G$ and $H$. The disjoint union of $k$ copies of $G$ is written as $kG$. The join $G \vee H$ of $G$ and $H$ is the graph obtained from $G + H$ by joining each vertex of $G$ to each vertex of $H$.

Let $G$ be a graph. The complement of $G$, denoted by $\overline{G}$, is the graph with the same vertex set as $G$, but where two vertices are adjacent if and only if they are not adjacent in $G$. The line graph $L(G)$ of $G$ is the graph whose vertex set is $E(G)$ in which two vertices are adjacent if and only if they are adjacent in $G$. The total graph $T(G)$ of $G$ is the graph whose vertex set is $V(G) \cup E(G)$ in which two vertices are adjacent if and only if they are adjacent or incident in $G$. Wu and Meng [8] generalized the concept of total graphs to total transformation graph $G^{xyz}$ with $x, y, z \in \{-, +\}$, where $G^{+++}$ is exactly the total graph of $G$, and $G^{---}$ is the complement of $G^{+++}$. Formally, $G^{---}$ is the graph with $V(G) \cup E(G)$ as its vertex set in which two vertices $x$ and $y$ are joined by an edge if one of the following situations occurs: $(i)$ $x, y \in V(G)$, $x$ and $y$ are not adjacent in $G$; $(ii)$ $x, y \in E(G)$, $x$ and $y$ are not adjacent in $G$; $(iii)$ one of $x$ and $y$ is in $V(G)$ while the other one is in $E(G)$, and they are not incident in $G$.

Fleischner and Hobbs [5] showed that $G^{+++}$ is hamiltonian if and only if $G$ contains an EPS-subgraph. We refer readers to [10] for the more results and problems on $G^{xyz}$. Harary and Nash-Williams [7] proved that $L(G)$ is hamiltonian if and only if $G$ has a closed trail incident with each edge. We call the complement of $L(G)$ as the jump graph $J(G)$, by adopting the notion in [3]. Wu and Meng [9] give a simple sufficient and necessary condition for a graph $G$ to have the hamiltonian jump graph $J(G)$. Motivated from the results above, in this paper we prove that $\kappa(G^{---}) \geq \delta(G^{---}) - 1$ and further establish the following.

**Theorem 1.** *Let $G$ be a graph. Then $G^{---}$ is hamiltonian if and only if $G$ is not isomorphic to any graphs in $\{K_{1,r} \mid r \geq 1\} \cup \{K_{1,s} + K_1 \mid s \geq 1\} \cup \{K_{1,t} + e \mid t \geq 2\} \cup \{K_2 + 2K_1, K_3 + K_1, K_3 + 2K_1, K_4\}$.*

## 2   Preliminary

We start with some simple observations. Let $G$ be a graph of order $p$ and size $q$. Then the order of $G^{---}$ is $p + q$. If $e = uv \in E(G)$ and $w \in V(G)$, then clearly, $d_{G^{---}}(e) = p + q - d(u) - d(v) - 1$, $d_{G^{---}}(w) = p + q - 2d(w) - 1$. So $\delta(G^{---}) = p + q - 2\Delta(G) - 1$.

**Lemma 1.** *For a graph $G$, we have $\alpha(G^{---}) = 3$ if $\Delta(G) = 1$, and $\alpha(G^{---}) = \Delta(G) + 1$, otherwise.*

*Proof.* If $\Delta(G) = 0$, then $G^{---}$ is a complete graph, so $\alpha(G^{---}) = 1$.

If $\Delta(G) = 1$, then $G$ consists of some isolated vertices and isolated edges. Let $e = uv \in E(G)$. Obviously, $\{u, v, e\}$ is an independent set of $G^{---}$. On the other hand, let $S$ be an independent set of $G^{---}$. Since $\Delta(G) = 1$, any two edges of $G$ are adjacent in $G^{---}$ and for a vertex $w$ of $G$, at most one vertex of $V(G) \setminus \{w\}$ is not adjacent to $w$ in $G^{---}$. It follows that $|S \cap E(G)| \leq 1$ and $|S \cap V(G)| \leq 2$, and so $|S| \leq 3$. Hence $\alpha(G^{---}) = 3$.

Next we assume that $\Delta(G) \geq 2$. First note that for a vertex $v \in V(G)$, $\{v\} \cup I_G(v)$ is an independent set of $G^{---}$, and since $|I_G(v)| = d(v)$, we have $\alpha(G^{---}) \geq \Delta(G) + 1$. Now let $S$ be an independent set of $G^{---}$. Note that each vertex of $S \cap V(G)$ is an end vertex of each edge of $S \cap E(G)$ in $G$. So, if $|S \cap V(G)| = 2$, then $|S \cap E(G)| \leq 1$, and thus $|S| \leq 3 \leq \Delta(G) + 1$; if $|S \cap V(G)| > 2$, then $|S \cap E(G)| = 0$. Thus $S$ is a clique of $G$ and $|S| \leq \Delta(G) + 1$; if $|S \cap V(G)| = 1$, and say $v \in V(G) \cap S$, then $S \cap E(G) \subseteq I_G(v)$, thus $|S| \leq \Delta(G) + 1$. In any case, we obtain $|S| \leq \Delta(G) + 1$. The result then follows.     □

For simplicity, if a graph $G$ is isomorphic to $H$, we write $G \cong H$, and if it is not, $G \ncong H$. For a graph $G$ and a set $A$ of graphs, we denote by $G \in A$ that $G$ is isomorphic to a graph in $A$, and $G \notin A$, otherwise.

Wu and Meng [8] proved that if a graph $G$ is isomorphic to neither a star nor a triangle, then $G^{---}$ is connected.

**Theorem 2.** *For any graph $G$, $\kappa(G^{---}) \geq \delta(G^{---}) - 1$, and equality holds if and only if $G$ contains a triangle and $\Delta(G) = 2$.*

*Proof.* Assume that $S$ is a minimum cut of $G^{---}$ with $|S| < \delta(G^{---})$. We shall show that $|S| = \delta(G^{---}) - 1$, and furthermore, $\Delta(G) = 2$ and $G$ contains a triangle. Since $|S| < \delta(G^{---})$, each component of $G^{---} - S$ is nontrivial. We say that a component $H$ of $G^{---} - S$ is of type-1 (respectively, type-2, or type-3) if $V(H) \subseteq V(G)$ (respectively, $V(H) \subseteq E(G)$, or $V(H) \cap V(G) \neq \emptyset$ and $V(H) \cap E(G) \neq \emptyset$).

*Fact 1.* All the components of $G^{---} - S$ are of the same type.

*Proof of Fact 1.* Let $H_i$ be a component of type-i for $i = 1, 2, 3$. Let $e_2$ and $e_2'$ be two adjacent vertices of $H_2$, and $v_k \in V(H_k) \cap V(G)$ with $k = 1$ or $3$. By the definition of $G^{---}$, as edges of $G$, $e_2$ and $e_2'$ are not adjacent, but both of them are incident with $v_k$ in $G$ (equivalently, they are adjacent in $G$), a contradiction. It implies that $G^{---} - S$ does not contain type-2 components and type-k components with $k = 1$ or $3$ at the same time. It remains to show that type-1 and type-3 components does not exist at the same time in $G^{---} - S$. Let $x$ and $y$ be two adjacent vertices of $H_1$, and $e_3 \in V(H_3) \cap E(G)$. Then $u$ and $v$ are not adjacent, but $e_3$ is incident with both $x$ and $y$ in $G$ since $e$ and $x$ (or $y$) are in a different components of $G^{---} - S$, a contradiction. The result is true.     □

Now let $F_1, F_2, \ldots, F_\omega$ be all the components of $G^{---} - S$.

*Fact 2.* All the components of $G^{---} - S$ are of type-3.

*Proof of Fact 2.* By Fact 1, first suppose all $F_i's$ are of type-1. Since $|S| \leq \delta(G^{---}) - 1 = p + q - 2\Delta - 2$, $|V(G^{---} - S)| \geq p + q - (p + q - 2\Delta - 2) = 2\Delta + 2$. On the other hand, let $u_i \in V(F_i)$ for $i = 1, 2$. Then for each $i = 1$ and $2$, $u_i$ is adjacent to all vertices of $\cup_{j \neq i} V(F_j)$ in $G$, and

$$\sum_{j \neq i} |V(F_j)| \leq d_G(u_i) \leq \Delta(G).$$

It follows that $|V(F_1)| \leq \Delta(G)$ and $|V(G^{---} - S)| = \sum_{j=2} |V(F_j)| + |V(F_1)| \leq 2\Delta(G)$, a contradiction.

Now suppose all the components of $G^{---} - S$ are of type-2. We claim that $\omega \leq 3$, and $|V(F_i)| = 2$ for each $i = 1, \ldots, \omega$. By contradiction, suppose $e_1$, $e_2$, $e_3$, and $e_4$ are four vertices taken from the different components. Then $G[\{e_1, e_2, e_3, e_4\}] \cong K_{1,4}$, and let $u$ be the common vertex of $e_1$, $e_2$, $e_3$ and $e_4$ in $G$. Take a vertex $e_1'$ adjacent to $e_1$ in $G^{---} - S$. Then $e_1$ and $e_1'$ are not adjacent in $G$. Clearly, $G[\{e_1', e_2, e_3, e_4\}] \cong K_{1,4}$, it implies that $u$ must be an end vertex of $e_1'$ in $G$ as well. So $e_1$ and $e_1'$ are not adjacent in $G^{---}$. This contradiction shows that $\omega \leq 3$. Next, suppose that $|V(F_1)| \geq 3$. For each $i = 1, 2$, we take two adjacent vertices $e_i$ and $e_i'$ from $F_i$. Then $G[\{e_1, e_1', e_2, e_2'\}] \cong C_4$. Since $F_1$ has order at least 3, there exists a vertex $e \in V(F_1)$ which is adjacent to at least one of $e_1$ and $e_1'$ in $F_1$ (equivalently, $e$ is not adjacent to at least one of $e_1$ and $e_1'$ in $G$). But, $e$ is adjacent to both $e_2$ and $e_2'$ in $G$, and since $G[\{e_1, e_1', e_2, e_2'\}] \cong C_4$, it must be adjacent to both $e_1$ and $e_1'$ in $G$. Again it is a contradiction. This proves the claim.

From the claim above, if $\omega = 2$, $|S| = p + q - 4$ and if $\omega = 3$, $|S| = p + q - 6$. Combining these with $|S| \leq p + q - 2\Delta(G) - 2$, we have $\Delta(G) \leq 1$ or $\Delta(G) \leq 2$ based on $\omega = 2$ or 3, respectively. However, if $\omega = 2$, and since $G[F_1 \cup F_2] \cong C_4$, then $\Delta(G) \geq 2$, and if $\omega = 3$ and since $G[F_1 \cup F_2 \cup F_3] \cong K_4$, then $\Delta(G) \geq 3$, a contradiction.

Therefore all components of $G^{---} - S$ are of type-3.  □

*Fact 3.* All components of $G^{---} - S$ are of type-3 and $\omega = 3$.

*Proof of Fact 3.* By Fact 2, it remains to show $\omega = 3$. For any $e \in V(G^{---} - S) \cap E(G)$, by the definition of $G^{---}$, $e$ is adjacent to all vertices of $G$ except for its two end vertices. Note that each component of $G^{---} - S$ contains a vertex of $G$ since it is of type-3. It implies $\omega \leq 3$, and if $\omega = 2$, then for each $i \in \{1, 2\}$, $|V(F_i) \cap V(G)| \leq 2$ since otherwise one can find find an edge in $V(F_j) \cap E(G)$ has three vertices of $V(F_i) \cap V(G)$ as its end vertices in $G$, where $\{i, j\} = \{1, 2\}$, a contradiction. Suppose $\omega = 2$. We consider three cases.

*Case 1.* $|V(F_i) \cap V(G)| = 2$, for each $i = 1, 2$.

Let $V(F_i) \cap V(G) = \{v_i, v_i'\}$ for $i = 1, 2$. Then $V(F_1) \cap E(G) = \{v_2 v_2'\}$ and $V(F_2) \cap E(G) = \{v_1 v_1'\}$. Obviously, $v_1 v_1'$ and $v_2 v_2'$ is adjacent in $G^{---}$, which contradicts that $v_1 v_1'$ and $v_2 v_2'$ are in different components of $G^{---} - S$.

*Case 2.* $|V(F_1) \cap V(G)| = 2$ and $V(F_2) \cap V(G) = 1$.

Let $V(F_1) \cap V(G) = \{v_1, v_1'\}$. Then $V(F_2) = \{v_2, v_1 v_1'\}$, where $v_2 \in V(G)$. Note that each element of $V(F_1) \cap E(G)$ has $v_2$ and one of $v_1$ and $v_1'$ as its end vertices. It follows that $V(F_1) \cap E(G) \subseteq \{v_2 v_1, v_2 v_1'\}$, and thus $F_1$ is not connected, a contradiction.

By symmetry, if $|V(F_2) \cap V(G)| = 2$ and $V(F_1) \cap V(G) = 1$, One can derive a contradiction similarly.

*Case 3.* $|V(F_i) \cap V(G)| = 1$, for each $i = 1, 2$.

In this case, we have $|V(F_i) \cap E(G)| = 1$, for each $i = 1, 2$. If it is not, suppose $|V(F_1) \cap E(G)| \geq 2$, without loss of generality. We take $e_1, e'_1 \in E(G) \cap V(F_1)$, and $e_2 \in E(G) \cap V(F_1)$. Let $w$ be the common end vertex of $e_1$ and $e_2$ in $G$. Then $w$ is also an end vertex of $e'_1$, and $e_1 = vw = e'_1$, a contradiction. So $|V(F_i)| = 2$ for each $i = 1, 2$. Since $4 = p + q - |S| \geq p + q - (p + q - 2\Delta(G) - 2) = 2\Delta(G) + 2$, $\Delta(G) \leq 1$. But, it is not difficult to check that if $\Delta(G) = 1$, $\kappa(G^{---}) = \delta(G^{---})$, a contradiction.

This proves $\omega = 3$, Fact 3 is true.                                                        □

Now we are ready to show that $|S| = \delta(G^{---}) - 1$, $\Delta(G) = 2$, and $G$ contains a triangle. From Fact 3, $|V(F_i) \cap V(G)| = 1$ for each $i = 1, 2, 3$. Let $V(F_i) \cap V(G) = \{x_i\}$. Then $V(F_i) \cap E(G_i) = \{x_j x_k\}$ for each $i = 1, 2, 3$, where $\{i, j, k\} = \{1, 2, 3\}$. So $|(V(G) \cup E(G)) \setminus S| = 6$, and since

$$p + q - 6 = |S| \leq \delta(G^{---}) - 1 = p + q - 2\Delta(G) - 2, \quad (*)$$

we have $\Delta(G) \leq 2$. From $G[\{u_1, u_2, u_3\}] \cong K_3$, $\Delta(G) \geq 2$. Hence, $\Delta(G) = 2$, and together with $(*)$, it follows that $|S| = \delta(G^{---}) - 1$. So

$$\kappa(G^{---}) \geq \delta(G^{---}) - 1, \quad (**)$$

and if the equality holds, $G$ must contains a triangle and $\Delta(G) = 2$. Now let $G$ be a graph with $\Delta(G) = 2$ and containing a triangle $H$. Let $S = (V(G) \setminus V(H)) \cup (E(G) \setminus E(H))$. It is obvious that $G^{---} - S \cong K_3^{---} \cong 3K_2$, and thus $S$ is a vertex cut of $G^{---}$ with cardinality $|S| = p + q - 6 = p + q - 2\Delta(G) - 2 = \delta(G^{---}) - 1$. Combining with $(**)$, one obtains $\kappa(G^{---}) = \delta(G^{---}) - 1$.

The proof is complete.                                                                           □

The following classical result is due to Chvátal and Erdös [4].

**Theorem 3.** *If $\alpha(G) \leq \kappa(G)$ for a graph $G$ of order at least three, then $G$ is hamiltonian.*

**Corollary 1.** *Let $G$ be a graph of order $p$ and size $q$. If $p + q \geq 3\Delta(G) + 2$ and $G \notin \{2K_1, K_2 + 2K_1, K_3 + 2K_1\}$, then $G^{---}$ is hamiltonian.*

*Proof.* Assume $G$ is a graph as given in the hypotheses of the corollary. We consider three cases in view of Lemma 1 and Theorem 2.

*Case 1.* $\Delta(G) = 0$ or 1. If $\Delta(G) = 0$, then $q = 0$ and $p = p + q \geq 2$. Since $G \ncong 2K_1$, $p \geq 3$. Thus $G^{---} \cong K_p$ and $G^{---}$ is hamiltonian. If $\Delta(G) = 1$, and $G \ncong K_2 + 2K_1$, $p + q \geq 6$. By Lemma 1 and Theorem 2, $\kappa(G^{---}) = \delta(G^{---}) = p + q - 2\Delta(G) - 1 \geq 3 = \alpha(G^{---})$. By Theorem 3, $G^{---}$ is hamiltonian.

*Case 2.* $\Delta(G) > 2$, or $\Delta(G) = 2$ and $G$ contains no triangles.

By Lemma 1 and Theorem 2, $\alpha(G^{---}) = \Delta(G) + 1$ and $\kappa(G^{---}) = \delta(G^{---})$. Since $\delta(G^{---}) = p + q - 2\Delta(G) - 1$, and $p + q \geq 3\Delta(G) + 2$, we have $\kappa(G^{---}) \geq \Delta(G) + 1 = \alpha(G^{---})$. Thus $G^{---}$ is hamiltonian by Theorem 3.

*Case 3.* $\Delta(G) = 2$ and $G$ contains a triangle.

Since $p+q \geq 3\Delta(G)+2 = 8$ and $G \not\cong K_3+2K_1, p+q \geq 9$. So, by Lemma 1 and Theorem 2, $\alpha(G^{---}) = 3$ and $\kappa(G^{---}) = \delta(G^{---}) - 1 = p+q - 3\Delta(G) - 2 \geq 3 = \alpha(G^{---})$. By Theorem 3, $G^{---}$ is hamiltonian.                               □

## 3   On $G^{---}$ of Some Special Graphs $G$

We need a useful observation. Let $K^*_{r,r}$ be a bipartite graph with two partite sets $\{e_1, e_2, \ldots, e_r\}$ and $\{v_1, v_2, \ldots, v_r\}$, in which $e_i v_j \in E(K^*_{r,r})$ if and only if $i \neq j$. Note that if $r \geq 3$, then for any edge $e_i v_j$, there exists a Hamilton cycle of $K^*_{r,r}$ containing $e_i v_j$.

**Lemma 2.** *If $G$ is isomorphic to a graph in $\{K'_{1,r}, K_{1,r} + \overline{K_t} \mid r, t \geq 2\} \cup \{(K_{1,s} + 2e)_1 \mid s \geq 3\} \cup \{(K_{1,t} + 2e)_2 \mid t \geq 4\} \cup \{K_p \mid p \geq 5\} \cup \{K_s \vee \overline{K_t} \mid s, t \geq 2\}$, then $G^{---}$ is hamiltonian.*

*Proof.* For an integer $r \geq 2$, let $S_r \cong K_{1,r}$ with $V(S_r) = \{u, v_1, v_2, \ldots, v_r\}$ and $E(S_r) = \{e_1, e_2, \ldots, e_r\}$, where $d_{S_r}(u) = r$ and $e_i = uv_i$ for each $i = 1, 2, \ldots, r$. Then $S_r^{---} \cong K^*_{r,r} + K_1$.

First let us consider $G \cong K'_{1,r}$. Let $V(G) = V(S_r) \cup \{v_{r+1}\}$ and $E(G) = E(S_r) \cup \{e_{r+1}\}$, where $e_{r+1} = v_r v_{r+1}$. If $r = 2$, then $G \cong P_4$ and $ue_3 e_1 v_2 v_1 e_2 v_3 u$ is a Hamilton cycle of $G^{---}$. For the case $r \geq 3$, let $C$ be a Hamilton cycle of $K^*_{r,r}$ containing $e_1 v_2$. We can obtain a Hamilton cycle of $G^{---}$ from $C$ by replacing the edge $e_1 v_2$ with the path $e_1 e_{r+1} u v_{r+1} v_2$.

Suppose $G \cong K_{1,r} + \overline{K_t}$ with $r \geq 2$ and $t \geq 2$. Let $V(G) = V(S_r) \cup \{u_1, u_2, \ldots, u_t\}$ and $E(G) = E(S_r)$. Let $C$ be a Hamilton cycle of $K^*_{r,r}$ containing $e_1 v_2$. By replacing the edge $e_1 v_2$ of $C$ with the path $e_1 u_1 u u_2 \cdots u_t$, one obtain a Hamilton cycle of $G^{---}$.

Now we consider $G \cong (K_{1,s} + 2e)_1$ with $s \geq 3$. Let $V(G) = V(H_s)$ and $E(G) = E(H_s) \cup \{e, f\}$, where $e = v_1 v_2$ and $f = v_2 v_3$. Let $C$ be a Hamilton cycle of $K^*_{r,r}$ containing $v_1 e_3$. By replacing the path $v_1 e_3$ of $C$ with $v_1 f u e e_3$, one obtain a Hamilton cycle of $G^{---}$.

Suppose $G \cong (K_{1,t} + 2e)_2$ with $t \geq 4$. Let $V(G) = V(H_s)$ and $E(G) = E(H_s) \cup \{e, f\}$, where $e = v_1 v_2$ and $f = v_3 v_4$. Let $C$ be a Hamilton cycle of $K^*_{r,r}$ containing $v_1 e_3$. By replacing the path $v_1 e_3$ of $C$ with $v_1 f u e e_3$, one obtain a Hamilton cycle of $G^{---}$.

If $G \cong K_p$ with $p \geq 5$, we have

$$p+q - 3\Delta(G) - 2 = p + \frac{p(p-1)}{2} - 3(p-1) - 2 = \frac{p^2 - 5p + 2}{2} > 0.$$

Thus $G^{---}$ is hamiltonian by Corollary 1.

Finally we consider the case $G \cong K_s \vee \overline{K_t}$ with $s \geq 2, t \geq 2$. Since $p = s + t$, $q = st + \frac{s(s-1)}{2}$, and $\Delta(G) = s + t - 1$, we have

$$p + q - 3\Delta(G) - 2 = \frac{s^2}{2} - \frac{5s}{2} + t(s-2) + 1$$

$$\geq \frac{s^2}{2} - \frac{5s}{2} + 2(s-2) + 1$$

$$= \frac{1}{2}s(s-1) - 3.$$

Therefore, if $s \geq 3$ then $\Delta(G) \geq 3$, and $G^{---}$ is hamiltonian by Corollary 1. For $s = 2$, let $V(G) = \{u, u'\} \cup \{v_1, v_2, \ldots, v_t\}$, $E(G) = \{e\} \cup \{e_1, e_2, \ldots, e_t\} \cup \{e'_1, e'_2, \ldots, e'_t\}$, where $e = uu'$, $e_i = uv_i$ and $e'_i = u'v_i$ for each $i = 1, 2, \cdots, t$. Then for $t \geq 3$, we can find a Hamilton cycle $C$ of $G^{---}$:

$$C = \begin{cases} e_1 e'_2 e_3 e'_4 \cdots e_{t-2} e'_{t-1} u e'_1 e_t u' e_2 e'_3 \cdots e_{t-1} e'_t v_1 e v_2 v_3 \cdots v_t e_1, & \text{if } t \text{ is odd}; \\ e_1 e'_2 e_3 e'_4 \cdots e_{t-1} u' e_2 e'_t u e'_1 e_4 e'_3 \cdots e_t e'_{t-1} v_1 e v_2 v_3, \cdots v_t e_1, & \text{if } t \text{ is even}. \end{cases}$$

If $t = 2$, $G \cong K_4^-$ and $ev_1 e'_2 u e'_1 e_2 u' e_1 v_2 e$ is a Hamilton cycle of $(K_4^-)^{---}$.    □

Next we introduce a definition. Let $G$ be a graph of order $p$ and size $q$, $e = uv \in E(G)$. We call $e$ removable if $d(u) + d(v) < \frac{p+q-1}{2}$, equivalently, $d_{G^{---}}(e) > \frac{p+q-1}{2}$. We will use the following result.

**Lemma 3. (Bondy and Chvátal [1])** *Let $G$ be a graph with order $p$ and let $u$ and $v$ be nonadjacent vertices in $G$ such that $d(u) + d(v) \geq p$. Then $G$ is hamiltonian if and only if $G + uv$ is hamiltonian.*

**Corollary 2.** *For a graph $G$, if an edge $e$ of $G$ is removable and $(G - e)^{---}$ is hamiltonian, then $G^{---}$ is hamiltonian.*

*Proof.* Let $u$ and $v$ be the two end vertices of $e$ in $G$. Note that $u$ and $v$ are not adjacent in $G^{---}$, but they are adjacent in $(G - e)^{---}$. Since $e$ is removable, $d(u) + d(v) < \frac{p+q-1}{2}$ and so

$$d_{G^{---}}(u) + d_{G^{---}}(v) = (p + q - 2d(u) - 1) + (p + q - 2d(v) - 1)$$

$$= 2(p + q) - 2 - 2(d(u) + d(v))$$

$$\geq 2(p + q) - 2 - (p + q) + 2$$

$$= p + q.$$

Hence, from Lemma 3, to prove $G^{---}$ is hamiltonian, it suffices to prove that $H$ is hamiltonian, where $H$ is the graph obtained from $G^{---}$ by adding an edge between $u$ and $v$. Observe that $H - e = (G - e)^{---}$. Let $C$ be a Hamilton cycle of $(G - e)^{---}$. Since the order of $(G - e)^{---}$ is $p + q - 1$ and $d_{G^{---}}(e) > \frac{p+q-1}{2}$, $e$ must be adjacent to two consecutive vertices of $C$ in $H$. It follows that $H$ is hamiltonian.    □

All graphs of order 4 and their corresponding transformation graphs $G^{---}$ are given in Figure 1. It is straightforward to check that $3K_1$ is the unique graph $G$ of order no more than 3 such that $G^{---}$ is hamiltonian, and $4K_1$, $2K_2$, $P_4$, $C_4$, $K_4^-$ are all the graphs $G$ of order 4 such that $G^{---}$ is hamiltonian.

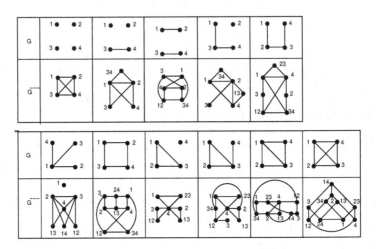

**Fig. 1.**

There are 34 graphs of order 5 (see [6]), which are treated in the next lemma.

**Lemma 4.** *Let $G$ be a graph of order 5. If $G$ is not isomorphic to $K_{1,4}$, $K_{1,4} + e$, $K_{1,3} + K_1$, or $K_3 + 2K_1$, then $G^{---}$ is hamiltonian.*

*Proof.* Let $p$ and $q$ be the order and the size of $G$, respectively. If $\Delta(G) = 0$, then $G^{---}$ is the complete graph of order 5, and is hamiltonian. If $\Delta(G) = 1$, then $p+q \geq 6$, $G^{---}$ is hamiltonian by Corollary 1. As for $\Delta(G) = 2$, clearly $p+q \geq 7$. If $p + q = 7$, then $G \cong P_3 + 2K_1$. One can easily check that $(P_3 + 2K_1)^{---}$ is hamiltonian. If $p + q \geq 8$, and since $G \not\cong K_3 + 2K_1$, then by Corollary 1 $G^{---}$ is hamiltonian.

Now assume $\Delta(G) = 3$. If $q \geq 6$, then $p + q \geq 11$, and by Corollary 1 $G^{---}$ is hamiltonian. Next we consider the case $3 \leq q \leq 5$. If $G$ is not connected, and since $\Delta(G) = 3$, it follows that $G \cong G_1 + K_1$, where $G_1$ is a connected graph of order 4. Since $G \not\cong K_{1,3} + K_1$, $G_1 \in \{K_{1,3} + e, K_4 - e, K_4\}$. One can easily check that $G^{---}$ is hamiltonian.

If $G$ is connected, then $q \geq 5 - 1 = 4$. If $q = 4$, then $G \cong K'_{1,3}$, and by Lemma 2 $(K'_{1,3})^{---}$ is hamiltonian. If $q = 5$, since $\Delta(G) = 3$, there exists a vertex $u \in V(G)$ with $d_G(u) = 1$. Let $v$ be the neighbor of $u$ in $G$. So $d_G(u) + d_G(v) = 1 + d_G(v) \leq 4 < \frac{9}{2} = \frac{p+q-1}{2}$, or $uv$ is removable. Then $G - uv$ is not connected, and we have just seen $(G - uv)^{---}$ is hamiltonian. By Corollary 2, $G^{---}$ is hamiltonian.

Finally let $\Delta(G) = 4$. Then, since $G \not\cong K_{1,4}$, $K_{1,4} + e$, $G$ is connected and $q \geq 6$. If $q \geq 9$, then $G^{---}$ is hamiltonian by Corollary 1. Next we consider $6 \leq q \leq 8$.

Let $w \in V(G)$ with $d(w) = 4$, and $G' = G - w$. Then $G'$ is a graph of order 4 and size between 2 and 4. So $G' \in \{2K_2, P_3 + K_1, P_4, K_{1,3}, K_3 + K_1, K_{1,3} + e, C_4\}$, and therefore $G$ must be one of graphs depicted in Figure 2 below. One can check that $G^{---}$ is hamiltonian for each $G$.                                                   □

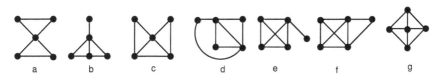

Fig. 2.

## 4   The Proof of Theorem 1

First, we prove the necessity. If $G$ is isomorphic to one of $K_{1,r}$, $K_{1,s} + K_1$, and $K_{1,t} + e$ for some $r \geq 1$, $s \geq 1$, and $t \geq 2$, it is trivial to check that $G^{---}$ is not hamiltonian. Also, for every graph $G$ in $\{K_3, K_3 + K_1, K_3 + 2K_1, K_4\}$, $G^{---}$ is not hamiltonian.

Next, we prove the sufficiency of the Theorem 1 by contradiction. Suppose that there exists a graph $G$ not isomorphic to any graph in the hypothesis of Theorem 1, such that $G^{---}$ is not hamiltonian. Furthermore, we assume that $G$ is such a graph of minimum $p + q$. By the results in Section 3, we have $p \geq 6$. Hence, if $\Delta(G) \leq 2$, then $p + q \geq 6 + \Delta(G) \geq 3\Delta(G) + 2$. Since $G \not\cong K_3 + 2K_1$, by Corollary 1, $G^{---}$ is hamiltonian. Thus $\Delta(G) \geq 3$.

*Claim 1.* For any $e \in E(G)$, $(G - e)^{---}$ is hamiltonian.

*Proof of Claim 1.* Suppose, on the contrary, $e_0 \in E(G)$ such that $(G - e_0)^{---}$ is not hamiltonian. Then by the minimality of $G$ and $p \geq 6$, $G - e_0$ must be isomorphic to either $K_{1,p-2} + K_1$ or $K_{1,p-1} + e$. Therefore, $G$ is isomorphic to one of the following graphs in $\{K'_{1,p-2}, (K_{1,p-1} + 2e)_1, (K_{1,p-1} + 2e)_2\}$. By Lemma 2, $G^{---}$ is hamiltonian, a contradiction.                                                   □

*Claim 2.* For any $uv \in E(G)$, $d(u) + d(v) \geq \frac{p+q-1}{2}$.

*Proof of Claim 2.* Suppose, on the contrary, that there exists an edge $e_0 = xy \in E(G)$ such that $d(x) + d(y) < \frac{p+q-1}{2}$. Thus $e_0$ is removable, and since $(G - e_0)^{---}$ is hamiltonian by Claim 1, it follows that $G^{---}$ is hamiltonian Corollary 2, a contradiction.                                                   □

*Claim 3.* There exist two edges with no common vertex in $G$, say $uv$ and $xy$, such that $G[\{u, v, x, y\}]$ is isomorphic to neither $K_4^-$ nor $K_4$.

*Proof of Claim 3.* If $G$ has at least two nontrivial components, then we are done by simply taking two edges from two different components. Moreover, since $G \not\cong \overline{K_p}$, next we consider the case that $G$ has just one nontrivial component, say $H$. Denotes the order of $H$ as $s$. Then $G \cong H + \overline{K_{p-s}}$ with $s \geq 2$. We shall show that $H$ is not a complete graph. Otherwise, $H$ is a complete graph of order $s \geq 2$. By Lemma 2, $H^{---} = K_s^{---}$ is hamiltonian for $s \geq 5$. Applying this to the fact $G^{---} = H^{---} \vee K_{p-s}$, it implies that $G^{---}$ is also hamiltonian, a contradiction. If $s = 2, 3$, or $4$, then $\Delta(G) = 1, 2$, or $3$, respectively. Since $p \geq 6$, it is easy to check that $G \notin \{2K_1, K_2 + 2K_1, K_3 + 2K_1\}$ and $p + q \geq 3\Delta(G) + 2$. By Corollary 1, $G^{---}$ is hamiltonian, a contradiction.

So $H$ contains two nonadjacent vertices, say $u$ and $x$. If $N_H(u) \neq N_H(x)$, without loss of generality $N_H(u) \backslash N_H(x) \neq \emptyset$, then we take $v \in N_H(u) \backslash N_H(x)$ and $y \in N_H(x)$. It is easy to see that $uv, xy \in E(G)$ are exactly those as we required. Next we assume that for any two nonadjacent vertices $u$ and $x$, $N_H(u) = N_H(x)$. Then we define a relation $\sim$ on $V(H)$ as follows: $u \sim x$ if and only if $N_H(u) = N_H(x)$. It is easy to see that $\sim$ is an equivalent relation on $V(H)$, which partitions $V(H)$ into equivalent classes $V_1, V_2, \ldots, V_r$. It is not difficult to see that two vertices of $H$ are adjacent if and only if they are in different classes. Thus $H$ is a complete $r$-partite graph. If there are two sets $V_i$ and $V_j$, $i \neq j$, with their cardinalities at least two, we are done, since we can take the desired four vertices, two from $V_i$ and the other two from $V_j$.

Thus, it remains to consider $H \cong K_{s-t} \vee \overline{K_t}$ for some $t \geq 2$ with $t < s$. In this case, $G \cong (K_{s-t} \vee \overline{K_t}) + \overline{K_{p-s}}$. Since $G$ is not a star, the equalities $p = s$ and $s - t = 1$ do not hold at the same time. If $s - t \geq 2$, then $(K_{s-t} \vee \overline{K_t})^{---}$ is hamiltonian by Lemma 2, and thus $G^{---}$ is also hamiltonian. If $s - t = 1$, then $p \neq s$. Moreover, since $G \not\cong K_{1,p-2} + K_1$, we have $G \cong K_{1,t} + \overline{K_{p-s}}$ with $t \geq 2$ and $p - s \geq 2$. By Lemma 2, $G^{---}$ is hamiltonian, a contradiction.    □

By Claim 2 and Claim 3, we have $q \geq d(u) + d(v) - 1 + d(x) + d(y) - 1 - 2 = d(u) + d(v) + d(x) + d(y) - 4 \geq p + q - 1 - 4 = p + q - 5$. Then $p \leq 5$, which contradicts $p \geq 6$. The proof of Theorem 1 is complete.    □

# Acknowledgements

The authors are grateful to the referees for their careful reading and helpful comments.

# References

1. Bondy, J.A., Chvátal, V.: A Method in Graph Theory. Discrete Math. **15** (1976) 111–135
2. Bondy, J.A., Murty, U.S.R.: Graph Theory with Applications. American Elsevier, New York, Macmillan, London (1976)
3. Chartrand, G., Hevia, H., Jarrett, E.B., Schultz, M.: Subgraph Distances in Graphs Defined by Edge Transfers. Discrete Math. **170** (1997) 63–79
4. Chvátal, V., Erdös, P.: A Note on Hamiltonian Circuits. Discrete Math. **2** (1972) 111–113
5. Fleischner, H., Hobbs, A.M.: Hamiltonian Total Graphs. Math. Notes **68** (1975) 59–82
6. Harary, F.: Graph Theory. Addison-Wesley (1971)

7. Harary, F., Nash-Williams, C.St.J.A.: On Eulerian and Hamiltonian Graphs and Line Graphs. Canad. Math. Bull. **8**(1965) 701–709

8. Wu, B., Meng, J.: Basic Properties of Total Transformation Graphs. J. Math. Study **34** (2)(2001) 109–116

9. Wu, B., Meng, J.: Hamiltonian Jump Graphs. Discrete Math. **289** (2004) 95–106

10. Wu, B., Zhang, L., Zhang, Z.: The Transformation Graph $G^{xyz}$ when $xyz = -++$. Discrete Math. **296** (2005) 263–270

# Isolated Toughness and Existence of $f$-Factors[*]

Yinghong Ma[1] and Qinglin Yu[2,3]

[1] School of Management, Shandong Normal University, Jinan, Shandong, P.R. China
[2] Center for Combinatorics, LPMC, Nankai University, Tianjin, P.R. China
[3] Department of Mathematics and Statistics, Thompson Rivers University,
Kamloops, BC, Canada

**Abstract.** Let $G$ be a graph with vertex set $V(G)$ and edge set $E(G)$. The *isolated toughness* of $G$ is defined as $I(G) = min\{|S|/i(G-S) \mid S \subseteq V(G), i(G-S) \geq 2\}$ if $G$ is not complete; otherwise, set $I(G) = |V(G)| - 1$. Let $f$ and $g$ be two nonnegative integer-valued functions defined on $V(G)$ satisfying $a \leq g(x) \leq f(x) \leq b$ . The purpose in this paper are to present sufficient conditions in terms of the isolated toughness and the minimum degree for graphs to have $f$-factors and $(g, f)$-factors $(g < f)$. If $g(x) \equiv a < b \equiv f(x)$, the conditions can be weakened.

**Keywords:** isolated toughness, $(g, f)$-factor, $f$-factor.

**AMS Subject Classifications:** 05C70.

## 1 Introduction

All graphs considered in this paper are simple and undirected. Let $G = (V(G), E(G))$ be a graph, where $V(G)$ and $E(G)$ denote the vertex set and the edge set of $G$, respectively. For any $S \subseteq V(G)$, the subgraph of $G$ induced by $S$ is denoted by $G[S]$ and we write $G - S$ for $G[V(G)\backslash S]$. We use $i(G - S)$ to denote the number of isolated vertices of $G - S$. For $S \subseteq V(G)$ and $T \subseteq V(G)$, let $E(S, T) = \{uv \in E(G) \mid u \in S, v \in T\}$ and $e(S, T) = |E(S, T)|$. Other notation and terminology not defined in this paper can be found in [1].

Let $g$ and $f$ be two nonnegative integer-valued functions defined on $V(G)$ and let $H$ be a spanning subgraph of $G$. We call $H$ a $(g, f)$-*factor* of $G$ if $g(x) \leq d_H(x) \leq f(x)$ holds for each $x \in V(G)$. Similarly, $H$ is an $f$-*factor* of $G$ if $g(x) = f(x)$ for each $x \in V(G)$. If $g(x) \equiv a$ and $f(x) \equiv b$ for each $x \in V(G)$, where $a, b$ are positive integers, then a $(g, f)$-factor is called an $[a, b]$-*factor*.

For any function $f(x)$ and a vertex subset $S$, we define $f(S) = \sum_{x \in S} f(x)$.

The well-known necessary and sufficient condition for a graph $G$ to have an $f$-factor was given by Tutte [9].

**Tutte's $f$-Factor Theorem ([9]).** *A graph $G$ has an $f$-factor if and only if*

$$f(S) - f(T) + d_{G-S}(T) - o(G - (S \cup T)) \geq 0$$

[*] This work is supported by Taishan Scholar Project of Shandong Province, NSFC of China (grant number 10471078), Natural Sciences and Engineering Research Council of Canada (grant number OGP0122059).

J. Akiyama et al. (Eds.): CJCDGCGT 2005, LNCS 4381, pp. 120–129, 2007.
© Springer-Verlag Berlin Heidelberg 2007

*for any pair of disjoint subsets $S$ and $T$ of $V(G)$, where $o(G-(S\cup T))$ denotes the number of components $C$ of $G - (S\cup T)$ such that $e_G(V(C), T) + \sum_{x\in V(C)} f(x)$ is odd.*

For convenience, we denote $\delta(S, T; f) = f(S) - f(T) + d_{G-S}(T) - o(G - (S \cup T))$. So a graph $G$ has an $f$-factor if and only if $\delta(S, T; f) \geq 0$ for any pair of disjoint $S$ and $T$. Furthermore, he noticed that $\delta(S, T; f) \equiv \sum_{x\in V(G)} f(x)$ ( mod 2).

Lovász generalized Tutte's $f$-factor theorem to $(g, f)$-factors by minor change in the notion $\delta(S, T; f)$.

**Lovász's $(g, f)$-Factor Theorem ([7]).** *Let $G$ be a graph and $g$, $f$ be integer-valued functions defined on $V(G)$ such that $g(x) \leq f(x)$ for any $x \in V(G)$. Then $G$ has a $(g, f)$-factor if and only if*

$$f(S) - g(T) + d_{G-S}(T) - o(G - (S \cup T)) \geq 0$$

*for any pair of disjoint sets $S, T \subseteq V(G)$, where $o(G - (S \cup T))$ denotes the number of components $C$ of $G-(S\cup T)$ such that $g(x) = f(x)$ for any $x \in V(C)$ and $e(V(C), T) + \sum_{x\in V(C)} f(x)$ is odd.*

For $g(x) < f(x)$, Heinrich *et al.* [4] simplified Lovász's $(g, f)$-factor theorem and obtained the following necessary and sufficient condition for the existence of $(g, f)$-factors.

**Lemma 1 (Heinrich et al., [4]).** *Let $g$ and $f$ be nonnegative integer-valued functions defined on $V(G)$. If either one of the following conditions holds*

**(i)** $g(x) < f(x)$ *for every* $x \in V(G)$;
**(ii)** $G$ *is bipartite;*

*then $G$ has a $(g, f)$-factor if and only if for any set $S$ of $V(G)$*

$$g(T) - d_{G-S}(T) \leq f(S),$$

*where* $T = \{x \,|\, x \in V(G) - S, \ d_{G-S}(x) \leq g(x)\}$.

Through the effort of many researchers, there have been many sufficient conditions for the existence of $f$-factors or $(g, f)$-factors. For example, the toughness conditions for the existence of some factors are obtained by Katerinis [5] and Chvátal [2]. In particular, Chvátal conjectured that $G$ has $k$-factors if $G$ is $k$-tough. This conjecture is confirmed by Enomoto *et al.* [3] and generalized to the following version in [5].

**Katerinis' Generalization.** *Let $G$ be a graph and $a \leq b$ be two positive integers.*

**(1)** *Suppose that $t(G) \geq \frac{(b+a)^2+2(b-a)}{4a}$ when $b \equiv a\,(\mathrm{mod}\ 2)$ and $t(G) \geq \frac{(b+a)^2+2(b-a)+1}{4a}$ when $b \not\equiv a\,(\mathrm{mod}\ 2)$. If $f$ is an integer-valued function such that $a \leq f(x) \leq b$ and $\sum_{x\in V(G)} f(x) \equiv 0\,(\mathrm{mod}\ 2)$, then $G$ has an $f$-factor;*

**(2)** *If $t(G) \geq (a-1) + \frac{a}{b}$ and $a|V(G)|$ is even when $a = b$, then $G$ has an $[a, b]$-factor.*

The *isolated toughness* was first introduced by Ma and Liu [8] and is motivated from Chvátal's toughness by replacing $c(G - S)$ with $i(G - S)$ in the definition, defined as $I(G) = min\{|S|/i(G - S) \,|\, S \subseteq V(G), \; i(G - S) \geq 2\}$ if $G$ is not complete; otherwise, set $I(G) = |V(G)| - 1$. Clearly, $I(G) \geq t(G)$ for any graph and $I(G) \leq \frac{|V(G)| - \alpha(G)}{\alpha(G)}$, where $\alpha(G)$ is the size of an independent set.

In this paper, we take advantage of the notion of isolated toughness $I(G)$ to obtain several sufficient conditions for the existence of $f$-factors and $(g, f)$-factors. The main purpose is to present the following results. Some of them are more general than Katerinis' results.

**Theorem 1.** *Let $G$ be a $K_{1,n}$-free graph and $f$ an integer-valued function on $V(G)$ satisfying $a \leq f(x) \leq b$ for any $x \in V(G)$ and $\sum_{x \in V(G)} f(x) \equiv 0 \,(\bmod \; 2)$. If $\delta(G) \geq \frac{(a+b-1)^2 + 4(b+n-1)}{4(a-n+1)}$ and $I(G) \geq \frac{(a+b-1)^2 + 4(b+n-1)}{4(a-n+1)}$, where $a, b$ are positive integers satisfying $2 \leq n - 1 \leq a \leq b$, then $G$ has an $f$-factor.*

When the condition $g(x) < f(x)$ for each $x \in V(G)$ is posted, we have the following.

**Theorem 2.** *Let $G$ be a graph and $f$, $g$ be two nonnegative integer-valued functions with $a \leq g(x) < f(x) \leq b$. If $\delta(G) \geq \frac{(a+b)^2 + 2(b-a) + 1}{4a}$ and $I(G) \geq \frac{(a+b)^2 + 2(b-a) + 1}{4a}$, where $a \leq b$ are two positive integers, then $G$ has a $(g, f)$-factor.*

For $[a, b]$-factors, the isolated toughness condition in Theorem 2 can be weakened. Since its proof is very similar to that of Theorem 2, we choose to state the theorem only.

**Theorem 3.** *Let $a$ and $b$ be integers with $2 \leq a < b$ and let $G$ be a graph. If $\delta(G) \geq a$ and $I(G) \geq (a-1) + \frac{a}{b}$, then $G$ has an $[a, b]$-factor.*

Let $a = 1 < b$ in Theorem 3, then the isolated toughness condition becomes a necessary and sufficient condition for $G$ having $[1, b]$-factors in terms of $I(G)$. This can be derived easily from the criterion of star-factor due to Vergnas [6].

**Proposition 1.** *Let $G$ be a graph with $\delta(G) \geq 1$ and $b > 1$ be a positive integer. Then $G$ has a $[1, b]$-factor if and only if $I(G) \geq \frac{1}{b}$*

## 2    Proof of Theorem 1

A subset $I$ of $V(G)$ is an *independent set* if no two vertices of $I$ are adjacent in $G$ and a subset $C$ of $V(G)$ is a *covering set* if every edge of $G$ has at least one end in $C$. It is easy to verify that a set $I \subseteq V(G)$ is an independent set of $G$ if and only if $V(G) - I$ is a covering set of $G$.

To prove the main theorems, we need the following result from Katerinis [5].

**Lemma 2 (Katerinis, [5]).** *Let $H$ be a graph and $S_1, S_2, \ldots, S_{k-1}$ be a partition of $V(H)$ such that $x \in S_j$ if and only if $d_H(x) \leq j$. Then there exist an independent set $I$ and a covering set $C$ of $V(H)$ such that*

$$\sum_{j=1}^{k-1}(k-j)c_j \leq \sum_{j=1}^{k-1}j(k-j)i_j,$$

*where $|I \cap S_j| = i_j$ and $|C \cap S_j| = c_j$ for every $j = 1, 2, \ldots, k-1$.*

Now we are ready to prove Theorem 1.

*Proof of Theorem 1.* Suppose, by the contrary, that there exists an integer-valued function $f$ which satisfies all the conditions in the theorem, but $G$ has no $f$-factors. Then, by Tutte's $f$-factor Theorem, there exists a pair of disjoint subsets of $V(G)$, say $S$ and $T$, such that

$$0 > \delta(S, T; f). \tag{1}$$

Recall $\delta(S, T; f) = f(S) - f(T) + d_{G-S}(T) - o(G - (S \cup T))$. We choose $S$ and $T$ such that $\delta(S, T; f)$ is the *minimum* and then $|S \cup T|$ is as *large* as possible.

First, we consider the case of $T = \emptyset$. If $S = \emptyset$, then $o(G) = 0$ (since $\sum_{x \in V(G)} f(x)$ is even) and $f(T) - d_{G-S}(T) = 0$. Thus $0 > \delta(S, T; f) = 0$, a contradiction. If $S \neq \emptyset$, then $o(G - S) \leq (n-1)|S|$ since $G$ is $K_{1,n}$-free. Then $(n-1)|S| \geq o(G-S) > f(S) \geq a|S|$ by (1) and thus $a < n-1$, which contradicts to the condition given in the theorem.

So, we may assume that $T \neq \emptyset$. Next we prove the following two claims.

*Claim 1.* $i(G - (S \cup T)) = 0$.

If $i(G - (S \cup T)) \neq 0$, then there exists an isolated vertex, say $v$, in $G - (S \cup T)$. If $e_G(v, T) > f(v)$, then set $S' = S \cup \{v\}$ and we have

$$\delta(S', T; f) = f(S') + d_{G-S'}(T) - f(T) - o(G - (S' \cup T))$$

$$\leq f(S) + f(v) + d_{G-S}(T) - e_G(v, T) - f(T) - (o(G - (S \cup T)) - 1)$$

$$= \delta(S, T; f) + f(v) + 1 - e_G(v, T)$$

$$\leq \delta(S, T; f),$$

which contradicts to the maximum of $|S \cup T|$ with respect to the minimum of $\delta(S, T; f)$. If $e_G(v, T) \leq f(v)$, then set $T' = T \cup \{v\}$ and we have

$$\delta(S, T'; f) = f(S) + d_{G-S}(T') - f(T') - o(G - (S \cup T'))$$

$$\leq f(S) + d_{G-S}(T) + e_G(v, T) - f(T) - f(v) - (o(G - (S \cup T)) - 1)$$

$$= \delta(S, T; f) + e_G(v, T) - f(v) + 1$$

$$\leq \delta(S, T; f) + 1.$$

Since $\delta(S, T'; f) \equiv \sum_{x \in V(G)} f(x) \equiv \delta(S, T; f) \pmod 2$, we have $\delta(S, T'; f) \leq \delta(S, T; f)$. Again this is a contradiction to the maximum of $|S \cup T|$ with respect to the minimum of $\delta(S, T; f)$. Therefore, $i(G - (S \cup T)) = 0$.

*Claim 2.* $d_{G-S}(x) \leq b + n - 1$ for any $x \in T$.

For any $x \in T$, let $T' = T - \{x\}$. By the minimum of $\delta(S, T; f)$, we have $\delta(S, T'; f) \geq \delta(S, T; f)$. Since $G$ is a $K_{1,n}$-free graph, $x$ is adjacent to at most $n - 1$ components of $G - (S \cup T)$ or $o(G - (S \cup T')) \geq o(G - (S \cup T)) - (n - 1)$. Therefore

$$\delta(S, T; f) \leq \delta(S, T'; f) = f(S) - f(T') + d_{G-S}(T') - o(G - (S \cup T'))$$

$$\leq f(S) - f(T) + f(x) + d_{G-S}(T) - d_{G-S}(x)$$

$$-(o(G - (S \cup T)) - (n - 1))$$

$$= \delta(S, T; f) + f(x) - d_{G-S}(x) + n - 1.$$

Thus $d_{G-S}(x) \leq b + n - 1$ as $f(x) \leq b$.

Let $T^j = \{x \mid x \in T, d_{G-S}(x) = j\}$, $t_j = |T^j|$ for every $j = 0, 1, 2, \ldots, b+n-1$ and $H = G[T^1 \cup T^2 \cup \cdots \cup T^{b+n-1}]$. Then $T^0$ is the set of the isolated vertices and $\{T^j \mid j = 1, 2, \ldots, b + n - 1\}$ is a vertex partition of $H$. Applying Lemma 2 with $k = b + n$, then there exist an independent set $I$ and a covering $C$ of $V(H)$ such that

$$\sum_{j=1}^{b+n-1} (b + n - 1 - j)c_j \leq \sum_{j=1}^{b+n-1} j(b + n - 1 - j)i_j, \tag{2}$$

where $|I \cap T^j| = i_j$ and $|C \cap T^j| = c_j$ for every $j = 1, 2, \ldots, b + n - 1$. Clearly, Lemma 2 holds for any independent set $I' \supseteq I$ and the cover set $C$. So, without loss of generality, we may assume that $I$ is a maximal independent set.

Set $W = G - (S \cup T)$ and $U = S \cup C \cup (N_G(I) \cap V(W))$. Then

$$|U| \leq |S| + \sum_{j=1}^{b+n-1} j i_j, \tag{3}$$

$$i(G - U) \geq \sum_{j=1}^{b+n-1} i_j + t_0. \tag{4}$$

*Case 1.* $i(G - U) \geq 2$.

Since $i(G - U) \geq 2$, by the definition of $I(G)$,

$$|U| \geq i(G - U)I(G). \tag{5}$$

Combining (3), (4) and (5), we have

$$|S| \geq \sum_{j=1}^{b+n-1} (I(G) - j)i_j + I(G)t_0. \tag{6}$$

Since $a \leq f(x) \leq b$ for each $x \in V(G)$, so $o(G - (S \cup T)) > f(S) - f(T) + d_{G-S}(T) \geq a|S| - b|T| + d_{G-S}(T)$. On the other hand, since $G$ is a $K_{1,n}$-free graph, we have $o(G - (S \cup T)) \leq (n-1)(|S| + |T|)$. Thus $(n-1)(|S| + |T|) > a|S| - b|T| + d_{G-S}(T)$ and this implies that

$$(b + n - 1)|T| - d_{G-S}(T) > (a - n + 1)|S|. \tag{7}$$

However, $(b + n - 1)|T| - d_{G-S}(T) = \sum_{j=0}^{b+n-1}(b + n - 1 - j)t_j \leq \sum_{j=1}^{b+n-1}(b + n - 1 - j)i_j + \sum_{j=1}^{b+n-1}(b + n - 1 - j)c_j + (b + n - 1)t_0$, since $t_j \leq c_j + i_j$ in $T$. Thus

$$\sum_{j=1}^{b+n-1}(b+n-1-j)i_j + \sum_{j=1}^{b+n-1}(b+n-1-j)c_j + (b+n-1)t_0 > (a-n+1)|S|. \tag{8}$$

Combining (8) and (6), we have

$$\sum_{j=1}^{b+n-1}(b + n - 1 - j)c_j > \sum_{j=1}^{b+n-1}[(a - n + 1)(I(G) - j) - (b + n - 1 - j)]i_j \tag{9}$$

$$+ [(a - n + 1)I(G) - (b + n - 1)]t_0.$$

Notice that $(a - n + 1)I(G) - (b + n - 1) > 0$ since $I(G) \geq \frac{(a+b-1)^2 + 4(b+n-1)}{4(a-n+1)}$. Therefore, (9) implies that

$$\sum_{j=1}^{b+n-1}(b + n - 1 - j)c_j > \sum_{j=1}^{b+n-1}[(a - n + 1)(I(G) - j) - (b + n - 1 - j)]i_j. \tag{10}$$

By (10) and (2), we have

$$\sum_{j=1}^{b+n-1} j(b + n - 1 - j)i_j > \sum_{j=1}^{b+n-1}[(a - n + 1)(I(G) - j) - (b + n - 1 - j)]\,i_j.$$

Hence there exists some $j \in \{1, 2, \ldots, b+n-1\}$ such that $j(b+n-1-j) > (a-n+1)(I(G)-j)-(b+n-1-j)$, that is, $j(b+a-j-1) > (a-n+1)I(G)-b-n+1$.

Let $h(j) = j(b + a - 1 - j)$. The maximum value of $h(j)$ is $\frac{(a+b-1)^2}{4}$ when $j = \frac{a+b-1}{2} \leq b+n-1$. But $(a-n+1)I(G)-b-n+1 \geq \frac{(a+b-1)^2}{4} \geq j(b+a-j-1)$ for any $j \in \{1, 2, \ldots, b+n-1\}$ since $I(G) \geq \frac{(a+b-1)^2 + 4(b+n-1)}{4(a-n+1)}$, a contradiction.

Case 2. $i(G - U) = 0$.

By (4), we have $\sum_{j=1}^{b+n-1} i_j + t_0 \leq 0$ or $t_0 = i_j = 0$ for all $j = 1, 2, \ldots, b+n-1$. Since $I$ is an maximal independent set, we have $T = \emptyset$, a contradiction to our assumption that $T \neq \emptyset$.

*Case 3.* $i(G - U) = 1$.

Then, by (4), we have $\sum_{j=1}^{b+n-1} i_j + t_0 \leq 1$.

If $t_0 = i_j = 0$ for all $j = 1, 2, \ldots, b+n-1$, it is exactly Case 2.

If $t_0 = 1$ and $i_j = 0$ for all $j \in \{1, \ldots, b+n-1\}$, then $T$ is an isolated vertex, say $v$. Therefore, by $o(G - (S \cup T)) \leq (n-1)(|S| + |T|) = (n-1)(|S| + 1)$ and $o(G - (S \cup T)) > f(S) - f(T) + d_{G-S}(T) \geq a|S| - b$, it yields

$$(a - n + 1)|S| < b + n - 1. \tag{11}$$

On the other hand,

$$\delta(G) \leq d_G(v) = e(v, S) \leq |S|. \tag{12}$$

Thus, by (11) and (12), we have $\delta(G)(a - n + 1) < b + n - 1$. But this is impossible because $\delta(G)(a - n + 1) - (b + n - 1) \geq \frac{(b+a-1)^2}{4} > 0$ since $\delta(G) \geq \frac{(a+b-1)^2 + 4(b+n-1)}{4(a-n+1)}$.

If there exists some $j_0 \in \{1, 2, \ldots, b+n-1\}$ such that $i_{j_0} = 1$ and $i_j = t_0 = 0$ for all $j \in \{1, 2, \ldots, b+n-1\} \backslash j_0$, then the maximality of $I$ implies that $H$ is a complete graph. Let $I = \{u\}$ for some vertex $u \in V(H)$. Then $d_G(u) \leq |S| + j_0$ and so $|S| \geq \delta(G) - j_0$. By (8), we have

$$\sum_{j=1}^{b+n-1} (b + n - 1 - j)c_j > (a - n + 1)(\delta(G) - j_0) - (b + n - 1 - j_0). \tag{13}$$

Combining (2) and (13), we get $j_0(b + n - 1 - j_0) > (a - n + 1)(\delta(G) - j_0) - (b + n - 1 - j_0)$. The maximum value of $j_0(b + n - 1 - j_0) + (a - n + 1)j_0 - j_0 = j_0(b + a - j_0 - 1)$ is $\frac{(a+b-1)^2}{4}$, but $(a - n + 1)\delta(G) - (b + n - 1) \geq \frac{(a+b-1)^2}{4}$ since $\delta(G) \geq \frac{(a+b-1)^2 + 4(b+n-1)}{4(a-n+1)}$, a contradiction again.

In all the cases, we derive a contradiction and thus complete the proof.    □

## 3    Proof of Theorem 2

In this section, we provide a proof for Theorem 2.

*Proof of Theorem 2.* Suppose that there exist two functions $g$ and $f$ which satisfy the conditions of the theorem but $G$ has no $(g, f)$-factors. Then, by Lemma 1, there exists a vertex set $S \subset V(G)$ such that

$$g(T) - d_{G-S}(T) > f(S), \tag{14}$$

where $T = \{x \mid x \in V(G) - S, \ d_{G-S}(x) \leq g(x)\}$.

Choose $T$ such that $T$ is minimal subject to (14). Suppose that there exists $x \in T$ such that $d_G(x) = g(x)$. Then the sets $S$ and $T - \{x\}$ satisfy (14), which contradicts to the choice of $T$. Hence we have $d_G(x) \leq g(x) - 1$ for all $x \in T$.

We assume that $S \neq \emptyset$. Otherwise, since $\delta(G) \geq \frac{(a+b)^2 + 2(b-a) + 1}{4a} > b \geq g(x)$ for every $x \in V(G)$, thus $T = \emptyset$ and (14) does not hold. Without loss of generality, we use $T = \{x \mid x \in V(G) - S, \ d_{G-S}(x) \leq b - 1\}$ instead of $T = \{x \mid x \in V(G) - S, \ d_{G-S}(x) \leq g(x) - 1\}$ since $g(x) \leq b$ for every $x \in V(G)$.

For each $0 \leq i \leq b - 1$, let $T^i = \{x \mid x \in T, \ d_{G-S}(x) = i\}$ and $t_i = |T^i|$ (we allow $T^i = \emptyset$ for some $i$), then $T^0$ is the set of the isolated vertices. Let $H = G\left[T^1 \cup T^2 \cup \cdots \cup T^{b-1}\right]$, then $d_H(x) \leq i$ for each $x \in T^i$ and $\{T^i \mid i = 1, 2, \ldots, b-1\}$ is a vertex partition of $H$. By Lemma 2, there exist an independent set $I$ and a covering set $C$ of $V(H)$ such that

$$\sum_{j=1}^{b-1}(b-j)c_j \leq \sum_{j=1}^{b-1} j(b-j)i_j, \tag{15}$$

where $|I \cap T^j| = i_j$ and $|C \cap T^j| = c_j$ for every $j = 1, 2, \ldots, b - 1$.

Without loss of generality, we may choose $I$ to be a maximal independent set of $H$. Set $W = G - (S \cup T)$ and $U = S \cup C \cup (N_{G-S}(I) \cap V(W))$. Then

$$|U| \leq |S| + \sum_{j=1}^{b-1} j i_j, \tag{16}$$

$$i(G - U) \geq t_0 + \sum_{j=1}^{b-1} i_j. \tag{17}$$

*Case 1.* $i(G - U) \geq 2$.

By the definition of $I(G)$,

$$|U| \geq i(G - U)I(G). \tag{18}$$

Combining (16), (17) and (18), we have

$$|S| \geq \sum_{j=1}^{b-1}(I(G) - j)i_j + I(G)t_0. \tag{19}$$

On the other hand, since $g(x) \leq b$ for every $x \in V(G)$, $g(T) - d_{G-S}(T) \leq b|T| - d_{G-S}(T) = \sum_{j=0}^{b-1}(b-j)t_j \leq \sum_{j=1}^{b-1}(b-j)i_j + \sum_{j=1}^{b-1}(b-j)c_j + bt_0$. Since $f(x) \geq a$ for every $x \in V(G)$, we have $f(S) \geq a|S|$. From (14) and (19), we obtain

$$\sum_{j=1}^{b-1}(b-j)i_j + \sum_{j=1}^{b-1}(b-j)c_j > a|S| \geq a\sum_{j=1}^{b-1}(I(G) - j)i_j + (aI(G) - b)t_0$$

$$\geq a\sum_{j=1}^{b-1}(I(G) - j)i_j,$$

this implies that

$$\sum_{j=1}^{b-1}(b-j)c_j > \sum_{j=1}^{b-1}(aI(G) - aj - b + j)i_j. \tag{20}$$

By (20) and (15), we have

$$\sum_{j=1}^{b-1} j(b-j)i_j > \sum_{j=1}^{b-1}(aI(G) - aj - b + j)i_j.$$

Hence, there exists some $j \in \{1, 2, \ldots, b-1\}$ such that $j(b-j) > aI(G) - aj - b + j$. But $j(b-j) + aj - j = -j^2 + (a+b-1)j \leq \frac{(a+b-1)^2}{4}$ and $aI(G) - b \geq \frac{(b+a)^2+2(b-a)+1}{4} - b = \frac{(b+a-1)^2}{4}$ due to $I(G) \geq \frac{(b+a)^2+2(b-a)+1}{4a}$, a contradiction.

*Case 2.* $i(G - U) = 0$.

By (17), we have $0 \geq t_0 + \sum_{j=1}^{b-1} i_j$ or $t_0 = i_j = 0$ for all $j = 1, 2, \ldots b-1$. Since $I$ is maximal, it yields $T = \emptyset$. Hence $g|T| - d_{G-S}(T) = 0 > g(S) > 0$, a contradiction.

*Case 3.* $i(G - U) = 1$ or $1 \geq t_0 + \sum_{j=1}^{b-1} i_j$ from (17).

If $t_0 = i_j = 0$ for all $j = 1, 2, \ldots, b-1$, it is exactly Case 2.

If $t_0 = 1$, then for all $j = 1, 2, \ldots, b-1$, $i_j = 0$ and $T$ is an isolated vertex. Let $T = \{v\}$. Since $a \leq g(x) < f(x) \leq b$ for every $x \in V(G)$, we have $g(T) - d_{G-S}(T) = g(v) \leq b$ and $f(S) \geq a|S| \geq ad_G(v) \geq a\delta(G) \geq \frac{(b+a)^2+2(b-a)+1}{4} \geq b$, a contradiction to (14).

Suppose there exists some $j_0 \in \{1, 2, \ldots, b-1\}$ such that $i_{j_0} = 1$ and $i_j = 0$ for all $j \in \{1, 2, \ldots, b-1\} - \{j_0\}$. Since $I$ is maximal, then $H$ is a complete graph. From (16), we have $|U| \leq |S| + j_0$. On the other hand, $|U| \geq |S| + d_{G-S}(v) \geq \delta(G)$ and it yields

$$|S| \geq |U| - j_0 \geq \delta(G) - j_0. \tag{21}$$

On the other hand, $g(T) - d_{G-S}(T) \leq b|T| - d_{G-S}(T) = \sum_{j=1}^{b-1}(b-j)t_j \leq \sum_{j=1}^{b-1}(b-j)i_j + \sum_{j=1}^{b-1}(b-j)c_j = (b-j_0) + \sum_{j=1}^{b-1}(b-j)c_j$. Since $f(x) \geq a$ for every $x \in V(G)$, by (21), we have $f(S) \geq a|S| \geq a(\delta(G) - j_0)$. These inequalities imply

$$\sum_{j=1}^{b-1}(b-j)c_j + (b-j_0) > a(\delta(G) - j_0), \tag{22}$$

and by (15),

$$\sum_{j=1}^{b-1}(b-j)c_j \leq j_0(b-j_0). \tag{23}$$

Thus, from (22) and (23), we have

$$j_0(b-j_0) > a(\delta(G) - j_0) - (b-j_0)$$

$$\geq a\left(\frac{(a+b)^2 + 2(b-a) + 1}{4a} - j_0\right) - (b-j_0)$$

or

$$\frac{(a+b)^2 + 2(b-a) + 1}{4} - b < -j_0^2 + (a+b-1)j_0. \tag{24}$$

However, for any $j_0$ $(1 \leq j_0 \leq b-1)$, it is not hard to see that $(a+b)^2+2(b-a)+1-4b \geq (a+b-1)^2 - (2j_0 - (a+b-1))^2$ or equivalently $\frac{(a+b)^2+2(b-a)+1}{4} - b \geq \frac{(a+b-1)^2}{4} - \left(j_0 - \frac{a+b-1}{2}\right)^2 = -j_0^2 + (a+b-1)j_0$, a contradiction to (24).

The proof is complete. $\qquad\qquad\qquad\qquad\qquad\qquad\qquad\qquad\qquad\qquad\square$

Although all the graphs considered are simple, the theorems in this paper can be extended to graphs with multiple edges as well (but without loops). To see this, one needs only to notice that a graph with multiple edges has the same isolated toughness as its underlying graph.

## Acknowledgements

The authors are indebted to Prof. Guizhen Liu and Dr. Jiansheng Cai for their constructive discussion and comments. The gratitude also goes to the anonymous referees for their detailed corrections and suggestions.

## References

1. Bollobás, B.: Modern Graph Theory. Springer-Verlag New York (1998)
2. Chvátal, V.: Tough Graphs and Hamiltonian Circuits. Discrete Math. **5** (1973) 215–228
3. Enomoto, H., Jackson, B., Katerinis, P., Saito, A.: Toughness and the Existence of $k$-Factors. J. Graph Theory **9** (1985) 87–95
4. Heinrich, K., Hell, P., Kirkpatrick, D., Liu, G.Z.: A Simple Existence Criterion for $(g, f)$-Factors. Discrete Math. **85** (1990) 313–317
5. Katerinis, P.: Toughness of Graphs and the Existence of Factors. Discrete Math. **80** (1990) 81–92
6. Las Vergnas, M.: An Extension of Tutte's 1-Factors of a Graph. Discrete Math. **2** (1972) 241–255
7. Lovász, L.: Subgraphs with Prescribed Valencies. J. Combinatorial Theory Ser. B **8** (1970) 391–416
8. Ma, Y.H., Liu, G.Z.: Isolated Toughness and the Existence of Fractional Factors in Graphs. Acta Appl. Math. Sinica (in Chinese) **26** (2003) 133–140
9. Tutte, W.T.: The Factor of Graphs. Canadian J. Math. **4** (1952) 314–328

# A Note on the Integrity of Middle Graphs

Aygul Mamut and Elkin Vumar[*]

College of Mathematics and System Sciences, Xinjiang University,
Urumqi 830046, P.R. China
vumar@xju.edu.cn

**Abstract.** The integrity $I(G)$ of a noncomplete connected graph $G$ is a measure of network invulnerability and is defined by $I(G) = \min\{|S| + m(G - S)\}$, where $S$ and $m(G - S)$ denote the the subset of $V$ and the order of the largest component of $G - S$, respectively. In this paper, we determine the integrity and some other parameters of middle graphs of some classes of graphs.

**Keywords:** integrity, covering number, independence number, vertex dominating number.

## 1 Introduction

Throughout this paper we consider only simple, finite and undirected graph $G = (V, E)$. For the notation and terminology not defined here we refer to Bondy & Murty [5]. The symbols $\delta(G)$, $\kappa(G)$, $\kappa'(G)$ and $\beta(G)$ denote the minimum degree, the vertex connectivity (connectivity), the edge-connectivity and the independence number of $G$, respectively. For a vertex $v \in V(G)$, we use $N(v)$ to denote the set of vertices adjacent to $v$, and $N[v] = N(v) \cup \{v\}$. The join $G \vee H$ of two graphs $G$ and $H$ is a graph obtained by joining every vertex of $G$ to every vertex of $H$.

There are several measures of the invulnerability of a communication network. One is connectivity, measures the "vulnerability" of a graph or network. A communication network can be considered to be highly vulnerable to disruption if the destruction of a few elements can result in very few members being able to communicate. But it was found that neither vertex nor the edge-connectivity measures how easy it is to break the graph into small pieces. In the analysis of vulnerability of a communication network we often consider the following quantities:

**1** the number of members of largest remaining group within mutual communication can still occur,
**2** the number of elements that are not functioning.

To estimate these quantities, the concept of integrity was introduced by Barefoot, Entringer and Swart [3] as a measure of the stability of a graph.

---

[*] Corresponding author. The project sponsored by SRF for ROCS, REM.

The *integrity* $I(G)$ of a graph $G$ is defined by $I(G) = \min\{|S| + m(G-S): S \subset V(G)\}$, where $m(G - S)$ denotes the order of a largest component of $G - S$. An *I-set* of $G$ is any (strict) subset $S$ of $V(G)$ for which $I(G) = S + m(G - S)$ [1]. A subset $S$ of $V$ is called a *cut set* of a graph $G$, if $\omega(G - S) > \omega(G)$ or $G - S$ is an isolated vertex, where $\omega(G)$ is the number of components of $G$.

Note that the definition of integrity does not require that a graph be connected. Moreover the original definition permitted the removal of all vertices. As immediate consequences of the definition, we have that if $G$ is a graph of order $n$, then $1 \le I(G) \le n$, and for any subgraph $H$ of $G$, $I(H) \le I(G)$. For a given graph $G$ and an integer $k$, the problem of deciding whether the integrity of $G$ is at most $k$ is $NP$-complete, even for planar graphs [6]. However, it is possible to determine the integrity of large classes of graphs. For more results on integrity see [1].

Now we give the definition of middle graph.

For a graph $G$, the *middle graph* $M(G)$ is the graph with vertex set $V(G) \cup E(G)$ in which the vertices $x$ and $y$ are joined by an edge if one of the following conditions holds: (i) $x$, $y \subset E(G)$, and $x$ and $y$ are adjacent in $G$, (ii) one of $x$ and $y$ is in $V(G)$ and the other is in $E(G)$, and they are incident in $G$.

We give the following definitions for later use.

**Definition 1.** *A* covering set *of $G$ is a subset $S$ of $V(G)$ such that every edge of $G$ has at least one end in $S$. The number of vertices in a minimum covering set of $G$ is called the* covering number *of $G$ and is denoted by $\alpha(G)$.*

**Definition 2.** *A vertex dominating set for a graph $G$ is a set $S$ of vertices such that every vertex of $G$ belongs to $S$ or is adjacent to a vertex of $S$. The cardinality of a minimum vertex dominating set in a graph $G$ is called the* vertex dominating number *of $G$ and is denoted by $\sigma(G)$.*

Obviously, $\sigma(G) \le \beta(G)$ for every graph $G$. In the next section we shall determine the integrity and some other parameters of the middle graphs of some classes of graphs.

## 2   Integrity of Some Classes of Graphs

Let $P_n$, $C_n$ and $K_n$ denote a path, a cycle and a complete graph with $n$ vertices, respectively. A star with $n + 1$ vertices is the complete bipartite graph $K_{1,n}$, while a wheel with $n + 1$ vertices, denoted by $W_{1,n}$, is the join of $K_1$ and $C_n$, i.e., $W_{1,n} = k_1 \vee C_n$.

We first cite some known results.

**Lemma 1.** [1] *If $S$ is an I-set of $G$, then*

(a) $I(G - S) = m(G - S)$;
(b) *$S$ is a cut set of $G$ unless $G$ is complete;*
(c) *if $S$ is minimal, then each vertex $v$ of $S$ is a cut vertex of $G - (S - \{v\})$.*

**Lemma 2.** [4] *For any graph $G$*

$$\kappa(M(G)) = \kappa'(G) \text{ and } \kappa'(M(G)) = \delta(M(G)) = \delta(G).$$

**Lemma 3.** [1] *For any graph $G$,*

(a) $\delta(G) + 1 \leq I(G) \leq \alpha(G) + 1$;
(b) $I(G \vee H) = \min\{I(G) + |H|, I(H) + |G|\}$, *in particular,* $I(G \vee K_r) = I(G) + r$;
(c) $I(G) \geq (\nu(G) - \kappa(G))/\beta(G) + \kappa(G)$.

**Theorem 1.** [1] $I(K_n) = n$, $I(K_{m,n}) = 1 + \min\{m, n\}$, $I(K_n^c) = 1$, $I(P_n) = \lceil 2\sqrt{n+1} \rceil - 2$, $I(C_n) = \lceil 2\sqrt{n} \rceil - 1$, $I(W_{1,n}) = 1 + I(C_n) = \lceil 2\sqrt{n} \rceil$.

Note that $M(G)$ has a covering set of cardinality $\varepsilon(G)$ and an independent set of cardinality $\nu(G)$. Hence by Lemmas 2 and 3 we have

**Theorem 2.** *For any graph $G$,*

(a) $\alpha(M(G)) = \varepsilon(G)$;
(b) $\beta(M(G)) = \nu(G)$;
(c) $\delta(G) + 1 \leq I(M(G)) \leq \varepsilon(G) + 1$;
(d) $I(M(G)) \geq 1 + \varepsilon(G)/\nu(G) + (1 - 1/\nu(G))\kappa'(G) \geq 1 + \varepsilon(G)/\nu(G)$.

*Proof.* We only prove $(a)$. The proof of $(b)$ is similar, and $(c)$ and $(d)$ are consequences of $(a)$, $(b)$ and Lemma 3.

Without loss of generality we may suppose that $G$ is not empty. Let $S$ be a covering set of $M(G)$. If $S$ is not $E(G)$, then $S \cap V(G) \neq \emptyset$ and $S \cap E(G) \neq \emptyset$. By the definition, $E(G) - S$ is an independent set in $M(G)$, and there is no edge connecting $E(G) - S$ and $V(G) - S$. Since in $M(G)$ each vertex in $E(G) - S$ is adjacent exactly to two vertices in $S \cap V(G)$ and since any two different vertices in $E(G) - S$ have the different neighborhoods in $S \cap V(G)$, we deduce that $|S \cap V(G)| \geq |E(G) - S|$, and this in turn implies that $|S| \geq |E(G)| = \varepsilon(G)$. Hence $E(G)$ is a minimum covering set of $M(G)$.     □

Now we present some parameters for the middle graphs of some special classes of graphs.

**Theorem 3.** *Let $P_n$, $C_n$, $W_{1,n}$ and $K_n$ be the graphs defined as above. Then*

(a) $\alpha(M(P_n)) = n - 1$, $\beta(M(P_n)) = n$, $\sigma(M(P_n)) = \lceil n/2 \rceil$ *and* $I(M(P_n)) = \lceil 2\sqrt{2n} \rceil - 2 = I(P_{2n-1})$;
(b) $\alpha(M(C_n)) = n$, $\beta(M(C_n)) = n$, $\sigma(M(C_n)) = \lceil n/2 \rceil$ *and* $I(M(C_n)) = \lceil 2\sqrt{2n} \rceil - 1 = I(C_{2n})$;
(c) $\alpha(M(K_{1,n})) = n$, $\beta(M(K_{1,n})) = n + 1$, $\sigma(M(K_{1,n})) = n$ *and* $I(M(K_{1,n})) = n + 1$;
(d) $\alpha(M(W_{1,n})) = 2n$, $\beta(M(W_{1,n})) = n + 1$, $\sigma(M(W_{1,n})) = \lceil n/2 \rceil + 1$ *and* $I(M(W_{1,n})) = \lceil 2\sqrt{2n} \rceil = I(C_{2n}) + 1$.

(e) $\alpha(M(K_n)) = n(n-1)/2$, $\beta(M(K_n)) = n$, $\sigma(M(K_n)) = \lceil n/2 \rceil$, $\kappa'(M(K_n))$
$= \kappa(M(K_n)) = \delta(M(K_n)) = \delta(K_n) = n-1$ and $I(M(K_n)) \leq n(n-1)/2+1$.

*Proof.* We only prove $I(M(P_n)) = \lceil 2\sqrt{2n} \rceil - 2$ and $\sigma(M(K_n)) = \lceil n/2 \rceil$, since the other results are consequences of the previous lemma or can be proved in the similar manner.

Let $S$ be a cut set of $M(P_n)$ with $|S| = s$. Note that $M(P_n)$ has a Hamilton path, and hence $M(G) - S$ has $s+1$ or fewer components. This implies

$$I(M(P_n)) \geq \min\{s + (2n - 1 - s)/(s+1)\}.$$

Set $f(x) = x + (2n - 1 - x)/(x + 1)$. Then $f(x)$ has the minimum value $f_{\min} = 2\sqrt{2n} - 2$. The integrity is integer value, and so we round this up to get a lower bound. Since this value can be achieved for each $n$, we have $I(M(P_n)) = \lceil 2\sqrt{2n} \rceil - 2$. If we label $V(K_n) = \{v_i : i = 1, \ldots, n\}$ and $E(K_n) = \{e_{i,j} : e_{i,j} = v_i v_j \in K_n\}$, then it is not difficult to prove that the set $\{e_{1,2}, e_{3,4}, e_{5,6}, \ldots\}$ is a minimum covering set of $M(K_n)$, and consequently $\sigma(M(K_n)) = \lceil n/2 \rceil$.  □

**Theorem 4.** *If $T$ is a tree with $n$ vertices, then $I(M(T)) \leq I(M(K_{1,n-1})) = n$.*

*Proof.* By induction on covering number $\alpha(T)$.

If $\alpha(T) = 1$, then $T$ is a star and $I(M(T)) = n$. Assume that the result holds for $\alpha(T) \leq k$. Now suppose $\alpha(T) = k+1$ and let $V' = \{v_1, v_2, \ldots, v_{k+1}\}$ be a covering set of $T$.

*Case 1.* $V' = \{v_1, v_2, \ldots, v_{k+1}\}$ is an independent set in $T[V']$.

Then there exists a vertex $u$ that is adjacent to $v_1$ in $T$ and $V' - \{v_1\}$ is a covering set in $(T - N[v_1]) \cup \{u\}$. Note that $|V' - \{v_1\}| = k$ and $|(T - N[v_1]) \cup \{u\}| = n - d(v_1)$. For an $I$-set $S'$ in $G' := M((T - N[v_1]) \cup \{u\})$ we add the vertices adjacent to $v_1$ in $G := M(T)$ and obtain another cut-set $S$ in $M(T)$ such that $|S| = |S'| + d(v_1)$. Hence

$$I(M(T)) \leq m(G - S) + S = m(G' - S') + |S'| + d(v_1) \leq n - d(v_1) + d(v_1) = n.$$

*Case 2.* $V' = \{v_1, v_2, \ldots, v_{k+1}\}$ is not an independent set in $T[V']$.

Since $T$ is a tree, there exists a vertex $v_i \in V'$ such that $d(v_i) = 1$ in $T[V']$. Without loss of generality we may assume that $d_{T[V']}(v_1) = 1$, and the vertex adjacent to $v_1$ is $v_2$ in $T[V']$. Note that $|V' - \{v_1\}| = k$, and $V' - \{v_1\}$ is a covering set in $(T - N[v_1]) \cup \{v_2\}$. Using the same method as that in Case 1, one can obtain $I(M(T)) \leq n$.  □

*Remark 1.* The preceding Theorem 4 shows that the $M(K_{1,n-1})$ has the maximum integrity among the middle graphs of trees with $n$ vertices. On the other hand, it seems that the minimum integrity of the middle graphs of trees with $n$ vertices would be achieved in $M(P_n)$. Another problem is to determine the integrity of the middle graph of a complete graph $K_n$.

# References

1. Bagga, K., Beineke,L., Goddard, W., Lipman, L., Pippert, R.: A Survey of Integrity. Discrete Applied Math. **37/38** (1992) 13–28
2. Bagga, K., Beineke, L., Lipman, M., Pippert, R.: Edge-Integrity: A Survey. Discrete Math. **124** (1994) 3–12
3. Barefoot, C.A., Entringer, R., Swart, H.: Vulnerability in Graphs—A Comparative Survey. J.Combin. Math. Combin. Comput. **1** (1987) 13–21
4. Bauer, B., Tindell, R.: The Connectivities of Middle Graph. J. Combin. Inform. Sys. Sci. **7(1)** (1982) 54–55
5. Bondy, J.A., Murty, U.S.R.: Graph Theory with Applications. Macmillan London and Elsevier. New York (1976)
6. Clark, L.H., Entringer, R.C., Fellows, M.: Computational Complexity of Integrity. J. Combin. Math. Combin. Comput. **2** (1987) 179–191
7. Goddard, W., Swart, H.: Integrity in Graphs: Bounds and Basics. J. Combin. Math. Combin.Comput. **7** (1990) 139–151
8. Hamada, T., Yushimura, I.: Tranversability and Connectivity of the Middle Graph of a Graph. Discrete Math. **14** (1976) 247–255
9. Kratsch, D., Kloks, T., Müller, H.: Measuring the Vulnerability for Classes of Intersection Graphs. Discrete Applied Math. **77** (3) (1997) 259–270
10. Dündar, P., Aytac, A.: Integrity of Total Graphs via Certain Parameters. Mathematical Notes Vol.76. No.5 (2004) 665–672
11. Vince, A.: The Integrity of a Cubic Graph. Discrete Applied Math. **140** (2004) 223–239

# Indecomposable Coverings*

János Pach[1], Gábor Tardos[2], and Géza Tóth[3]

[1] City College, CUNY and Courant Institute of Mathematical Sciences,
New York University, New York, NY 10012, USA
pach@cims.nyu.edu
[2] School of Computer Science, Simon Fraser University,
Burnaby, BC, Canada, V5A 1S6 and
Rényi Institute of the Hungarian Academy of Sciences,
H-1364 Budapest, P.O.B. 127, Hungary
tardos@cs.sfu.ca
[3] Rényi Institute of the Hungarian Academy of Sciences,
H-1364 Budapest, P.O.B. 127, Hungary
geza@renyi.hu

**Abstract.** We prove that for every $k > 1$, there exist $k$-fold coverings of
the plane (1) with strips, (2) with axis-parallel rectangles, and (3) with
homothets of any fixed concave quadrilateral, that cannot be decomposed
into two coverings. We also construct, for every $k > 1$, a set of points $P$
and a family of disks $\mathcal{D}$ in the plane, each containing at least $k$ elements
of $P$, such that no matter how we color the points of $P$ with two colors,
there exists a disk $D \in \mathcal{D}$, all of whose points are of the same color.

## 1  Multiple Arrangements: Background and Motivation

The notion of multiple packings and coverings was introduced independently by
Davenport and László Fejes Tóth. Given a system $\mathcal{S}$ of subsets of an underlying
set $X$, we say that they form a *k-fold packing (covering)* if every point of $X$
belongs to *at most (at least)* $k$ members of $\mathcal{S}$. A 1-fold packing (covering) is
simply called a *packing (covering)*. Clearly, the union of $k$ packings (coverings)
is always a $k$-fold packing (covering). Today there is a vast literature on this
subject [FTG83], [FTK93].

Many results are concerned with the determination of the maximum den-
sity $\delta^k(C)$ of a $k$-fold packing (minimum density $\theta^k(C)$ of a $k$-fold covering)
with congruent copies of a fixed convex body $C$. The same question was studied for multiple *lattice packings (coverings)*, giving rise to the parameter $\delta_L^k(C)$
$(\theta_L^k(C))$. Throughout this paper, it is always assumed that the geometric arrangements, packings, and coverings under consideration are *locally finite*, that is, any
bounded region intersects only finitely many members of the arrrangement.

* János Pach has been supported by NSF Grant CCF-05-14079, and by grants from
NSA, PSC-CUNY, Hungarian Research Foundation OTKA, and BSF. Gábor Tardos
has been supported by OTKA T-046234, AT-048826, and NK-62321. Géza Tóth has
been supported by OTKA T-038397.

J. Akiyama et al. (Eds.): CJCDGCGT 2005, LNCS 4381, pp. 135–148, 2007.

Because of the strongly combinatorial flavor of the definitions, it is not surprising that combinatorial methods have played an important role in these investigations. For instance, Erdős and Rogers [ER62] used the "probabilistic method" to show that $\mathbf{R}^d$ can be covered with congruent copies (actually, with translates) of a convex body so that no point is covered more than $e(d \ln d + \ln \ln d + 4d)$ times (see [PA95], and [FuK05] for another combinatorial proof based on Lovász' Local Lemma). Note that this easily implies that there exist positive constants $\theta_d, \delta_d$, depending only on $d$, such that

$$k \leq \theta^k(C) \leq k\theta(C) \leq k\theta_d,$$
$$k\delta_d \leq k\delta(C) \leq \delta^k(C) \leq k.$$

Here $\delta(C)$ and $\theta(C)$ are shorthands for $\delta^1(C)$ and $\theta^1(C)$.

To establish almost tight density bounds, at least for lattice arrangements, it would be sufficient to show that any $k$-fold packing (covering) splits into roughly $k$ packings (coverings), or into about $k/l$ disjoint $l$-fold packings (coverings) for some $l < k$. The initial results were promising. Blundon [Bl57] and Heppes [He59] proved that for unit disks $C = B^2$, we have

$$\theta_L^2(C) = 2\theta_L(C), \quad \delta_L^k(C) = k\delta_L(C) \text{ for } k \leq 4,$$

and these results were extended to arbitrary centrally symmetric convex bodies in the plane by Dumir and Hans-Gill [DuH72] and by G. Fejes Tóth [FTG77], [FTG84]. In fact, there was a simple reason for this phenomenon: It turned out that every 3-fold lattice packing of the plane can be decomposed into 3 packings, and every 4-fold lattice packing into *two* 2-fold ones. This simple scheme breaks down for larger values of $k$. As $k$ tends to infinity, Cohn [Co76] and Bolle [Bo89] proved that

$$\lim_{k \to \infty} \frac{\theta_L^k(C)}{k} = \lim_{k \to \infty} \frac{\theta^k(C)}{k} = 1 \leq \theta(C),$$
$$\lim_{k \to \infty} \frac{\delta_L^k(C)}{k} = \lim_{k \to \infty} \frac{\delta^k(C)}{k} = 1 \geq \delta(C).$$

For convex bodies $C$ with a "smooth" boundary, the inequalities on the right-hand side are strict [Sch61], [Fl78].

The situation becomes slightly more complicated if we do not restrict our attention to *lattice* arrangements. In reply to a question raised by László Fejes Tóth, the senior author noted [P80] that any 2-fold packing of homothetic copies of a plane convex body splits into 4 packings. Furthermore, any $k$-fold packing $\mathcal{C}$ with not too "elongated" convex sets splits into at most $9\lambda k$ packings, where

$$\lambda := \max_{C \in \mathcal{C}} \frac{(\text{circumradius}(C))^2 \pi}{\text{area}(C)}.$$

(Here the constant factor $9\lambda$ can be improved. See also [Ko04].)

One would expect that similar results hold for coverings rather than packings. However, in this respect we face considerable difficulties. For any $k$, it is easy to construct a $k$-fold covering of the plane with not too elongated convex sets (of different shapes but of roughly the same size) that cannot be decomposed even to *two* coverings [P80]. The problem is far from being trivial even for coverings with congruent disks. In an unpublished manuscript, P. Mani-Levitska and the (then junior and now) senior author have shown that every 33-fold covering of the plane with congruent disks splits into two coverings [MP87]. Another positive result was established in [P86].

**Theorem 1.1.** *For any centrally symmetric convex polygon $P$, there exists a constant $k = k(P)$ such that every $k$-fold covering of the plane with translates of $P$ can be decomposed into two coverings.*

At first glance, one may believe that approximating a disk by centrally symmetric polygons, the last theorem implies that any sufficiently thick covering with congruent disks is decomposable. The trouble is that, as we approximate a disk with polygons $P$, the value $k(P)$ tends to infinity. Nevertheless, it follows from Theorem 1.1 that if $k = k(\varepsilon)$ is sufficiently large, then any $k$-fold covering with disks of radius 1 splits into a covering and an "almost covering" in the sense that it becomes a covering if we replace each of its members by a concentric disk whose radius is $1 + \varepsilon$.

Recently, Tardos and Tóth [TaT06] have managed to extend Theorem 1.1 to any (not necessarily centrally symmetric) convex polygon $P$. Here the assumption that $P$ is convex cannot be dropped.

Surprisingly, the analogous decomposition result is false for multiple coverings with balls in *three* and higher dimensions.

**Theorem 1.2. [MP87]** *For any $k$, there exists a $k$-fold covering of $\mathbf{R}^3$ with unit balls that cannot be decomposed into two coverings.*

Somewhat paradoxically, it is the very heavily covered points that create problems. Pach [P80], [AS00] (p. 68) noticed that by the Lovász Local Lemma we obtain

**Theorem 1.3. [AS00]** *Any $k$-fold covering of $\mathbf{R}^3$ with unit balls, no $c2^{k/3}$ of which have a point in common, can be decomposed into two coverings. (Here $c$ is a positive constant.)*

Similar theorems hold in $\mathbf{R}^d$ ($d > 3$), except that the value $2^{k/3}$ must be replaced by $2^{k/d}$.

# 2    Cover-Decomposable Families: Statement of Results

These questions can be reformulated in a slightly more general combinatorial setting.

**Definition 2.1.** *A family $\mathcal{F}$ of sets in $\mathbf{R}^d$ is called* cover-decomposable *if there exists a positive integer $k = k(\mathcal{F})$ such that any $k$-fold covering of $\mathbf{R}^d$ with members from $\mathcal{F}$ can be decomposed into two coverings.*

In particular, Theorem 1.1 above can be rephrased as follows. The family consisting of all translates of a given centrally symmetric convex polygon in the plane is cover-decomposable. Theorem 1.2 states that the translates of a unit ball is 3-space is not cover-decomposible. These results are valid for both *open* and *closed* polygons and balls.

Note that Theorem 1.1 has an equivalent "dual" form. Given a system $\mathcal{S}$ of translates of $P$, let $C(\mathcal{S})$ denote the set of centers of all members of $\mathcal{S}$. Clearly, $\mathcal{S}$ forms a $k$-fold covering of the plane if and only if every translate of $P$ contains at least $k$ elements of $C(\mathcal{S})$. Recall that, by assumption, $\mathcal{S}$ is a locally finite arrangement. Therefore, any bounded region contains only *finitely many* points of $C(\mathcal{S})$. We call such a point set *locally finite*.

The fact that the family of translates of $P$ is cover-decomposable can be expressed by saying that there exists a positive integer $k$ satisfying the following condition: any locally finite set $C$ of points in the plane such that $|P' \cap C| \geq k$ for all translates $P'$ of $P$ can be partitioned into two disjoint subsets $C_1$ and $C_2$ with

$$|C_1 \cap P'| \neq \emptyset \text{ and } |C_2 \cap P'| \neq \emptyset \text{ for every translate } P' \text{ of } P.$$

We can think of $C_1$ and $C_2$ as "color classes."

This latter condition, in turn, can be reformulated as follows. Let $H(C)$ denote the (infinite) hypergraph whose vertex set is $C$ and whose (hyper)edges are precisely those subsets of $C$ that can be obtained by taking the intersection of $C$ by a translate of $P$. By assumption, every hyperedge of $H(C)$ is of size at least $k$. The fact that $C$ can be split into two color classes $C_1$ and $C_2$ with the above properties is equivalent to saying that $H(C)$ is *two-colorable*.

**Definition 2.2.** *A hypergraph is* two-colorable *if its vertices can be colored by two colors such that no edge is monochromatic.*

*A hypergraph is called* two-edge-colorable *if its edges can be colored by two colors such that every vertex is contained in edges of both colors.*

Obviously, a hypergraph $H$ is two-edge-colorable if and only if its *dual hypergraph* $H^*$ is two-colorable. (By definition, the vertex set and the edge set of $H^*$ are the edge set and the vertex set of $H$, respectively, with the containment relation reversed.)

Summarizing, Theorem 1.1 can be rephrased in two equivalent forms. For any centrally symmetric convex polygon $P$ in the plane, there is a $k = k(P)$ such that

1. any $k$-fold covering of $\mathbf{R}^2$ with translates of $P$ (regarded as an infinite hypergraph on the vertex set $\mathbf{R}^2$) is two-edge-colorable;
2. for any locally finite set of points $C \subset \mathbf{R}^2$ with the property that each translate of $P$ covers at least $k$ elements of $C$, the hypergraph $H(C)$ whose edges are the intersections of $C$ with all translates of $P$ is two-colorable.

Clearly, the above two statements are also equivalent for translates of *any* set $P$, that is, we do not have to assume here that $P$ is a polygon or that it is convex or connected. However, if instead of *translates*, we consider congruent, similar, or homothetic copies of $P$, then assertions 1 and 2 do not necessarily remain equivalent.

The aim of this paper is to give various geometric constructions showing that certain families of sets in the plane are not cover-decomposable.

Let $T_k$ denote a rooted $k$-ary tree of depth $k-1$. That is, $T_k$ has $1 + k + k^2 + k^3 + \ldots + k^{k-1} = \frac{k^k - 1}{k-1}$ vertices. The only vertex at level 0 is the root $v_0$. For $0 \le i < k-1$, each vertex at level $i$ has precisely $k$ children. The $k^{k-1}$ vertices at level $k-1$ are all leaves.

**Definition 2.3.** *For any rooted tree $T$, let $H(T)$ denote the hypergraph on the vertex set $V(T)$, whose hyperedges are all sets of the following two types:*
**1.** Sibling *hyperedges: for each vertex $v \in V(T)$ that is not a leaf, take the set $S(v)$ of all children of $v$;*
**2.** Descendent *hyperedges: for each leaf $v \in V(T)$, take all vertices along the unique path from the root to $v$.*

Obviously, $H_k =: H(T_k)$ is a $k$-uniform hypergraph with the following property. No matter how we color the vertices of $H_k$ by two colors, red and blue, say, at least one of the edges will be monochromatic. In other words, $H_k$ is not two-colorable. Indeed, assume without loss of generality that the root $v_0$ is red. The children of the root form a sibling hyperedge $S(v_0)$. If all points of $S(v_0)$ are blue, we are done. Otherwise, pick a red point $v_1 \in S(v_0)$. Similarly, there is nothing to prove if all points of $S(v_1)$ are blue. Otherwise, there is a red point $v_2 \in S(v_1)$. Proceeding like this, we either find a sibling hyperedge $S(v_i)$, all of whose elements are blue, or we construct a red descendent hyperedge $\{v_0, v_1, \ldots, v_{k-1}\}$.

**Definition 2.4.** *Given any hypergraph $H$, a* planar realization *of $H$ is defined as a pair $(P, \mathcal{S})$, where $P$ is a set of points in the plane and $\mathcal{S}$ is a system of sets in the plane such that the hypergraph obtained by taking the intersections of the members of $\mathcal{S}$ with $P$ is isomorphic to $H$.*

*A realization of the dual hypergraph of $H$ is called a* dual realization *of $H$.*

In the sequel, we show that for any rooted tree $T$, the hypergraph $H(T)$ defined above has both a planar and a dual realization, in which the members of $\mathcal{S}$ are open strips (Lemmas 3.1–4.1). In particular, the hypergraph $H_k = H(T_k)$ permits such realizations for every positive $k$. These results easily imply the following

**Theorem 2.5.** *The family of open strips in the plane is not cover-decomposable.*

Indeed, fix a positive integer $k$, and assume that we have shown that $H_k = H(T_k)$ has a dual realization with strips (see Lemma 4.1). This means that the set of vertices of $T_k$ can be represented by a collection $\mathcal{S}$ of strips, and the set of (sibling and descendent) hyperedges by a point set $P \subset \mathbf{R}^2$ whose every element is covered by the corresponding $k$ strips. Recall that $H_k$ is not two-colorable, hence its dual hypergraph $H_k^*$ is not two-edge-colorable. In other words, no matter how we color the strips in $\mathcal{S}$ with two colors, at least one point in $P$ will be covered only by strips of the same color. Add now to $\mathcal{S}$ all open strips that do not contain any element of $P$. Clearly, the resulting (infinite) family of strips,

$S'$, forms a $k$-fold covering of the plane, and it does not split into two coverings. This proves Theorem 2.5.

In fact, a "degenerate" version of Theorem 2.5 is also true, in which strips are replaced by straight-lines (that is, by "strips of width zero").

**Theorem 2.6.** *The family of straight lines in the plane is not cover-decomposable.*

We prove this theorem in Section 4. It implies the following generalization of Theorem 2.5: The family of open strips of *unit width* in the plane is not cover-decomposable.

Lemma 5.1 was originally established in [MP87]. For completeness, here we include a somewhat simpler proof (see Section 5). Lemma 5.1 easily implies that, for any $d \geq 3$, the family of open unit balls in $\mathbf{R}^d$ is not cover-decomposable, for any $d \geq 3$ (Theorem 1.2).

In Section 6, we show that the hypergraph $H_k = H(T_k)$ permits a dual realization in the plane with axis-parallel rectangles, for every positive $k$ (Lemma 6.1). This implies, in exactly the same way as outlined in the paragraph below Theorem 2.5, that the following theorem is true.

**Theorem 2.7.** *The family of axis-parallel open rectangles in the plane is not cover-decomposable.*

We cannot decide whether $H_k$ permits a planar realization. However, it can be shown [CPST06] that a sufficiently large randomly and uniformly selected point set $P$ in the unit square, say, with large probability has the following property. No matter how we color the points of $P$ with two colors, there is an axis-parallel rectangle containing at least $k$ elements of $P$, all of the same color.

Recall that the family of translates of any convex polygon $Q$ is cover-decomposable (see Theorem 1.1 and [TaT06]). The next result shows that this certainly does not hold for some *concave* polygons $Q$.

**Theorem 2.8.** *The family of all translates of a given (open) concave quadrilateral is not cover-decomposable.*

The proofs presented in the next five sections also yield that Theorems 2.5, 2.7, and 2.8 remain true for *closed* strips, rectangles, and quadrilaterals. Most arguments follow the same general inductive scheme, but the subtleties require separate treatment.

## 3   Planar Realization with Strips

A *strip* is an open set $S$ in the plane, bounded by two parallel lines. The counterclockwise angle $\alpha$   $(-\frac{\pi}{2} < \alpha \leq \frac{\pi}{2})$ from the $x$-axis to these lines is called the *direction* or *slope* of $S$.

**Lemma 3.1.** *For any rooted tree $T$, the hypergraph $H(T)$ permits a planar realization with strips. That is, there is a set of points $P$ and a set of strips $\mathcal{S}$ in the plane such that the hypergraph on the vertex set $P$ whose hyperedges are the sets $S \cap P$ $(S \in \mathcal{S})$ is isomorphic to $H(T)$.*

**Proof:** We prove the lemma by induction on the number of vertices of $T$. The statement is trivial if $T$ has only one vertex. Suppose that $T$ has $n$ vertices and that the statement has been proved for all rooted trees with fewer vertices. Let $v_0$ be the root of $T$, and let $v_0 v_1 \ldots v_m$ be a path of maximum length starting at $v_0$. Let $U = \{u_1, u_2, \ldots u_k\}$ be the set of children of $v_{m-1}$. Each member of $U$ is a leaf of $T$, and one of them is $v_m$. Delete the members of $U$ from $T$, and let $T'$ denote the resulting rooted tree. Clearly, $v_{m-1}$ is a leaf of $T'$. By the induction hypothesis, there is a planar realization $(P, \mathcal{S})$ of $H(T')$ with open strips. We can assume without loss of generality that no element of $P$ lies on the boundary of any strip in $\mathcal{S}$, otherwise we could slightly decrease the widths of some strips without changing the containment relation.

Let $S \in \mathcal{S}$ be the strip representing the descendent hyperedge $\{v_0, v_1, \ldots, v_{m-1}\}$, i.e., a strip that contains precisely the points corresponding to these vertices of $T'$. (See Definition 2.3.) Rotating $S$ through very small angles, the resulting strips $S^1, S^2, \ldots, S^k$ contain the same points of $P$ as $S$ does. Moreover, we can make sure that the new strips are not parallel to each other or to any old strip. Hence, we can choose a line $\ell$, not passing through any element of $P$, such that $S^1, S^2, \ldots, S^k$ intersect $\ell$ in pairwise disjoint intervals that are also disjoint from all members of $\mathcal{S}$. For each $i$, $1 \le i \le k$, pick a point $p^i$ in $\ell \cap S^i$, and add these points to $P$. Replace $S$ in $\mathcal{S}$ by the strips $S^1, S^2, \ldots, S^k$, and add another member to $\mathcal{S}$: a very narrow strip $\bar{S}$ around $\ell$, which contains all $p^i$, but no other point of $P$.

In this way, we obtain a planar realization of $H(T)$, where $p_1, p_2, \ldots, p_k$ represent the vertices (leaves) $u_1, u_2, \ldots u_k \in V(T)$, the strip $\bar{S}$ represents the sibling hyperedge $U = \{u_1, u_2, \ldots u_k\}$ of $H(T)$, while $S^1, S^2, \ldots, S^k$ represent the descendent hyperedges, corresponding to the paths from $v_0$ to $u_1, u_2, \ldots u_k$, respectively. □

A hypergraph is *k-uniform* if all of its hyperedges have precisely $k$ vertices.

**Corollary 3.2.** *For any $k \ge 2$, there exists a k-uniform hypergraph which is not two-colorable and which permits a planar realization by open strips.* □

## 4 Dual Realization with Strips: Proofs of Theorems 2.5 and 2.6

Recall that a *dual* realization of a hypergraph $H$ is a planar realization of its dual $H^*$. That is, given a tree $T$, a dual realization of $H(T)$ is a pair $(P, \mathcal{S})$, where $P$ is a set of points in the plane representing the (sibling and descendent) hyperedges of $H(T)$, and $\mathcal{S}$ is a system of regions representing the vertices of $T$ such that a region $S \in \mathcal{S}$ covers a point $p \in P$ if and only if the vertex corresponding to $S$ is contained in the hyperedge corresponding to $p$.

**Lemma 4.1.** *For any rooted tree $T$, the hypergraph $H(T)$ permits a dual realization with strips.*

**Proof:** Most of our proof is identical to the proof of Lemma 3.1. We establish the statement by induction on the number of vertices of $T$. The statement is trivial if $T$ has only one vertex. Suppose that $T$ has $n$ vertices and that the statement has been proved for all rooted trees with fewer than $n$ vertices. Let $v_0$ be the root of $T$, and let $v_0 v_1 \ldots v_m$ be a path of maximum length starting at $v_0$. Let $U = \{u_1, u_2, \ldots u_k\}$ denote the set of children of $v_{m-1}$. Clearly, each element of $U$ is a leaf of $T$, one of them is $v_m$, and $U$ is a sibling hyperedge of $H(T)$. Let $T'$ denote the tree obtained by deleting from $T$ all elements of $U$. The vertex $v_{m-1}$ is then a leaf of $T'$.

By the induction hypothesis, $H(T')$ permits a dual realization $(P, \mathcal{S})$ with open strips. We can assume without loss of generality that no element of $P$ lies on the boundary of any strip in $\mathcal{S}$, otherwise we could slightly decrease the widths of some strips without changing the containment relation.

Let $p \in P$ be the point corresponding to the descendent hyperedge $\{v_0, v_1, \ldots, v_{m-1}\}$ of $H(T')$. Let $p_1, p_2, \ldots, p_k$ be distinct points so close to $p$ that they are contained in exactly the same strips from $\mathcal{S}$ as $p$ (namely, in the ones corresponding to $v_0, v_1, \ldots, v_{m-1}$). The point $p_i$ will correspond to the descendent hyperedge of $T$ containing $v_i$. Choose a point $q$ such that all lines $p_i q$ for $1 \le i \le k$ are distinct and they do not pass through any element of $P$. This point will correspond to the sibling hyperedge $\{u_1, \ldots, u_k\}$ of $T$. For $1 \le i \le k$, let $S^i$ be an open strip around the line $p_i q$ that is narrow enough so that it does not contain any element of $P$ or any point $p_j$ with $j \ne i$. This strip represents the vertex $u_i$ of $T$.

Add $S^1, S^2, \ldots, S^k$ to $\mathcal{S}$. Delete $p$ from $P$, and add $p_1, \ldots, p_k$, and $q$. The resulting configuration is a dual realization of $H(T)$ with open strips, so we are done. □

**Proof of Theorem 2.6:** Let $C_k^n$ be a $k \times k \times \ldots \times k$ piece of the $n$-dimensional integer grid, that is,

$$C_k^n = \{(x_1, x_2, \ldots, x_n) : x_i \in \{0, 1, \ldots, k-1\}\}.$$

A $k$-*line* is a set of $k$ collinear points of $C_k^n$. Denote by $H_k^n$ the $k$-uniform hypergraph on the vertex set $C_k^n$, whose hyperedges are the $k$-lines. The following statement is a direct consequence of the Hales-Jewett theorem.

**Lemma 4.2. [HaJ63]** *The hypergraph $H_k^n$ is not two-colorable.*

Our goal is to construct an indecomposable covering of the plane by (continuously many) straight lines such that every point is covered at least $k$ times. Project $C_k^n$ to a "generic" plane so that no two elements of $C_k^n$ are mapped into the same point and no three noncollinear points become collinear.

Applying a duality transformation, we obtain a family $\mathcal{L}$ of $k^n$ lines and a set $P$ of so-called $k$-*points*, dual to the $k$-lines, such that each $k$-point belongs to precisely $k$ members of $\mathcal{L}$. It follows from Lemma 4.2 that for any two-coloring of the members of $\mathcal{L}$, there is a $k$-point $p \in P$ such that all lines passing through $p$ are of the same color.

It remains to extend the family $\mathcal{L}$ into a $k$-fold covering of the whole plane with lines without destroying the last property. This can be achieved by simply adding to $\mathcal{L}$ all straight lines that do not pass through any point in $P$.     □

## 5  Planar Realization with Disks

In this section, all *disks* are assumed to be open. A pair $(P, \mathcal{D})$ consisting of a point set $P$ and a system of disks $\mathcal{D}$ in the plane is said to be in *general position*, if no element of $P$ lies on the boundary of a disk $D \in \mathcal{D}$, no two members of $\mathcal{D}$ are tangent to each other, and no three circles bounding members of $\mathcal{D}$ pass through the same point.

In order to facilitate the induction, we prove a slightly stronger lemma than what we need.

**Lemma 5.1.** *For any rooted tree $T$, the hypergraph $H(T)$ permits a planar realization $(P, \mathcal{D})$ with disks in general position such that every disk $D \in \mathcal{D}$ has a point on its boundary that does not belong to the closure of any other disk $D' \in \mathcal{D}$.*

**Proof:** By induction on the number of vertices of $T$. The statement is trivial if $T$ has only one vertex. Suppose that $T$ has $n$ vertices and that the statement has already been proved for all rooted trees with fewer than $n$ vertices. Let $v_0$ denote the root of $T$, and let $v_0 v_1 \ldots v_m$ be a path of maximum length starting at $v_0$. Let $U = \{u_1, u_2, \ldots u_k\}$ be the set of children of $v_{m-1}$. Each element of $U$ is a leaf of $T$, and one of them is $v_m$. Remove all elements of $U$ from $T$, and let $T'$ denote the resulting rooted tree. Clearly, $v_{m-1}$ is a leaf of $T'$. By the induction hypothesis, $H(T')$ permits a planar realization $(P, \mathcal{D})$ with disks satisfying the conditions in the lemma.

Let $D$ denote the disk representing the descendent hyperedge $\{v_0, v_1, \ldots, v_{m-1}\}$ of $H(T')$. Let $v$ be a point on the boundary of $D$, which does not belong to the closure of any other disk $D' \in \mathcal{D}$. Choose a small neighborhood $N(v, \varepsilon)$ of $v$, which still disjoint from any disk $D' \in \mathcal{D}$ other than $D$.

To obtain a planar realization of $H(T)$, we have to add $k$ new points to $P$ that will represent the vertices $u_1, u_2, \ldots u_k \in V(T)$, and replace $D$ by $k$ new disks that will represent the descending hyperedges of $H(T)$, corresponding to the paths connecting the root to $u_1, u_2, \ldots u_k$. We also add a disk representing the sibling hyperedge $U = \{u_1, u_2, \ldots u_k\}$ of $H(T)$. This can be achieved, as follows.

Let $\ell$ denote the straight line connecting the center of $D$ to $v$, and let $w$ be the point on $\ell$, outside of $D$, at distance $\varepsilon/2$ from $v$. Let $D(1), D(2), \ldots, D(k)$ be $k$ disks obtained from $D$ by rotating it about the point $w$ through very small angles, so that $D(i) \cap P = D \cap P$ holds for any $1 \le i \le k$. Further, let $D'$ denote the disk of radius $\varepsilon/2$, centered at $w$. Then $D(i)$ and $D$ are tangent to each other; let $p(i)$ denote their point of tangency $(1 \le i \le k)$. Add the points $p(1), p(2), \ldots, p(k)$ to $P$; they will represent $u_1, u_2, \ldots, u_k \in V(T)$, respectively. Remove $D$ from $\mathcal{D}$, and replace it by the disks $D(1), D(2), \ldots, D(k)$ and $D'$.

Now we are almost done: the new pair $(P, \mathcal{D})$ is almost a planar realization of $H(T)$, with the disk $D'$ representing the sibling hyperedge $\{u_1, u_2, \ldots u_k\}$ of $H(T)$. The only problem is that the points $p(i)$ lie on the boundaries of $D(i)$ and $D'$, rather than in their interiors. This can be easily fixed by increasing the radii of the disks $D(i)$ $(1 \le i \le k)$ and $D'$ by a very small positive number $\delta < \varepsilon/2$, so that the enlarged $D'$ contains $p(1), p(2), \ldots, p(k)$, but no other points in $P$.

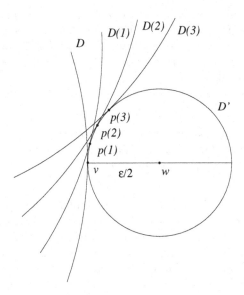

**Fig. 1.** *Replace $D$ by $D(1), D(2), \ldots, D(k)$*

It remains to verify that the new realization $(P, \mathcal{D})$ meets the extra requirements stated in the lemma: it is in general position and each disk $D \in \mathcal{D}$ has a boundary point that does not belong to the closure of any other disks in $\mathcal{D}$. However, these conditions are automatically satisfied if $\delta$ is sufficiently small. For instance, each disk $D(i)$ has point on its boundary, very close to $p(i)$, which is not covered by any other disk in $\mathcal{D}$. To see that the same property holds for $D'$, notice that *any* boundary point of $D'$, "sufficiently far" from $p(1), p(2), \ldots, p(k)$, will do. This completes the induction step, and hence the proof of the lemma. $\square$

**Corollary 5.2.** *For any $k \ge 2$, there exists a k-uniform hypergraph which is not two-colorable and which permits a planar realization by open disks.* $\square$

## 6   Dual Realization with Axis-Parallel Rectangles

All rectangles in this section are assumed to be *closed*, but our results and proofs also apply to open rectangles.

**Lemma 6.1.** *For any rooted tree $T$, the hypergraph $H(T)$ permits a dual realization with axis-parallel rectangles.*

**Proof:** Let $\sigma_0$ and $\sigma_1$ denote the segments $y = x$, $1 \leq x \leq 2$ and $y = x + 2$, $0 \leq x \leq 1$. First, consider the sub-hypergraph $H'$ of $H(T)$, consisting of all descendent hyperedges. We claim that it permits a dual realization with closed intervals and points of $\sigma_1$. To see this, choose an arbitrary interval in $\sigma_1$ to represent the root of $T$. If an interval $I$ represents a vertex $v$ of $T$ and $v$ has $k \geq 1$ children, choose any $k$ pairwise disjoint sub-intervals of $I$ to represent them. Finally, for every leaf $v$, pick any point of the interval representing $v$ to represent the descendent hyperedge of $H'$ that contains $v$. It is straightforward to check that the resulting system is indeed a dual realization of $H'$.

Now we construct a dual realization of $H(T)$ with axis-parallel rectangles. Let the descendent hyperedges be represented by the same point in $\sigma_1$ as in the construction above. For the sibling hyperedges, we choose distinct points of $\sigma_0$ to represent them. Let any vertex $x$ of $T$ be represented by the axis-parallel rectangle whose lower right corner is the point that represents the sibling hyperedge containing $x$, and whose intersection with $\sigma_1$ is the interval that represented $x$ in the previous construction. (Note that the root of $T$ is not contained in any sibling hyperedge. Therefore, if $x$ is the root, we have to modify the above definition. In this case, let the lower right vertex of the corresponding rectangle be any point of $\sigma_0$ that does not represent any sibling hyperedge.) Clearly, the resulting system of points and rectangles is a dual representation of $H(T)$.   □

# 7 Planar and Dual Realizations with Concave Quadrilaterals

The aim of this section is to prove Theorem 2.8. For the proof, it is irrelevant whether we consider closed or open quadrilaterals.

One of the two diagonals of a concave quadrilateral $Q$ is inside $Q$, the other is outside $Q$. We call the line of the diagonal outside $Q$ the *supporting line of $Q$*.

**Lemma 7.1** *For any rooted tree $T$ and for any concave quadrilateral $Q$, the hypergraph $H(T)$ permits both planar and dual realizations with translates of $Q$. Moreover, we can achieve that all translates of $Q$ used in the planar realization can be obtained from $Q$ by translations parallel to its supporting line, while all points used in the dual realization lie on the supporting line.*

**Proof:** The two realizations are dual to each other, so it is enough to prove the existence of a *planar* realization. Let the vertices of $Q$ be $a$, $b$, $c$, and $d$ in this order, and assume $b$ is the concave vertex. The supporting line of $Q$ is the line $ac$. We start with a planar realization $(P, \mathcal{S})$, in which each member of $\mathcal{S}$ is a translate parallel to $ac$ of one of the two infinite wedges $W_a, W_c$. Here the sides of $W_a$ are the rays $ad$ and $ab$, while the sides of the $W_c$ are the rays $cd$ and $cb$. Once we have such a planar realization, we can shrink the point set so that the wedges can be replaced by $Q$, without changing the containment relation.

In our planar realization, all sibling hyperedges will be represented by translates of $W_a$, while all descendent hyperedges will be represented by translates of $W_c$. We construct the planar realization by induction on the depth of $T$, starting with the trivial case of depth 0.

For the inductive step, let $v_0$ be the root of $T$, let $v_1, \ldots, v_k$ denote its children, and let $T^i$ be the tree rooted at $v_i$, for $1 \le i \le k$. By the inductive hypothesis, for every $i$, $H(T^i)$ permits a *planar realization* $(P_i, \mathcal{S}_i)$, meeting the requirements. We assume that the following three additional conditions are also satisfied.

1. $W \cap P_j = \emptyset$, whenever $W \in \mathcal{S}_i$ and $i \ne j$.
2. $P_i \cap W_a = \emptyset$, for all $i$.
3. For any $i$, there exists a point $x_i \in W_a$ such that, for any $W \in \mathcal{S}_j$, we have $x_i \in W$ if and only if $i = j$ and $W$ is a translate of $W_c$.

To verify that one can make the above assumptions, note that $H(T^i)$ can also be realized by any translate of $(P_i, \mathcal{S}_i)$. Translating $(P_i, \mathcal{S}_i)$ through sufficiently fast increasing multiples of the vector $ac$, as $i$ increases, makes all of the above three properties satisfied.

It is easy to see that one can find a point $x$, common to all translates of $W_c$ in any of the families $\mathcal{S}_i$, with he property that $x$ is not contained in $W_a$ or in any of its translates considered. Let $y_i \in P_i$ denote the point representing the root $v_i$ of $T^i$.

Now we are in a position to define the pair $(P, \mathcal{S})$ realizing $T$: let

$$P = ((\cup_i P_i) \cup \{x_i | 1 \le i \le k\} \cup \{x\}) \setminus \{y_i | 1 \le i \le k\},$$

and let $S = (\cup_i \mathcal{S}_i) \cup \{W_a\}$. It is straightforward to check now that $(P, \mathcal{S})$ is a planar realization of $H(T)$, where sibling hyperedges are represented by translates of $W_a$ parallel to the line $ac$ and descendent hyperedges are represented by translates of $W_c$ parallel to the same line.     □

**Proof of Theorem 2.8:** Let $Q$ be a concave quadrilateral and let $k \ge 1$ arbitrary. We need to show that not all $k$-fold coverings of the plane by translates of $Q$ can be split into two coverings. Let us start with a dual realization $(P, \mathcal{S})$ of the $k$-uniform hypergraph $H_k = H(T_k)$ with translates of $Q$. We consider the set $\mathcal{S}'$ obtained from $\mathcal{S}$ by adding all translates of $Q$ disjoint from $P$. Clearly, $\mathcal{S}'$ cannot be split into two covering, as every point of $P$ can be covered only by members of $\mathcal{S}$, and we know that $H_k$ is not two-edge-colorable.

It remains to check that $\mathcal{S}'$ is a $k$-fold covering of the plane. For this, we use the fact that the dual realization $(P, \mathcal{S})$ of $H_k$, whose existence is guaranteed by Lemma 7.1, satisfies that all points of $P$ lie on the supporting line of $Q$. Clearly, any point that does not belong to this line is covered by infinitely many translates of $Q$ that are disjoint from the line. For a point $r \notin P$ that belongs to the supporting we can still find infinitely many translates of $Q$ which cover $r$ and which are disjoint from the finite set $P$. If $a$ is a vertex of $Q$ on the supporting line then any translation that carries a point $a' \ne a$ of $Q$ to $r$, where $a'$ is sufficiently close to $a$, will do here. Finally, each point of $P$ is covered by exactly $k$ members of $\mathcal{S}$, as $H_k$ is a $k$-uniform hypergraph.     □

The proof of Lemma 7.1 applies not only to concave quadrilaterals, but to many other concave polygons $Q'$, as well, implying that the families of translates of these polygons are not cover-decomposable. However, the statement is not true for *all* concave polygons. For instance, if $Q'$ can be expressed as a finite union of translates of a given convex polygon, then the family of translates of $Q'$ must be cover-decomposable. It would be interesting to find an exact criterion for deciding whether the family of translates of a polygon $Q'$ is cover-decomposable.

# References

[AS00] N. Alon and J.H. Spencer: *The Probabilistic Method* (2nd ed.), Wiley, New York, 2000.

[Bl57] W.J. Blundon: Multiple covering of the plane by circles, *Mathematika* **4** (1957), 7–16.

[Bo89] U. Bolle: On the density of multiple packings and coverings of convex discs, *Studia Sci. Math. Hungar.* **24** (1989), 119–126.

[BMP05] P. Brass, J. Pach, and W. Moser: *Research Problems in Discrete Geometry*, Springer, Berlin, 2005, p. 77.

[CPST06] X. Chen, J. Pach, M. Szegedy, and G. Tardos: Delaunay graphs of point sets in the plane with respect to axis-parallel rectangles, manuscript, 2006.

[Co76] M.J. Cohn: Multiple lattice covering of space, *Proc. London Math. Soc. (3)* **32** (1976), 117–132.

[DuH72] V.C. Dumir and R.J. Hans-Gill: Lattice double packings in the plane, *Indian J. Pure Appl. Math.* **3** (1972), 481–487.

[ER62] P. Erdős and C.A. Rogers: Covering space with convex bodies, *Acta Arith.* **7** (1961/1962), 281–285.

[FTG77] G. Fejes Tóth: A problem connected with multiple circle-packings and circle-coverings, *Studia Sci. Math. Hungar.* **12** (1977), 447–456.

[FTG83] G. Fejes Tóth: New results in the theory of packing and covering, in: *Convexity and its Applications* (P.M. Gruber, J.M. Wills, eds.), Birkhäuser, Basel, 1983, 318–359.

[FTG84] G. Fejes Tóth: Multiple lattice packings of symmetric convex domains in the plane, *J. London Math. Soc. (2)* **29** (1984), 556–561.

[FTK93] G. Fejes Tóth and W. Kuperberg: A survey of recent results in the theory of packing and covering, in: *New Trends in Discrete and Computational Geometry* (J. Pach, ed.), *Algorithms Combin.* **10**, Springer, Berlin, 1993, 251–279.

[Fl78] A. Florian: Mehrfache Packung konvexer Körper (German), *Österreich. Akad. Wiss. Math.-Natur. Kl. Sitzungsber. II* **186** (1978), 373–384.

[FuK05] Z. Füredi and J.-H. Kang: Covering Euclidean $n$-space by translates of a convex body, *Discrete Math.*, accepted.

[HaJ63] A.W. Hales and R.I. Jewett: Regularity and positional games, *Trans. Amer. Math. Soc.* **106** (1963), 222–229.

[He59] A. Heppes: Mehrfache gitterförmige Kreislagerungen in der Ebene (German), *Acta Math. Acad. Sci. Hungar.* **10** (1959), 141–148.

[Ko04] A. Kostochka: Coloring intersection graphs of geometric figures, in: *Towards a Theory of Geometric Graphs* (J. Pach, ed.), *Contemporary Mathematics* **342**, Amer. Math. Soc., Providence, 2004, 127–138.

[MP87] P. Mani-Levitska and J. Pach: Decomposition problems for multiple coverings with unit balls, manuscript, 1987.

[PA95] J. Pach and P.K. Agarwal: *Combinatorial Geometry*, Wiley, New York, 1995.

[P80] J. Pach: Decomposition of multiple packing and covering, *2. Kolloquium über Diskrete Geometrie*, Salzburg (1980), 169–178.

[P86] J. Pach: Covering the Plane with Convex Polygons, *Discrete and Computational Geometry* **1** (1986), 73–81.

[Sch61] W.M. Schmidt: Zur Lagerung kongruenter Körper im Raum (German), *Monatsh. Math.* **65** (1961), 154–158.

[TaT06] G. Tardos and G. Tóth: Multiple coverings of the plane with triangles, *Discrete and Computational Geometry*, to appear.

# The Decycling Number of Cubic Planar Graphs*

Narong Punnim

Department of Mathematics, Srinakharinwirot University, Sukhumvit 23, Bangkok
10110, Thailand
narongp@swu.ac.th

**Abstract.** Bau and Beineke [2] asked the following questions:

1 Which cubic graphs $G$ of order $2n$ have decycling number $\phi(G) = \lceil \frac{n+1}{2} \rceil$?

2 Which cubic planar graphs $G$ of order $2n$ have decycling number $\phi(G) = \lceil \frac{n+1}{2} \rceil$?

We answered the first question in [10]. In this paper we prove that if $\mathcal{P}(3^{2n})$ is the class of all connected cubic planar graphs of order $2n$ and $\phi\left(\mathcal{P}(3^{2n})\right) = \left\{\phi(G) \colon G \in \mathcal{P}(3^{2n})\right\}$, then there exist integers $a_n$ and $b_n$ such that there exists a graph $G \in \mathcal{P}(3^{2n})$ with $\phi(G) = c$ if and only if $c$ is an integer satisfying $a_n \leq c \leq b_n$. We also find all corresponding integers $a_n$ and $b_n$. In addition, we prove that if $\mathcal{P}\left(3^{2n}; \phi = \lceil \frac{n+1}{2} \rceil\right)$ is the class of all connected cubic planar graphs of order $2n$ with decycling number $\lceil \frac{n+1}{2} \rceil$ and $G_1, G_2 \in \mathcal{P}\left(3^{2n}; \phi = \lceil \frac{n+1}{2} \rceil\right)$, then there exists a sequence of switchings $\sigma_1, \sigma_2, \ldots, \sigma_t$ such that for every $i = 1, 2, \ldots, t - 1$, $G_1^{\sigma_1 \sigma_2 \cdots \sigma_i} \in \mathcal{P}\left(3^{2n}; \phi = \lceil \frac{n+1}{2} \rceil\right)$ and $G_2 = G_1^{\sigma_1 \sigma_2 \cdots \sigma_t}$.

## 1 Introduction

We consider only finite simple graphs. For the most part, our notation and terminology follow that of Bondy and Murty [3]. Let $G = (V, E)$ denote a graph with vertex set $V = V(G)$ and edge set $E = E(G)$. Let $V(G) = \{v_1, v_2, \ldots, v_n\}$ and $E(G) = \{e_1, e_2, \ldots, e_m\}$. We use $|S|$ to denote the cardinality of a set $S$ and define $n = \nu(G) = |V|$ to be the *order* of $G$ and $m = \varepsilon(G) = |E|$ the *size* of $G$. We write $e = uv$ for an edge $e$ that joins vertex $u$ to vertex $v$. A *path* of order $k$ in a graph $G$, denoted by $P_k$, is a sequence of distinct vertices $u_1 u_2 \cdots u_k$ of $G$ such that for all $i = 1, 2, \ldots, k - 1$, $u_i u_{i+1}$ is an edge of $G$. The *degree* of a vertex $v$ of a graph $G$ is defined as $d_G(v) = |\{e \in E \colon e = uv \text{ for some } u \in V\}|$. The maximum degree of a graph $G$ is usually denoted by $\Delta(G)$. If $S \subseteq V(G)$, the graph $G[S]$ is the subgraph induced by $S$ in $G$ and we also use the notation $\varepsilon(S)$ for $\varepsilon(G[S])$. For a graph $G$ and $X \subseteq E(G)$, we denote $G - X$ the graph obtained from $G$ by removing all edges in $X$. If $X = \{e\}$, we write $G - e$ for $G - \{e\}$. For a graph $G$ and $X \subseteq V(G)$, $G - X$ is the graph obtained from $G$ by removing all vertices in $X$ and all edges incident with vertices in $X$. For a graph $G$ and $X \subseteq E(\overline{G})$, we denote $G + X$ the graph obtained from $G$ by adding all edges in $X$. If $X = \{e\}$, we simply write $G + e$ for $G + \{e\}$. Two graphs $G$

* This work was carried out with financial support by The Thailand Research Fund.

J. Akiyama et al. (Eds.): CJCDGCGT 2005, LNCS 4381, pp. 149–161, 2007.

and $H$ are disjoint if $V(G) \cap V(H) = \emptyset$. For any two disjoint graphs $G$ and $H$, $G \cup H$ is defined by $V(G \cup H) = V(G) \cup V(H)$ and $E(G \cup H) = E(G) \cup E(H)$. We can extend this definition to a finite union of pairwise disjoint graphs. For a graph $G$ and $s \in V(G)$, the *neighborhood* of $s$ in $G$ is defined by $N(s) = \{v \in V(G) : sv \in E(G)\}$. If $S \subseteq V(G)$, then we define $N(S) = \bigcup_{s \in S} N(s)$. If $F \subseteq V(G)$, we write $N_F(S)$ for $N(S) \cap F$. A graph $G$ is called *r-regular* if all of its vertices have degree $r$. A 3-regular graph is called a *cubic graph*. Let $G$ be a graph of order $n$ and $V(G) = \{v_1, v_2, \ldots, v_n\}$ be the vertex set of $G$. The sequence $(d_G(v_1), d_G(v_2), \ldots, d_G(v_n))$ is called a *degree sequence* of $G$, and we simply write $(d(v_1), d(v_2), \ldots, d(v_n))$ if the underlying graph $G$ is clear from the context. A sequence $\mathbf{d} = (d_1, d_2, \ldots, d_n)$ of non-negative integers is a *graphic degree sequence* if it is a degree sequence of some graph $G$. In this case, $G$ is called a *realization* of $\mathbf{d}$. An algorithm for determining whether or not a given sequence of non-negative integers is graphic was independently obtained by Havel [6] and Hakimi [5]. We state their result in the following theorem.

**Theorem 1.** *Let* $\mathbf{d} = (d_1, d_2, \ldots, d_n)$ *be a non-increasing sequence of non-negative integers and denote the sequence*

$$(d_2 - 1, d_3 - 1, \ldots, d_{d_1+1} - 1, d_{d_1+2}, \ldots, d_n) = \mathbf{d'}.$$

*Then* $\mathbf{d}$ *is graphic if and only if* $\mathbf{d'}$ *is graphic.*

Let $G$ be a graph and $ab, cd \in E(G)$ be independent, where $ac, bd \notin E(G)$. Define

$$G^{\sigma(a,\, b;\, c,\, d)} = (G - \{ab, cd\}) + \{ac, bd\}.$$

The operation $\sigma(a, b; c, d)$ is called a *switching operation* on $G$. For a graph $G$, put $G_0 = G$. We define $G^{\sigma_1 \sigma_2 \cdots \sigma_t}$ recursively by for each $i = 1, 2, \ldots, t$, $G_i = G_{i-1}^{\sigma_i}$, where $\sigma_i$ is a switching on $G_{i-1}$. It is easy to see that the graph $G_i$ has the same degree sequence as $G$. The following theorem was proved by Havel [6] and later by Hakimi [5] independently.

**Theorem 2.** *Let* $\mathbf{d} = (d_1, d_2, \ldots, d_n)$ *be a graphic degree sequence. If* $G$ *and* $H$ *are any two realizations of* $\mathbf{d}$, *then* $H$ *can be obtained from* $G$ *by a finite sequence of switchings.*

As a consequence of Theorem 2, Eggleton and Holton [4] defined in 1978 *the graph* $\mathcal{R}(\mathbf{d})$ of realizations of $\mathbf{d}$ whose vertices are the set of all non-isomorphic graphs with degree sequence $\mathbf{d}$; two vertices being adjacent in the graph $\mathcal{R}(\mathbf{d})$ if one can be obtained from the other by a switching. They obtained the following theorem.

**Theorem 3.** *The graph* $\mathcal{R}(\mathbf{d})$ *is connected.*

The following theorem was obtained by Taylor [11] in 1980.

**Theorem 4.** *For a graphic degree sequence* $\mathbf{d}$, *let* $\mathcal{CR}(\mathbf{d})$ *be the set of all connected realizations of* $\mathbf{d}$. *Then the subgraph of the graph* $\mathcal{R}(\mathbf{d})$ *induced by* $\mathcal{CR}(\mathbf{d})$ *is connected.*

## 2    The Decycling Number

The problem of determining the minimum number of vertices whose removal eliminates all cycles in a graph $G$ is difficult even for some simply defined graphs. For a graph $G$, this minimum number is known as the *decycling number* of $G$, and is denoted by $\phi(G)$. Let $\mathbb{G}$ be the class of all graphs. A function $f\colon \mathbb{G} \to \mathbb{Z}$ is called a *graph parameter* if $G$, $H \in \mathbb{G}$ and $G \cong H$, then $f(G) = f(H)$. Let $\mathbb{J} \subseteq \mathbb{G}$. A graph parameter $f$ is called an *interpolation graph parameter with respect to* $\mathbb{J}$ or $f$ *interpolates over* $\mathbb{J}$ if $G$, $H \in \mathbb{J}$ and $a = f(G) < f(H) = b$, then for every integer $c$ between $a$ and $b$, there exists $K \in \mathbb{J}$ such that $f(K) = c$. If $f$ interpolates over $\mathbb{J}$, then $f(\mathbb{J}) := \{f(G)\colon G \in \mathbb{J}\}$ is uniquely determined by $\min(f, \mathbb{J}) := \min\{f(G)\colon G \in \mathbb{J}\}$ and $\max(f, \mathbb{J}) := \max\{f(G)\colon G \in \mathbb{J}\}$. Let $\mathbf{d} = r^n$ be a graphic regular degree sequence and let $f$ be a graph parameter. We use the following notation: $\min(f, r^n) = \min(f, \mathcal{R}(r^n))$, $\max(f, r^n) = \max(f, \mathcal{R}(r^n))$, $\mathrm{Min}(f, r^n) = \min(f, \mathcal{CR}(r^n))$, and $\mathrm{Max}(f, r^n) = \max(f, \mathcal{CR}(r^n))$.

We proved in [8] that for a graph $G$ and a switching $\sigma$ on $G$, $|\phi(G) - \phi(G^\sigma)| \leq 1$. Thus we have the following results.

**1** $\phi$ interpolates over $\mathcal{R}(\mathbf{d})$.

**2** $\phi$ interpolates over $\mathcal{CR}(\mathbf{d})$.

**3** Let $\mathbb{J} \subseteq \mathcal{R}(\mathbf{d})$. If the subgraph of the graph $\mathcal{R}(\mathbf{d})$ induced by $\mathbb{J}$ is connected, then $\phi$ interpolates over $\mathbb{J}$.

Let $G$ be a graph and $F \subseteq V(G)$. $F$ is called an *induced forest* of $G$ if $G[F]$ contains no cycle. An induced forest $F$ of $G$ is *maximal* if for every $v \in V(G-F)$, $G[F \cup \{v\}]$ contains a cycle. Let $t(G)$ be defined as

$$t(G) := \max\{|F|\colon F \text{ is an induced forest of } G\}.$$

It is clear that, for a graph $G$ of order $n$, $\phi(G) + t(G) = n$. Consequently, $t$ interpolates over $\mathbb{J} \subseteq \mathcal{R}(\mathbf{d})$ if and only if $\phi$ does.

## 3    Cubic Graphs

The interest of studying the decycling number of cubic graphs was motivated by the results of Zheng and Lu [12], Alon et al. [1], Liu and Zhao [7], and Bau and Beineke [2]. We found in [8] the values of $\min(\phi, 3^{2n})$ in all situations and proved that $\min(\phi, 3^{2n}) = \mathrm{Min}(\phi, 3^{2n})$.

The problem of finding upper bounds of $\phi(G)$, where $G$ runs over a class of cubic graphs, has been investigated in the literature. First observe that if $G$ is a cubic graph of order $n$ and $F$ is a maximum induced forest of $G$, then it is easy to see that $G - F$ is also a forest. Thus $|F| \geq \frac{n}{2}$. The bound is sharp if and only if $n$ is a multiple of 4. We proved in [8] that if $n = 4q + t$, $t = 0, 2$, then $\max(\phi, 3^n) = 2q$.

The problem of finding $\mathrm{Max}(\phi, 3^n)$ is more difficult. We have considered such problem in term of the graph parameter $t$ in [9]. A *cubic tree* is a tree whose vertices consisting of degree 1 or 3. Evidently, if $T$ is a cubic tree of order $n$, then $n = 2k + 2$, where $k$ is the number of vertices of degree 3 of $T$. Let $\mathbb{T}$ denote the family of cubic graphs obtained by taking cubic trees and replacing each vertex of degree 3 by a triangle and attaching a copy of $K_4$ with one subdivided edge at every vertex of degree 1.

A lower bound for the order of maximum induced forest in connected cubic graphs has been obtained by Liu and Zhao [7] as stated in the following theorem.

**Theorem 5.** *Let $G$ be a connected cubic graph of order $2n \geq 12$. Then $t(G) = \frac{5}{4}n - \frac{1}{4}$ if $G \in \mathbb{T}$ and $t(G) \geq \frac{5}{4}n$ if $G \notin \mathbb{T}$.*

By Theorem 5 and with the fact that a switching changes the value of $\phi$ by at most 1 in the resulting graph, we proved in [9] the following results.

**Theorem 6.** *Let $n$ be an integer with $n \geq 6$. Then*

$$\text{Max}(\phi, 3^{2n}) = \begin{cases} \frac{3}{4}n + \frac{1}{4}, & \text{if } n \equiv 1 \ (\text{mod } 4), \\ \lfloor \frac{3}{4}n \rfloor, & \text{otherwise.} \end{cases}$$

## 4   Cubic Planar Graphs

Note that $\phi$ interpolates over $\mathcal{CR}(3^{2n})$ with

$$\text{Min}(\phi, 3^{2n}) = \lceil \tfrac{n+1}{2} \rceil \text{ and } \text{Max}(\phi, 3^{2n}) = \begin{cases} \frac{3}{4}n + \frac{1}{4}, & \text{if } n \equiv 1 \ (\text{mod } 4), \\ \lfloor \frac{3}{4}n \rfloor, & \text{otherwise.} \end{cases}$$

For each $c \in \phi(3^{2n})$, let $\mathcal{CR}(3^{2n}; \phi = c) = \{G \in \mathcal{CR}(3^{2n}): \phi(G) = c\}$.

We answered Question 1 in [10] by determining all cubic graphs $G$ of order $2n$ and $\phi(G) = \lceil \frac{n+1}{2} \rceil$. In addition, we proved that the subgraph of $\mathcal{CR}(3^{2n})$ induced by $\mathcal{CR}\left(3^{2n}; \phi = \lceil \frac{n+1}{2} \rceil\right)$ is connected.

Let $\mathcal{P}(3^{2n})$ be the class of all connected cubic planar graphs of order $2n$. For small $n$, the difference between $\text{Max}(\phi, 3^{2n})$ and $\text{Min}(\phi, 3^{2n})$ is relatively small and it is easy to show by construction that $\phi(\mathcal{P}(3^{2n})) = \phi(\mathcal{CR}(3^{2n}))$, for all $n \leq 13$. We now consider the interpolation property of $\phi$ with respect to $\mathcal{P}(3^{34})$. Since $\text{Min}(\phi, 3^{34}) = \frac{17+1}{2} = 9$ and $\text{Max}(\phi, 3^{34}) = \frac{3 \cdot 17}{4} + \frac{1}{4} = 13$, $\mathcal{CR}(3^{34})$ can be partitioned into five classes $C_i (9 \leq i \leq 13)$, where $C_i = \{G \in \mathcal{CR}(3^{34}): \phi(G) = i\}$ and $C_i \neq \emptyset$, for all $i = 9, 10, 11, 12, 13$. Consider the following example.

*Example 1.* Planar graphs $G^1$, $G^2$, $G^3$, $G^4$, $G^5$

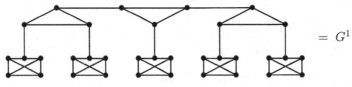

$G^1$ of order $34 = 8k + 10$ and $\phi(G^1) = 13 = \text{Max}(\phi, 3^{34})$.

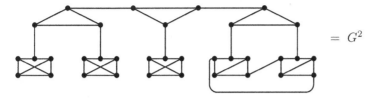

$G^2$ of order $34 = 8k + 10$ and $\phi(G^2) = 12 = \text{Max}(\phi, 3^{34}) - 1$.

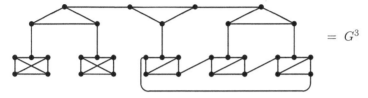

$G^3$ of order $34 = 8k + 10$ and $\phi(G^3) = 11 = \text{Max}\left(\phi, 3^{34}\right) - 2$.·

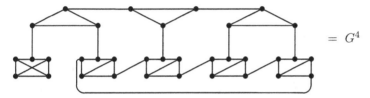

$G^4$ of order $34 = 8k + 10$ and $\phi(G^4) = 10 = \text{Max}(\phi, 3^{34}) - 3$.

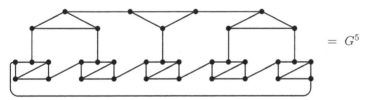

$G^5$ of order $34 = 8k + 10$ and $\phi(G^5) = 9 = \text{Max}(\phi, 3^{34}) - 4 = \text{Min}(\phi, 3^{34})$.

Thus for each $i = 9, 10, 11, 12, 13$, $C_i \cap \mathcal{P}(3^{34}) \neq \emptyset$. Therefore $\phi$ interpolates over $\mathcal{P}(3^{34})$. The same result can be obtained in $\mathcal{P}(3^{2n})$ for $n \in \{14, 15, 16\}$.

In general if we write $2n = 8k + 2i$, $i = 2, 3, 4, 5$ and $k \geq 4$, then we can construct a cubic planar graph $G_i$ of order $8k + 2i$ by a cubic tree $T_i$ of order $2k + 2$, where $k$ is the number of vertices of degree 3, replacing $k - 5 + i$ vertices of degree 3, by a triangle and attaching a copy of $K_4$ with one subdivided edge at every vertex of degree 1. Thus $G_i \in \mathcal{P}(3^{2n})$ and $\phi(G_i) = 3k - 1 + i = \text{Max}(\phi, 3^{2n})$. Hence $\max(\phi, \mathcal{P}(3^{2n})) = \text{Max}(\phi, 3^{2n})$. By forming appropriate sequence of switchings to $G_i$ as described in above example and with the fact that a switching changes the value of $\phi$ of the resulting graph by at most 1, we have the following theorem.

**Theorem 7.** $\phi$ *interpolates over* $\mathcal{P}(3^{2n})$, *with*

$$\min\{\phi(G)\colon G \in \mathcal{P}(3^{2n})\} = \mathrm{Min}(\phi, 3^{2n}), \text{ and}$$

$$\max\{\phi(G)\colon G \in \mathcal{P}(3^{2n})\} = \mathrm{Max}(\phi, 3^{2n}).$$

Let $G \in \mathcal{CR}(3^{2n}; \phi = \lceil\frac{n+1}{2}\rceil)$ and $S$ be a minimum decycling set of $G$. Let $F = V(G - S)$. We proved in [10] the following results.

**1** $\varepsilon(S) = 0$ if $n$ is odd and $\varepsilon(S) \leq 1$ if $n$ is even,
**2** if $n$ is odd, then $G[F]$ is a tree,
**3** if $n$ is even and $\varepsilon(S) = 1$, then $G[F]$ is a tree,
**4** if $n$ is even and $\varepsilon(S) = 0$, then $G[F]$ has 2 components.

**Lemma 1.** *There exists* $G \in \mathcal{P}(3^{2n})$ *such that* $\phi(G) = \lceil\frac{n+1}{2}\rceil$ *and* $G$ *contains a path of order* $\lfloor\frac{3n-1}{2}\rfloor$ *as its induced forest. Furthermore the induced subgraph of the corresponding decycling set of* $G$ *contains at most one edge and it contains no edge if and only if* $n$ *is odd.*

A proof of above lemma can be obtained easily by introducing the graphs $G_6$ and $G_8$ of order 6 and 8, respectively, as showed in the following figures and the definitions of *super subdivision.*

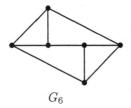

$G_6$

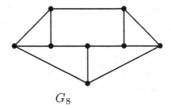

$G_8$

**Definition 1.** *Let* $G$ *be a connected cubic planar graph and let* $F$ *be an induced forest of* $G$. *If* $u, v \in F$ *and* $uv \in E(G)$, *define a super subdivision with respect to* $uv$ *of* $G$ *to be a graph* $G \cdot uv$ *obtained from* $G$ *by inserting three vertices* $x, y, z$ *on the edge* $uv$ *and joining* $x, y, z$ *to a new vertex* $w$.

**Definition 2.** *Let* $u \in F$, $d_F(u) = 1$. *Then there exists a unique vertex* $v \in F$ *that is adjacent to* $u$ *and there exist two vertices* $p, q \in V(G - F)$ *such that* $p$ *and* $q$ *are adjacent to* $u$. *Define a super subdivision with respect to* $u$ *of* $G$ *to be a graph* $G \cdot u$ *obtained from* $G$ *by inserting three vertices* $x, y, z$ *on the edge* $uv$ *to produce a path* $uxyzv$ *in* $F$, *then deleting* $qu$, *and adding the edge* $qz$, *and finally joining* $u, x, y$ *to a new vertex* $w$.

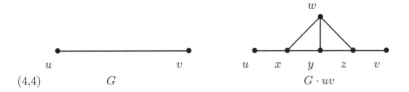

Super subdivision w. r. t. edge $uv$

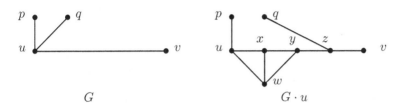

Super subdivision w. r. t. vertex $u$

Note that the concept of "super subdivision" may be defined by the concept of a contraction. A contraction that maps the free vertex, all three of its neighbors and a neighboring vertex $x$ on the tree to the vertex $x$.

Let $\mathcal{P}\left(3^{2n};\ \phi = \lceil\frac{n+1}{2}\rceil\right)$ be the class of cubic planar graphs $G$ of order $2n$ and $\phi(G) = \lceil\frac{n+1}{2}\rceil$. Let $G \in \mathcal{P}\left(3^{2n};\ \phi = \lceil\frac{n+1}{2}\rceil\right)$. Then for any minimum decycling set $S$ of $G$, $G-S$ is a tree or a union of two trees. Since $G$ is cubic, $\Delta(G-S) \le 3$. Let $n_3(G-S)$ be the number of vertices of degree 3 in $G-S$ and $n_3(G)$ be defined by

$$n_3(G) = \min\{n_3(G-S)\colon S \text{ is a minimum decycling set of } G\}.$$

For nonnegative integer $i$, put

$$\mathcal{P}_i(2n) = \left\{G \in \mathcal{P}\left(3^{2n};\ \phi = \lceil\frac{n+1}{2}\rceil\right) : n_3(G) = i\right\}.$$

We first consider the class $\mathcal{P}_0(2n)$ and prove that the subgraph of $\mathcal{CR}(3^{2n})$ induced by $\mathcal{P}_0(2n)$ is connected. In order to prove the result, we need to introduce some notation and terminology which will be used from now on.

Let $G \in \mathcal{P}_0(2n)$ and put $N = \lfloor\frac{3n-1}{2}\rfloor$. Thus there exists a minimum decycling set $S$ of $G$ such that $G - S = F_N = P_l \cup P_{N-l}$, where $l \ge N - l \ge 0$, $P_l = v_1 v_2 \cdots v_l$, and $P_{N-l} = v_{l+1} v_{l+2} \cdots v_N$. For a planar embedding of $G$, put $F_N$ in the horizontal direction and the vertices in $S$ are arranged either above or below $F_N$. Thus the vertices in $S$ can be partitioned into at most three subsets, $U, L, D$, according to the following rules.

1 For $s \in S$, $s \in U$ if $s$ lies above $F_N$ and all edges from $s$ to $F_N$ lie above $F_N$.
2 For $s \in S$, $s \in L$ if $s$ lies below $F_N$ and all edges from $s$ to $F_N$ lie below $F_N$.
3 Put $D = S - (U \cup L)$.

With a planar embedding we can also partition the edge set from $S$ to $F_N$ into $U'$ and $L'$ according to it is incident to a vertex of $F_N$ above or below $F_N$. It is clear that an edge from $s \in U$ to a vertex of $F_N$ is in $U'$ and likewise an edge from $s \in L$ to a vertex of $F_N$ is in $L'$ while an edge from $s \in D$ is either in $U'$ or $L'$ depending on the direction to its end. Note that for each planar embedding of $G \in \mathcal{P}_0(2n)$, the sets $U'$ and $L'$ are uniquely determined. Thus we can write $U'(G)$ and $L'(G)$ for $U'$ and $L'$ if we work with more than one graph.

**Definition 3.** *Let $G \in \mathcal{P}_i(2n)$. A switching $\sigma$ on $G$ is called a $\mathcal{P}_i$-switching with respect to $G$ if $G^\sigma \in \mathcal{P}_j(2n)$, for some $j \le i$. A sequence of switchings $\sigma_1, \sigma_2, \ldots, \sigma_t$ is called a sequence of $\mathcal{P}_i$-switchings with respect to $G$ if for all $k = 1, 2, \ldots, t$ there exists $j$ such that $j \le i$ and $G^{\sigma_1 \sigma_2 \cdots \sigma_k} \in \mathcal{P}_j(2n)$.*

A $\mathcal{P}_0$-switching plays a crucial role in proving the main results. We introduce a special type of $\mathcal{P}_0$-switching called *elementary switching*. Let $G \in \mathcal{P}_0(2n)$. We choose a planar embedding of $G$ as described above. Let $x$, $y$ be distinct vertices of $S$, $x$ is adjacent to $v_i$ and $y$ is adjacent to $v_{i+1}$, where $i \ne l$. If $xv_i \in U'$ and $yv_{i+1} \in L'$ or $xv_i \in L'$ and $yv_{i+1} \in U'$, then $\sigma(x, v_i; v_{i+1}, y)$ is a $\mathcal{P}_0$-switching with respect to $G$. This special type of $\mathcal{P}_0$-switching with respect to $G$ will be used throughout and we call it an *elementary switching with respect to $F_N$*. A sequence of elementary switchings can be similarly defined.

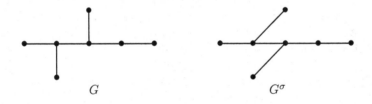

$$G \qquad\qquad G^\sigma$$

$\mathcal{P}_0$-switching: Elementary switching

By the definition of an elementary switching with respect to $F_N$ as described above, we assume that $x$ and $y$ are distinct. If $x = y$ and $x$ lies above $F_N$, then we can not apply the switching but we can redraw either an edge $xv_i$ or $xv_{i+1}$ so that the two edges lie above $F_N$. Similarly when $x$ lies below $F_N$. For a planar embedding of $G$, we want to avoid this situation. Thus for a planar embedding of $G$, if $x \in S$ is adjacent to $v_i$ and $v_{i+1}$ ($i \ne l$), then we draw $xv_i$, $xv_{i+1}$ and $x$ on the same side of $F_N$. Clearly if $\sigma$ is an elementary switching with respect to $F_N$, then the resulting graph $G^\sigma \in \mathcal{P}_0(2n)$ with $S$ as its minimum decycling set and $F_N$ as its induced forest. The following lemmas are useful and they are direct consequences of applying a sequence of elementary switchings.

**Lemma 2.** *Let $G \in \mathcal{P}_0(2n)$. Suppose that there are $k$ edges in $U'$ from $S$ to $P_l - \{v_1, v_l\}$ and there are $h$ edges in $L'$ from $S$ to $P_l - \{v_1, v_l\}$. Then there* *exists a sequence of elementary switchings* $\sigma_1, \sigma_2, \ldots, \sigma_t$ *with respect to* $F_N$ *such that* $G_t = G^{\sigma_1 \sigma_2 \cdots \sigma_t}$ *contains* $k$ *edges in* $U'(G_t)$ *from* $S$ *to* $\{v_2, v_3, \ldots, v_{k+1}\}$ *and* $h$ *edges in* $L'(G_t)$ *from* $S$ *to* $\{v_{k+2}, v_{k+3}, \ldots v_{k+h+1}\}$. *It is clear that* $k+h = l-2$.

The result in Lemma 2 is also true if $P_l$ is replaced by $P_{N-l}$. Moreover, it is also true that there exists a sequence of elementary switchings $\sigma_1, \sigma_2, \ldots, \sigma_t$ with respect to $F_N$ such that $G_t = G^{\sigma_1 \sigma_2 \cdots \sigma_t}$ contains $h$ edges in $L'(G_t)$ from $S$ to $\{v_2, v_3, \ldots, v_{h+1}\}$ and $k$ edges in $L'(G_t)$ from $S$ to $\{v_{h+2}, v_{h+3}, \ldots v_{h+k+1}\}$.

**Lemma 3.** *Let* $G \in \mathcal{P}_0(2n)$. *If* $e = sv_i \in U'$, $f = sv_j \in L'$, $1 \leq i < j \leq l$ *and* $s$ *is not adjacent to any internal vertex between* $v_i$ *and* $v_j$, *then there exists a sequence of elementary switchings* $\sigma_1, \sigma_2, \ldots, \sigma_t$ *with respect to* $F_N$ *such that* $s$ *is adjacent to consecutive vertices of* $P_l$ *in* $G^{\sigma_1 \sigma_2 \cdots \sigma_t}$.

*Proof.* We have already mentioned that if $e = sv_i \in U'$ and $f = sv_j \in L'$, $1 \leq i < j \leq l$, then $i + 2 \leq j$. Suppose that $j = i + 2$, there is a vertex $s' \in S$ such that $s \neq s'$ and $s'$ is adjacent to $v_{i+1}$. If $s'v_{i+1} \in U'$, then $\sigma(s, v_j; v_{i+1}, s')$ is an elementary switching and $s$ is adjacent to $v_i$ and $v_{i+1}$ in $G^{\sigma(s, v_j; v_{i+1}, s')}$. Similarly if $s'v_{i+1} \in L'$. Suppose $j = i + k$ with $k \geq 3$. Let $e_k$ denote an edge connecting with a vertex in $S$ and $v_k$, $k = i, i + 1, \ldots, j - 1$. Let $k$ be the first integer such that $i < k < j$ and $e_k \in L'$. Thus there exists a sequence of $k - i$ switchings $\sigma_1, \sigma_2, \ldots, \sigma_{k-i}$ with respect to $F_N$ such that $s$ is adjacent to $v_{i+1}$ and $v_j$ in $G^{\sigma_1 \sigma_2 \cdots \sigma_{k-i}}$. If $k = j$, then there exists a sequence of $j - i - 2$ switchings $\sigma_1, \sigma_2, \ldots, \sigma_{j-i-2}$ with respect to $F_N$ such that $s$ is adjacent to $v_i$ and $v_{i+1}$ in $G^{\sigma_1 \sigma_2 \cdots \sigma_{j-i-2}}$. Thus the proof is complete by continuing the process if necessary. □

The result in Lemma 3 is also true if $P_l$ is replaced by $P_{N-l}$. As a direct consequence of Lemma 3, if $s \in S$ is adjacent to $v_i$ and $v_j$, where $v_i$ and $v_j$ are in the same path, while $sv_i$ and $sv_j$ are in different sets, then there is a sequence of elementary switchings with respect to $F_N$ applying to $G$ and $s$ is adjacent to consecutive vertices of the path in the resulting graph. Thus we can redraw one of the edges so that $s$ and the two edges lie on the same side of $F_N$. It can happen that there is $s \in U$, $s$ is adjacent to $v_j$ and $sv_j \in U'$ but $sv_j$ lies above and below $F_N$.

**Lemma 4.** *Let* $G \in \mathcal{P}_0(2n)$. *If* $e = sv_i \in U'$, $f = sv_j \in U'$, $1 \leq i < j \leq l$ *but* $sv_j$ *lies above and below* $F_N$ *and* $s$ *is not adjacent to any internal vertex between* $v_i$ *and* $v_j$, *then there exists a sequence of elementary switchings* $\sigma_1, \sigma_2, \ldots, \sigma_t$ *with respect to* $F_N$ *such that* $s$ *is adjacent to consecutive vertices of* $P_l$ *in* $G^{\sigma_1 \sigma_2 \cdots \sigma_t}$.

*Proof.* If there is an integer $k$, $i < k < j$ such that $e_k \in L'$, then by Lemma 3 there exists a sequence of elementary switchings with respect to $F_N$ applying to $G$ so that $s$ is adjacent to $v_i$ and $v_{j'}$ with $i < j'$ and for all $k$, $i \leq k \leq j'$, $e_k \in U'$ in the resulting graph. Thus we can redraw the edge $sv_{j'}$ so that $sv_{j'} \in L'$. The proof follows by Lemma 3. □

Thus if $F_N = P_l \cup P_{N-l}$ and $N - l \geq 1$, then $n$ is even and for every $s \in S$, $s$ is adjacent to three vertices in $F_N$. Therefore for every $s \in S$, $s$ is adjacent to $P_l$ or $P_{N-l}$ by at least two vertices. If $s$ is adjacent to three vertices in $P_l$, then there exists a sequence of $\mathcal{P}_0$-switchings $\sigma_1, \sigma_2, \ldots, \sigma_t$ with respect to $F_N$ such that $s$ is adjacent to three consecutive vertices in $P_l$ of $G^{\sigma_1 \sigma_2 \cdots \sigma_t}$. Similarly if $s$ is adjacent to three vertices in $P_{N-l}$. If $s$ is adjacent to two vertices in $P_l$ and one in $P_{N-l}$, then there exists a sequence of $\mathcal{P}_0$-switchings $\sigma_1, \sigma_2, \ldots, \sigma_r$ with respect to $F_N$ such that $s$ is adjacent to two consecutive vertices in $P_l$ of $G_r = G^{\sigma_1 \sigma_2 \cdots \sigma_r}$. Thus there exists $i\,(1 \leq i < l)$ such that $s$ is adjacent to $v_i$ and $v_{i+1}$, and $s$, $sv_i$ and $sv_{i+1}$ lie in the same side of $F_N$. Suppose that $s$ lies above $F_N$ and $s$ is adjacent to $v_k$ for some $k\,(l + 1 \leq k \leq N)$. If $sv_k \in L'(G_r)$, then there exists a sequence of switchings $\sigma_{r+1}, \sigma_{r+2}, \ldots, \sigma_t$ with respect to $F_N$ such that for each $h\,(l + 1 \leq h < k)$, $e_k \in L'(G_t)$, where $G_t = G^{\sigma_1 \sigma_2 \cdots \sigma_t}$. Similarly if $sv_k \in U'(G_r)$ and $sv_k$ lies above and below $F_N$. Thus we have the following theorem.

**Theorem 8.** *Let $G \in \mathcal{P}_0(2n)$. If the planar embedding of $G$ is chosen in such a way that the corresponding set $D$ has minimum cardinality, then $D = \emptyset$ or there exists a sequence of elementary switchings $\sigma_1, \sigma_2, \ldots, \sigma_t$ with respect to $F_N$ such that $G^{\sigma_1 \sigma_2 \cdots \sigma_t}$ has a planar embedding with empty corresponding set $D$.*

**Definition 4.** *Let $G \in \mathcal{P}_0(2n)$. A vertex $s \in S$ is called a* free vertex *with respect to $F_N$ if $s$ is adjacent to three consecutive vertices in $F_N$.*

We have the following theorem.

**Theorem 9.** *If $G \in \mathcal{P}_0(2n)$, then there exists a sequence of switchings $\sigma_1, \sigma_2, \ldots, \sigma_t$ with respect to $F_N$ such that $G^{\sigma_1 \sigma_2 \cdots \sigma_t}$ contains a free vertex with respect to $F_N$.*

*Proof.* By Theorem 8 we may assume that there is a planar embedding of $G$ such that the corresponding set $D$ is empty.

*Case 1.* Suppose $\varepsilon(S) = 0$. Let $s \in U$ and $\mathbb{N}_G(s) = \{v_i, v_j, v_k\}$ with $1 \leq i < j < k \leq N$. We choose $s \in S$ so that $k - i$ is minimum. If there exists $x \in U$ such that $x$ is adjacent to $v_p$ for some $p$ with $i < p < j$ or $j < p < k$, then $\mathbb{N}_{F_N}(x) \subseteq \{v_{i+1}, v_{i+2}, \ldots, v_{j-1}\}$ or $\mathbb{N}_{F_N}(x) \subseteq \{v_{j+1}, v_{j+2}, \ldots, v_{k-1}\}$. This is a contradiction to the choice of $s$. Thus we can assume that for every $p$ with $i < p < j$ or $j < p < k$, $v_p$ is adjacent to a vertex in $L$. By Lemma 4, there exists a sequence of elementary switchings $\sigma_1, \sigma_2, \ldots, \sigma_t$ with respect to $P_N$, such that $G^{\sigma_1 \sigma_2 \cdots \sigma_t}$ contains $s$ as a free vertex with respect to $F_N$.

*Case 2.* Suppose $\varepsilon(S) = 1$ and $s_1 s_2 \in E(G)$. Thus $F_N = P_N$. Suppose $s_1 \in U$ and $s_1$ is adjacent to $v_i$ and $v_j$ with $i < j$ and if there exists $s \in U$ such that $s$ is adjacent to $v_p$ where $i < p < j$, then the result follows as in Case 1. Thus if $s \in S$ and $s \neq s_1$ and $s$ is adjacent to $v_p$, then $s \in L$. Thus we can assume that $s_1$ and $s_2$ are adjacent to consecutive vertices in $P_N$. Let $s \in S - \{s_1, s_2\}$ and $\mathbb{N}_G(s) = \{v_{i_1}, v_{i_2}, v_{i_3}\}$, where $i_1 < i_2 < i_3$. Choose $s$ in such a way that $i_3 - i_1$

is minimum. It follows by Case 1 that $s$ is a free vertex with respect to $P_N$ or there exists a sequence of elementary switchings $\sigma_1, \sigma_2, \ldots, \sigma_h$ with respect to $P_N$ such that $G_h = G^{\sigma_1 \sigma_2 \cdots \sigma_h}$ contains $s$ as a free vertex with respect to $P_N$.  □

**Lemma 5.** *Let $n$ be an even integer and $G \in \mathcal{P}_0(2n)$. Then $G$ contains $P_N$ as its induced forest or there exists a sequence of $\mathcal{P}_0$-switchings $\sigma_1, \sigma_2, \ldots, \sigma_t$ with respect to $G$ such that $G^{\sigma_1 \sigma_2 \cdots \sigma_t}$ contains $P_N$ as its induced forest.*

*Proof.* Let $S$ be a minimum decycling set of $G$ and let $G[S]$ be an empty graph. Let $F_N = G - S = P_l \cup P_{N-l}$, where $l \geq N - l \geq 1$, $P_l = v_1 v_2 \cdots v_l$, and let $P_{N-l} = v_{l+1} v_{l+2} \cdots v_N$ be an induced forest of $G$ of order $N = \frac{3n}{2} - 1$. By Theorem 8 we can assume that there is a planar embedding of $G$ with $F_N$ in a horizontal direction with empty corresponding set $D$. Further, by Theorem 9 we can assume that $G$ contains $s \in S$ as a free vertex with respect to $F_N$. Since $s$ is a free vertex with respect to $F_N$, it can freely belong to $U$ or $L$. Thus, without loss of generality, there exists a sequence of elementary switchings $\sigma_1, \sigma_2, \ldots, \sigma_h$ with respect to $F_N$ such that in $G_h = G^{\sigma_1 \sigma_2 \cdots \sigma_h}$ and $s$ is adjacent to $v_{l-2}, v_{l-1}, v_l$. Let $s_1 \in S$ where $s_1$ is adjacent to $v_l$. Thus $s \neq s_1$. We put $s$ and $s_1$ in different sets, say $s \in U$ and $s_1 \in L$. Let $p, q \in S$ that are adjacent to $v_{l+1}$ and at least one of which, say $p$, belongs to $U$. Then $\sigma(s, v_l; p, v_{l+1})$ is a $\mathcal{P}_0$-switching of $G_h$ and $G_h^{\sigma(s, v_l; p, v_{l+1})} \in \mathcal{P}_0(2n)$ having $P_N$ as its induced forest. Now suppose that $p, q \in L$ and $p$ lies on the left of $q$ in the planar embedding of $G_h$. Then $G_h^{\sigma} \in \mathcal{P}_0(2n)$ has $P_N$ as its induced forest, where $\sigma = \sigma(s_1, v_l; p, v_{l+1})$ if $s_1 \neq p$ and $\sigma = \sigma(s_1, v_l; q, v_{l+1})$ if $s_1 = p$.  □

**Theorem 10.** *Let $G_1, G_2 \in \mathcal{P}_0(2n)$. Then $G_1 \cong G_2$ or there exists a sequence of $\mathcal{P}_0$-switchings $\sigma_1, \sigma_2, \ldots, \sigma_t$ with respect to $G_1$ such that $G_2 = G_1^{\sigma_1 \sigma_2 \cdots \sigma_t}$.*

*Proof.* Up to isomorphism there is only one graph in $\mathcal{P}_0(4)$. Suppose that $n > 2$. Let $G_1, G_2 \in \mathcal{P}_0(2n)$. By Lemma 5 we can assume that both $G_1$ and $G_2$ have $P_N = v_1 v_2 \cdots v_N$ as their induced paths. By Theorem 9, we may assume that each of $G_1$ and $G_2$ contains a free vertex. Thus there exist $G_1'$ and $G_2'$ such that $G_1$ and $G_2$ are super subdivisions of $G_1'$ and $G_2'$ with either by edges or vertices, respectively. By induction on $n$, there exists a sequence of elementary switchings $\sigma_1, \sigma_2, \ldots, \sigma_t$ with respect to $P_{N-3}$ such that $G_2' = G_1'^{\sigma_1 \sigma_2 \cdots \sigma_t}$. Since a free vertex of $G \in \mathcal{P}_2(2n)$ with respect to $P_N$ can be in $U$ or $L$, the sequence of elementary switchings $\sigma_1, \sigma_2, \ldots, \sigma_t$ with respect to $P_{N-3}$ as described above can be regarded as a sequence of elementary switchings with respect to $P_N$ and $G_2 = G_1^{\sigma_1 \sigma_2 \cdots \sigma_t}$. This completes the proof.  □

We now consider the class $\mathcal{P}_i(2n)$ when $i \geq 1$. That is for a graph $G \in \mathcal{P}_i(2n)$, $G$ contains an induced forest $F_N$ of order $N = \lfloor \frac{3n-1}{2} \rfloor$ with $n_3(F_N) = i$. Note that $F_N$ is a tree if $n$ is odd or $n$ is even and $\varepsilon(S) = 1$, where $S = V(G - F_N)$, otherwise $F_N$ is a union of two trees.

**Theorem 11.** *Let $G \in \mathcal{P}_i(2n)$, $i \geq 1$ and let $F_N$ be an induced forest of $G$ of order $N = \lfloor \frac{3n-1}{2} \rfloor$ and $n_3(F_N) = i$. Then there exists a sequence of $\mathcal{P}_i$-switchings $\sigma_1, \sigma_2, \ldots, \sigma_t$ such that $G^{\sigma_1 \sigma_2 \cdots \sigma_t} \in \mathcal{P}_{i-1}$.*

*Proof.* Let $G \in \mathcal{P}_i(2n)$, $i \geq 1$ and let $F_N$ be an induced forest of $G$ of order $N = \lfloor \frac{3n-1}{2} \rfloor$ and $n_3(F_N) = i$. Let $P_k = v_1 v_2 \cdots v_k$ be a longest path of $F_N$ containing a vertex $v_l$ of degree 3 in the forest $F_N$. We choose a planar embedding of $G$ by placing $F_N$ in such a way that $v_1$ lies in the leftmost vertex of $F_N$ and placing the path as in the following figure.

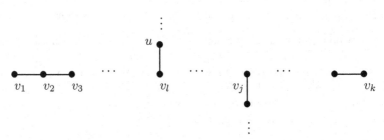

Suppose that $l$, $2 \leq l \leq k - 1$, is the least integer such that $d_{F_N}(v_l) = 3$. Suppose further that $v_l$ is adjacent to a vertex $u$ in $F_N$ that lies above $P_k$ as in above figure. Since $d_{F_N}(v_1) = 1$, there exist $p, q \in S = V(G - F_N)$ that are adjacent to $v_1$. Suppose further that $p, q$ lie above $P_k$ and $p$ lies on the left of $q$. We now consider the following cases.

*Case 1.* If $l = 2$, then $\sigma(p, v_1; v_2, u)$ is a $\mathcal{P}_i$-switching and $G^{\sigma(p, v_1; v_2, u)} \in \mathcal{P}_{i-1}(2n)$.

*Case 2.* Suppose that $l \geq 3$. Note that if there is a vertex $v \in S$ such that $v$ is adjacent to $v_j$ for some $j = 2, 3, \ldots, i - 1$ and ends under $P_k$. Choose $j$ to be maximum with this property, then we can analogously apply a sequence of elementary switchings with respect to $P_k$. Then we obtain the resulting graph $G'$ with decycling set $S$ and an induced forest $F_N = V(G' - S)$ containing $P_k$ as a longest path. Therefore, there exists $k$, $2 \leq k \leq l - 1$, such that $d_{F_N}(v_k) = 3$.

Thus we may assume that every vertex in $S$ that is adjacent to $v_j$ $j = 2, 3, \ldots, l - 1$, ends at $v_j$ above $P_k$.

If there exists $v \in S$, $v \neq p$, and $v \neq q$ such that $v$ is adjacent to $v_2$, then $\sigma(p, v_1; v_2, v)$ is a $\mathcal{P}_i$-switching, $G^{\sigma(p, v_1; v_2, v)} \in \mathcal{P}_i(2n)$, and $p$ is adjacent to $v_2$ under $P_k$. Thus there exists a sequence of $\mathcal{P}_i$-switchings that transforms $G$ to $G'$ with decycling set $S$, and an induced forest $F_N = V(G' - S)$ containing $P_k$ as its longest path. Therefore, there exists $k$, $2 \leq k \leq l - 1$, such that $d_{F_N}(v_k) = 3$.

Thus we can assume that $q$ is adjacent to $v_2$. Let $x, y \in F_N$ that are adjacent to $p$. Thus $\sigma(p, x; v_2, q)$ or $\sigma(p, y; v_2, q)$ is a $\mathcal{P}_i$-switching. If $\sigma(p, x; v_2, q)$ is a $\mathcal{P}_i$-switching, then $G^{\sigma(p, x; v_2, q)}$ contains an edge $pv_2$ which lies under $P_k$. Thus there exists a sequence of $\mathcal{P}_i$-switchings that transforms $G$ to $G'$ with decycling set $S$ and an induced forest $F_N = V(G' - S)$ containing $P_k$ as its longest path. Therefore, there exists $k$, $2 \leq k \leq i - 1$, such that $d_{F_N}(v_k) = 3$.

Therefore in any case there exists a sequence of $\mathcal{P}_i$-switchings $\sigma_1, \sigma_2, \ldots, \sigma_t$ such that $G^{\sigma_1 \sigma_2 \cdots \sigma_t} \in \mathcal{P}_{i-1}$. The proof is complete. $\quad\square$

The following corollaries are immediate consequence of the preceding theorem.

**Corollary 1.** *Let $G \in \mathcal{P}_i(2n)$, $i \geq 1$ and let $F_N$ be an induced forest of $G$ of order $N = \lfloor \frac{3n-1}{2} \rfloor$. Then there exists a sequence of $\mathcal{P}_i$-switchings $\sigma_1, \sigma_2, \ldots, \sigma_t$ such that $G^{\sigma_1 \sigma_2 \cdots \sigma_t} \in \mathcal{P}_0$.*

**Corollary 2.** *The subgraph $\mathcal{P}(3^{2n}; \phi = \lceil \frac{n+1}{2} \rceil)$ is connected.*

# Acknowledgements

The author would like to thank Prof. Jin Akiyama for his encouragement. The author is grateful to the referees whose valuable suggestions resulted in an improved paper.

# References

1. Alon, N., Mubayi, D., Thomas, R.: Large Induced Forests in Sparse Graphs. J. Graph Theory **38** (2001) 113–123
2. Bau, S., Beineke, L.W.: The Decycling Number of Graphs. Australas. J. Combin. **25** (2002) 285–298
3. Bondy, J.A., Murty, U.S.R.: Graph Theory with Applications. 1st Edition The MacMillan Press (1976)
4. Eggleton, R.B., Holton, D.A.: Graphic Sequences. Combinatorial Mathematics VI, (Proc. Sixth Austral. Conf., Univ. New England, Armidale, 1978) Lecture Notes in Math. **748** Springer Berlin (1979) 1–10
5. Hakimi, S.: On the Realizability of a Set of Integers as the Degree of the Vertices of a Graph. SIAM J. Appl. Math. **10** (1962) 496–506
6. Havel, V.: A Remark on the Existence of Finite Graphs (Czech). Časopis Pěst. Mat. **80** (1955) 477–480
7. Liu, J-P, Zhao, C.: A New Bound on the Feedback Vertex Sets in Cubic Graphs. Discrete Math. **184** (1996) 119–131
8. Punnim, N.: Decycling Regular Graphs. Australas. J. Combin. **32** (2005) 147–162
9. Punnim, N.: Decycling Connected Regular Graph. Australas. J. Combin. **35** (2006) 155–169
10. Punnim, N.: The Decycling Number of Cubic Graphs. Lecture Note in Computer Science (LNCS). **3330** (2005) 141–145
11. Taylor, R.: Constrained Switchings in Graphs. Combinatorial mathematics VIII (Geelong, 1980). Lecture Notes in Math. **884** Springer Berlin (1981) 314–336
12. Zheng, M., Lu, X.: On the Maximum Induced Forests of a Connected Cubic Graph without Triangles. Discrete Math. **85** (1990) 89–96

# Quasilocally Connected, Almost Locally Connected Or Triangularly Connected Claw-Free Graphs

Xiaoying Qu[1] and Houyuan Lin[2]

[1] Center for Combinatorics, LPMC, Nankai University,
Tianjin 300071, P.R. China
xiaoying0605@sohu.com
[2] School of Statistics and Mathematics Sciences, Shandong Economic University,
Jinan, 250014, P.R. China
houyuan_123@163.com

**Abstract.** The definitions of quasilocally connected graphs, almost locally connected graphs and triangularly connected graphs are introduced by Zhang, Teng and Lai et al. They are all different extensions of locally connected graphs. Many known results on the condition of local connectivity have been extended to these weaker conditions mentioned above. In this paper, we study the relations among these different conditions. In particular, we prove that every triangularly connected claw-free graph without isolated vertices is also quasilocally connected claw-free.

## 1 Introduction and Notations

In this paper, we consider only finite undirected graphs without loops and multiple edges. For terminologies and notations not defined here we refer to [1]. Let $\langle S \rangle$ denote the subgraph of $G$ induced by $S \subseteq V(G)$. A vertex $x \in V(G)$ is locally connected if $\langle N(x) \rangle$ is connected, where $N(x)$ stands for the open neighborhood of $x$. A graph $G$ is said to be locally connected if each vertex of $G$ is locally connected. A graph $G$ is said to be claw-free if it has no $K_{1,3}$ as its induced subgraph. For any two distinct vertices $x$ and $y$ in $V(G)$, $(x, y)$-path denotes a path with $x$, $y$ as its two end vertices. For $3 \leq l \leq |V(G)|$, $l$-cycle denotes a cycle of length $l$. A cycle of length $l$ containing the vertex set $S$ in $V(G)$ is called a $(S, l)$-cycle. For convenience, we write $(\{v\}, l)$-cycle as $(v, l)$-cycle. A graph $G$ is vertex pancyclic if every vertex of $G$ is contained in cycles of all lengths $l$ for $3 \leq l \leq |V(G)|$. A graph $G$ is called quasilocally connected if each vertex cut of $G$ contains a locally connected vertex. A graph $G$ is called almost locally connected if $B(G) = \{x \in V(G) : \langle N(x) \rangle$ is not connected$\}$ is independent and for any $x \in B(G)$, there is a vertex $y$ in $V(G) \backslash \{x\}$ such that $\langle N(x) \cup \{y\} \rangle$ is connected. A graph $G$ is triangularly connected if for every pair of edges $e_1$, $e_2 \in E(G)$, $G$ has a sequence of 3-cycles $C_1, C_2, \ldots, C_m$ such that $e_1 \in E(C_1)$, $e_2 \in E(C_m)$ and $E(C_i) \cap E(C_{i+1}) \neq \emptyset$, $(1 \leq i \leq m - 1)$.

Obviously, the condition of triangularly connected, quasilocally connected or almost locally connected is different extension of locally connected, respectively.

J. Akiyama et al. (Eds.): CJCDGCGT 2005, LNCS 4381, pp. 162–165, 2007.
© Springer-Verlag Berlin Heidelberg 2007

Clark [2] proved the following result.

**Theorem 1 (Clark, 1981).** *Every connected locally connected claw-free graph with $|V(G)| \geq 3$ is vertex pancyclic.*

Zhang introduced the definition of quasilocally connected graphs and obtained the following result in [3].

**Theorem 2 (Zhang, 1989).** *If the vertex $v$ of a quasilocally connected claw-free graph $G$ with $|V(G)| \geq 3$ is contained in a cycle of length $r$, then $v$ is contained in a cycle of length $h$ for any $h = r, r + 1, \ldots, n$.*

Some extensions of Theorem 2 are known.

**Theorem 3 (Veldman et al., 1990).** *Every quasilocally connected claw-free graph with $|V(G)| \geq 3$ is vertex pancyclic.*

**Theorem 4 (Teng et al., 2002).** *Every connected almost locally connected claw-free graph with $|V(G)| \geq 3$ is vertex pancyclic.*

**Theorem 5 (Lai et al., preprint).** *Every triangularly connected claw-free graph $G$ with $|E(G)| \geq 3$ is vertex pancyclic.*

In this paper, we study the relations among these different conditions and prove the following theorem.

**Theorem 6.** *Every triangularly connected claw-free graph without isolated vertices is also quasilocally connected claw-free.*

Note that we can easily conclude Theorem 5 by Theorem 3 and Theorem 6.

## 2   Claw-Free Graphs in Different Conditions

Note that $G_1$, $G_2$ and $G_3$ are showed in Fig. 1, Fig. 2 and Fig. 3, respectively. Obviously, $G_1$ is an almost locally connected claw-free graph. We can easily obtain that both of $v_1$ and $v_2$ are not locally connected vertices in $G_1$. However, $\{v_1, v_2\}$ is a vertex cut of $G_1$. Hence, $G_1$ is not a quasilocally connected claw-free graph. By the definition of triangularly connected, we know that $G_1$ is not a triangularly connected claw-free graph. Obviously, $G_2$ is a quasilocally connected claw-free graph but it is not a triangularly connected claw-free graph. Obviously, both of $v_1$ and $v_2$ are not locally connected vertex in $G_3$. Furthermore, we can not find a vertex $u$ in $V(G_3)\backslash\{v_1\}$ or $V(G_3)\backslash\{v_2\}$ such that $\langle N(v_1) \cup \{u\}\rangle$ or $\langle N(v_2) \cup \{u\}\rangle$ is connected. Hence, $G_3$ is not an almost locally connected claw-free graph. We can easily obtain that $G_3$ is a quasilocally connected claw-free graph and it is also a triangularly connected claw-free graph by the definition of quasilocally connected and triangularly connected.

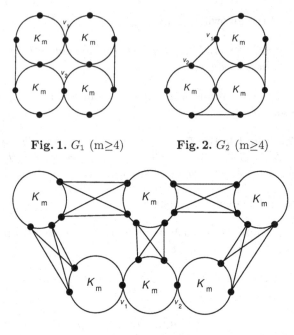

**Fig. 1.** $G_1$ (m≥4)          **Fig. 2.** $G_2$ (m≥4)

**Fig. 3.** $G_3$ (m≥6)

## 3   Proof of Theorem 6

Let $G$ be a triangularly connected claw-free graph without isolated vertices. Let $S$ be a vertex cut of $G$. It suffices to show that there is a locally connected vertex in $S$. Since $G$ is claw-free, if a vertex $u$ is not locally connected then $\langle N(u) \rangle$ is a disjoint union of two cliques. Hence, if $N(u)$ has a pair of nonadjacent vertices $x_1$, $x_2$ such that there is an $(x_1, x_2)$-path in $N(u)$, then $u$ should be a locally connected vertex. Let $A$ be a component of $G-S$ and write $\overline{A} = G-S \cup A$. Then there is an edge $e_1$ such that $V(e_1) \cap V(A) \neq \phi$, where $V(e_1)$ stands for the set of two end vertices of $e_1$. Also there is an edge $e_2$ such that $V(e_2) \cap V(\overline{A}) \neq \phi$. Since $G$ is triangularly connected, $G$ has a sequence of 3-cycles $C_1$, $C_2, \ldots, C_m$ such that $e_1 \in E(C_1)$, $e_2 \in E(C_m)$ and $E(C_i) \cap E(C_{i+1}) \neq \emptyset$, $(1 \leq i \leq m-1)$. Choose $e_1$, $e_2$ and $C_1$, $C_2, \ldots, C_m$ so that $m$ is as small as possible. Since $V(e_1) \cap V(A) \neq \phi$ and $V(e_2) \cap V(\overline{A}) \neq \phi$, we observe that $V(C_1) \neq V(C_m)$, which implies that $m \geq 2$. Let $V(C_1) = \{v_1, v_2, v_3\}$. Without loss of generality, we may assume $e_1 = v_1v_2$, $v_1 \in V(A)$ and $E(C_1) \cap E(C_2) = \{v_2v_3\}$. If $\{v_2, v_3\} \cap V(A) \neq \phi$, then we can take $v_2v_3$ as $e_1$ instead of $v_1v_2$ and we get a shorter sequence $C_2$, $C_3, \ldots, C_m$, which contradicts the choice of $e_1$, $e_2$ and $C_1$, $C_2, \ldots, C_m$. Hence $\{v_2, v_3\} \cap V(A) = \phi$, which means that $v_2$, $v_3 \in S$. Let $V(C_2) = \{v_2, v_3, v_4\}$. We show $v_1v_4 \notin E(G)$. Assume $v_1v_4 \in E(G)$, which implies $v_4 \notin \overline{A}$ and there is the next 3-cycle $C_3$. If $E(C_2) \cap E(C_3) = v_2v_3$, then we can skip $C_2$ and we get a shorter sequence $C_1$, $C_3, \ldots, C_m$, which contradicts the choice of $e_1$, $e_2$ and

$C_1, C_2, \ldots, C_m$. Hence, we know that either $E(C_2) \cap E(C_3) = \{v_2 v_4\}$ or $E(C_2) \cap E(C_3) = \{v_3 v_4\}$. By symmetry, we may assume that $E(C_2) \cap E(C_3) = \{v_3 v_4\}$. Then taking $e_1' = v_1 v_4$ instead of $e_1 = v_1 v_2$ and $V(C_1') = \{v_1, v_3, v_4\}$ instead of $C_1$, we again get a shorter sequence $C_1', C_3, \ldots, C_m$, which contradicts the choice of $e_1, e_2$ and $C_1, C_2, \ldots, C_m$. Now it is shown that $v_1 v_4 \notin E(G)$. Then, since $v_1 v_3 v_4$ is a $(v_1, v_4)$-path and $\{v_1, v_3, v_4\} \subseteq N(v_2)$, we observe that there is a pair of nonadjacent vertices $v_1, v_4$ in $N(v_2)$ such that $\langle N(v_2) \rangle$ has a $(v_1, v_4)$-path. This assures us that $v_2$ is a desired locally connected vertex and the proof of Theorem 6 is completed.

## Acknowledgements

The authors wish to express their appreciations to the referees, who made very helpful suggestions for the improvement of this paper.

## References

1. Bondy, J.A., Murty, U.S.R.: Graph Theory with Applications. Macmillan, London and Elsevier New York (1976)
2. Clark, L.: Hamiltonian Properties of Connected Locally Connected Graphs. Congr. Numer. **32** (1981) 199–204
3. Zhang, C.-Q.: Cycles of Given Length in Some $K_{1,3}$-Free Graphs. Discrete Math. **78** (1989) 307–313
4. Ainouche, A., Broersma, H.J., Veldman, H.J.: Remarks on Hamiltonian Properties of Claw-free Graphs. Ars Combin. **29C** (1990) 110–121
5. Teng, Y., You, H.: Every Connected Almost Locally Connected Quasi-Claw-Free Graph $G$ with $|V(G)| \geq 3$ is Fully Cycle Extendable. Journal of Shandong Normal University (Natural Science) **17** (4) (Dec. 2002) 5–8
6. Lai, H., Miao, L., Shao, Y., Wan, L.: Triangularly Connected Claw-free Graph. Preprint.

# Rotational Steiner Ratio Problem Under Uniform Orientation Metrics[*]

Songpu Shang[1], Xiaodong Hu[2], and Tong Jing[3]

[1] College of Mathematics And Information Sciences
North China University of Water Conservancy and Electric Power, Henan, China
[2] Institute of Applied Mathematics, Chinese Academy of Sciences
P.O. Box 2734, Beijing 100080, China
xdhu@amss.ac.cn
[3] Department of Computer Science and Technology, Tsinghua University
Beijing 100084, China

**Abstract.** Let $P$ be a set of $n$ points in a metric space. A Steiner Minimal Tree (SMT) on $P$ is a shortest network interconnecting $P$ while a Minimum Spanning Tree (MST) is a shortest network interconnecting $P$ with all edges between points of $P$. The Steiner ratio is the infimum over $P$ of ratio of the length of SMT over that of MST. Steiner ratio problem is to determine the value of the ratio. In this paper we consider the Steiner ratio problem in uniform orientation metrics, which find important applications in VLSI design. Our study is based on the fact that lengths of MSTs and SMTs could be reduced through properly rotating coordinate systems without increasing the number of orientation directions. We obtain the Steiner ratios with $|P| = 3$ when rotation is allowed and some bounds of Steiner ratios for general case.

**Keywords:** Steiner tree, Steiner ratio, Uniform orientation metrics, VLSI design.

## 1 Introduction

Let $P$ be a set of $n$ points in a metric space. A Steiner Minimal Tree (SMT) on $P$ is a shortest network interconnecting $P$, which is denoted by $\mathrm{SMT}(P)$ [10]. Since the problem to find a SMT has been shown to be NP-hard for $P$ in both Euclidean space and rectilinear space [6,7], it is important to study approximation algorithms for SMT problem.

A Minimum Spanning Tree (MST) is a shortest network interconnecting $P$ with all edges between points of $P$, which is denoted by $\mathrm{MST}(P)$. Since given $P$, an MST can be constructed in polynomial time and its length $L_m(P)$ is not too much longer than the length of SMT $L_s(P)$, MSTs thus could serve as yardsticks for the performances of SMT approximations. The *Steiner ratio* in metric space $M$ is defined as

$$r_M = \inf_{P \in M} \frac{L_s(P)}{L_m(P)}.$$

[*] Supported in part by NSF of China under Grant No. 70221001, 10401038, 10531070, and the Key Project of Chinese Ministry of Education under grant No. 106008.

J. Akiyama et al. (Eds.): CJCDGCGT 2005, LNCS 4381, pp. 166–176, 2007.

The *Steiner ratio problem* [4,10] is to determine the value of $r_M$ for $M$. Du and Hwang [5] proved that $r_M = \sqrt{3}/2$ for Euclidean plane, which settled the long-lasting conjecture by Gilbert and Pollak. Besides in the Euclidean plane, the Steiner ratio problem in other metric spaces also have received lots of attention for decades, for example, Hwang [9] proved that $r_M = 2/3$ for rectilinear plane while it was suspected that $r_M = (2 + \sqrt{2})/4$ for octilinear plane [12].

In this paper we study the variants of Steiner ratio problems in uniform orientation metrics, which find important applications in VLSI design [3,8,9,11-16]. Let $\lambda$ be a non-negative integer, $\lambda \geq 2$. Let $\omega = \frac{\pi}{\lambda}$, legal orientations are then defined by the straight lines passing through the origin and make angles $i\omega, i = 1, 2, \cdots, \lambda - 1$ with the $X$-axis. We call such a uniform orientation metrics $\lambda$-*geometry*. Rectilinear and octilinear planes correspond to $\lambda = 2$ and $\lambda = 4$, respectively, which are two of most important uniform orientation metrics in the design of VLSI.

There are two ways to reduce the lengths of SMTs and MSTs in $\lambda$-geometry, the first one is to increase value $\lambda$ so that routing could go more directions; the second one is to fix value $\lambda$ and rotate coordinate system so that yields shortcuts. In this paper, we focus on the second method. Obviously in the Euclidean plane rotating coordinate system will not reduce the lengths of MSTs and SMTs. But in the $\lambda$-geometry, as the coordinate system is rotated, that is, to rotate all legal orientations simultaneously for an angle $\theta \in [0, \omega]$, the lengths and topologies of MSTs and SMTs may change considerably [2].

In this paper we consider two different versions of Steiner ratio problem. Given $P$ in $\lambda$-geometry and $\theta \in [0, \omega]$, let $L_s(P, \theta)$ and $L_m(P, \theta)$ denote the lengths of SMT and MST of $P$, respectively, and let $\overline{L}_s(P) = \inf_\theta L_s(P, \theta)$ and $\overline{L}_m(P) = \inf_\theta L_m(P, \theta)$ denote the lengths of SMT and MST of $P$ over $\theta \in [0, \omega]$. Define two rotational Steiner ratios as follow:

$$\overline{r}_\lambda = \inf_P \frac{\overline{L}_s(P)}{\overline{L}_m(P)}, \quad \text{and} \quad \widehat{r}_\lambda = \inf_P \sup_\theta \frac{L_s(P, \theta)}{L_m(P, \theta)}.$$

In the above, the first ratio means, when the rotation of coordinate system is allowed, the maximum difference in lengths between SMT and MST (maybe in different coordinate system), while the second ratio means the maximum difference in lengths between SMT and MST in the same coordinate system. Moreover, when the number of points in $P$ are enough large or $\lambda$ goes sufficiently large, we will see that these two ratios could be acted as lower bounds for the classic Steiner ratio of $L_s(P)$ over $L_m(P)$.

The rest of this paper is organized as follows. In section 2, we give the definitions of rational Steiner ratios and compare these three Steiner ratios. In Section 3 and Section 4, we give some results on the first and second Steiner ratios under rotation for 3 points for $\lambda = 2, 4$. We also obtain the exact value of the first ratio for $\lambda = 6m$ and upper-bound for general $\lambda$. In section 5, we conclude the paper with some remarks.

## 2    Preliminary Results

The following two lemmas due to Brazil et al in [2] give some properties about MSTs and SMTs in $\lambda$-geometry when the coordinate system is rotated.

**Lemma 2.1.** *One of the edges in a rotationally optimal MST overlaps with one of the legal orientations.*

**Lemma 2.2.** *A rotationally optimal SMT in $\lambda$-geometry is a union of $\lambda$-FSTs such that at least one of them contains no bent edges, where FST is a full Steiner tree that all points in $P$ are leaf-nodes of the tree.*

**Theorem 2.3.** *For any $\lambda$-geometry, $\widehat{r}_\lambda \geq \overline{r}_\lambda \geq r_\lambda$.*

*Proof.* Consider the first inequality. For any $\epsilon > 0$, there exist a set $P_0$ of points and an angle $\theta_0$ such that

$$\frac{L_s(P_0, \theta_0)}{L_m(P_0, \theta_0)} - \widehat{r}_\lambda < \varepsilon, \quad \text{and} \quad \overline{L}_m(P_0) = L_m(P_0, \theta_0).$$

Thus we obtain

$$\overline{r}_\lambda \leq \frac{\overline{L}_s(P_0)}{\overline{L}_m(P_0)} \leq \frac{L_s(P_0, \theta_0)}{L_m(P_0, \theta_0)} < \widehat{r}_\lambda + \epsilon,$$

which implies $\overline{r}_\lambda \leq \widehat{r}_\lambda$.

Now consider the second inequality. For any $\epsilon > 0$, there exists a set of $P_1$ of points and an angle $\theta_1$ such that

$$\frac{\overline{L}_s(P_1)}{\overline{L}_m(P_1)} - \overline{r}_\lambda < \epsilon, \quad \text{and} \quad \overline{L}_s(P_1, \theta) = L_s(P_1, \theta_1).$$

Thus we obtain

$$r_\lambda \leq \min_\theta \frac{L_s(P_1, \theta)}{L_m(P_1, \theta)} \leq \frac{L_s(P_1, \theta_1)}{L_m(P_1, \theta_1)} \leq \frac{\overline{L}_s(P_1)}{\overline{L}_m(P_1)} < \overline{r}_\lambda + \epsilon,$$

which implies $r_\lambda \leq \overline{r}_\lambda$.    □

## 3    First Rotational Steiner Ratio on 3 Points

In this section we study the first rotational Steiner ratio $\overline{r}_\lambda$. For any three points $A, B$, and $C$, which make a triangle $\triangle ABC$. We denote $\angle A, \angle B, \angle C$ those three angles of the triangle, and $a, b, c$ the three edges of this triangle. We also denote $\mid BC \mid_{AB}$ the distance between $B$ and $C$ when $AB$ is a legal orientation, and so on. We denote $L_m(AB)$ the length of MST when $AB$ is a legal orientation, similar with $L_m(AC), L_m(BC)$. By Lemma 2.1, $\overline{L}_m$ is the smallest of $L_m(AB), L_m(AC)$, and $L_m(BC)$.

## 3.1  Rectilinear Plane ($\lambda = 2$)

We denote $L_s(AB)$ the length of SMT when $AB$ is an legal orientation, similar with $L_s(AC)$, $L_s(BC)$. By Lemma 2.2, $\overline{L}_s$ is the smallest of $L_s(AB), L_s(AC)$, and $L_s(BC)$. Suppose $a \leq b \leq c$, which implies $\angle A \leq \angle B \leq \angle C$.

**Lemma 3.1.** *For any* $\angle C \geq \pi/2$, $\overline{L}_s/\overline{L}_m \geq (\sqrt{7}+1)/4$.

*Proof.* Suppose $\angle C \geq \pi/2$, then $L_s(AB) = c + b \sin \angle A$, and $L_m(AC) = L_s(AC) = c(\sin \angle A + \cos \angle A), L_s(BC) = L_m(BC) = c(\sin \angle B + \cos \angle B)$. Because $L_s(BC) \geq L_s(AC)$, we only need to prove

$$\frac{\overline{L}_s}{\overline{L}_m} \geq \frac{L_s(AB)}{L_m(AC)} = \frac{c + b \sin \angle A}{c(\sin \angle A + \cos \angle A)} \equiv f(b) \geq \frac{\sqrt{7}+1}{4}.$$

Since $f'(b) \geq 0$, then $f$ reaches the minimum when $b$ is minimum, that is, when $a = b$ holds. In this case,

$$f = \frac{2a \cos \angle A + a \sin \angle A}{2a \cos \angle A (\cos \angle A + \sin \angle A)} \geq \frac{2 + \sin \angle A}{2(\cos \angle A + \sin \angle A)} \equiv g(\angle A).$$

When $\cos(\angle A + \pi/4) \geq \frac{\sqrt{2}}{4}$, we have $g'(A) \leq 0$; When $\cos(\angle A + \pi/4) \leq \frac{\sqrt{2}}{4}$, we have $g'(A) \geq 0$. Thus $g(\angle A)$ reaches the minimum when $\cos(\angle A + \pi/4) = \frac{\sqrt{2}}{4}$, and then we have

$$\sin \angle A = \frac{\sqrt{7}-1}{4}, \quad \cos \angle A = \frac{\sqrt{7}+1}{4}, \quad \text{and} \quad g \geq \frac{\sqrt{7}+1}{4},$$

which imply $\overline{L}_s/\overline{L}_m \geq f \geq g \geq (\sqrt{7}+1)/4$.  □

**Lemma 3.2.** *For any* $\angle C \leq \pi/2$, $\overline{L}_m = b + a(\sin \angle C + \cos \angle C)$ .

*Proof.* Suppose $\angle A \leq \angle B \leq \angle C \leq \pi/2$. We consider the following three cases.

Case 1. $AB$ is a legal orientation. In this case we have $\mid AB \mid_{AB} = c$, $\mid AC \mid_{AB} = b(\sin \angle A + \cos \angle A)$, $\mid BC \mid_{AB} = a(\sin \angle B + \cos \angle B)$. Thus $L_m(AB) = c + a(\sin \angle B + \cos \angle B)$.

Case 2. $AC$ is a legal orientation. In this case we have $\mid AB \mid_{AC} = c(\sin \angle A + \cos \angle A)$, $\mid AC \mid_{AC} = b$, $\mid BC \mid_{AC} = a(\sin \angle C + \cos \angle C)$. Thus $L_m(AC) = b + a(\sin \angle C + \cos \angle C)$.

Case 3. $BC$ is a legal orientation. In this case we have $\mid AB \mid_{BC} = c(\sin \angle B + \cos \angle B)$, $\mid AC \mid_{BC} = b(\sin \angle C - \cos \angle C)$, $\mid BC \mid_{BC} = a$. Thus $L_m(BC) = a + b(\sin \angle C + \cos \angle C)$.

We then have $\overline{L}_m = b + a(\sin \angle C + \cos \angle C)$ since $L_m(AC) \leq L_m(AB)$ and $L_m(AC) \leq L_m(BC)$.  □

**Lemma 3.3.** *For any* $\angle C \leq \pi/2$, $\overline{L}_s = a + b \sin \angle C$.

*Proof.* Suppose $\angle C \leq \pi/2$, then we have $L_s(AB) = c + a \sin \angle B$, $L_s(AC) = b + a \sin \angle C$, $L_s(BC) = a + c \sin \angle B$. Since $L_s(BC) - L_s(AB) = (a + c \sin \angle B) - (c + a \sin \angle B) \leq 0$ and $L_s(BC) - L_s(AC) = (a + b \sin \angle C) - (b + a \sin \angle C) \leq 0$, we then obtain $\overline{L}_s = L_s(BC) = b + a \sin \angle C$.  □

**Theorem 3.4.** *For three points in rectilinear plane, $\bar{r}_2 = (3 + \sqrt{3})/6$.*

*Proof.* For any $\angle C \geq \pi/2$, by Lemma 3.1, we have

$$\frac{\overline{L}_s}{\overline{L}_m} \geq \frac{\sqrt{7}+1}{4} \geq \frac{3+\sqrt{3}}{6}.$$

When $\angle A \leq \angle B \leq \angle C \leq \pi/2$, we define

$$f(\angle A) \equiv \frac{\overline{L}_s}{\overline{L}_m} = \frac{a + b\sin\angle C}{b + a(\sin\angle C + \cos\angle C)}$$

$$= \frac{\sin\angle A + \sin\angle(A + C)\sin\angle C}{\sin\angle(A + C) + \sin\angle A(\sin\angle C + \cos\angle C)}.$$

Since $f'(\angle A) < 0$, $f(\angle A)$ reaches the minimum when $\angle A$ approaches the maximum, that is, when $\angle A = \angle B$. Thus we obtain

$$\frac{\overline{L}_s}{\overline{L}_m} \leq \frac{1 + \sin C}{1 + \sin C + \cos C} \equiv g(\angle C).$$

Since $g'(C) > 0$, then $g(C)$ reaches the minimum when $\angle C$ approaches the minimum of $\angle C = \pi/3$. Therefore we have

$$\frac{\overline{L}_s}{\overline{L}_m} \geq g\left(\frac{\pi}{3}\right) = \frac{1 + \sqrt{3}/2}{1 + \sqrt{3}/2 + 1/2} = \frac{3 + \sqrt{3}}{6},$$

which implies $\bar{r}_2 \geq (3 + \sqrt{3})/6$. Since the equality holds if and only if $\angle A = \angle B = \angle C$, we have $\bar{r}_2 \leq (3 + \sqrt{3})/6$. Hence $\bar{r}_2 = (3 + \sqrt{3})/6$ for any three points. $\square$

### 3.2  Octilinear Case ($\lambda = 4$)

Suppose that $O$ is the Steiner point of rotationally optimal Steiner tree interconnecting points $A, B$ and $C$. By Lemma 2.2, $OA, OB, OC$ are all legal orientations, and we can suppose $\angle AOB = \pi/2$. $\angle AOC = \angle BOC = 3\pi/4$. Suppose $\angle BAO = \alpha$, and $\angle CAO = \beta$, $\angle BCO = \gamma$, and $\mid BO \mid \geq \mid AO \mid$ (see Fig. 3.1).

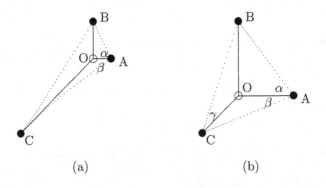

(a)                              (b)

**Fig. 3.1.** Illustration for the proofs of Lemma 3.5-6

**Lemma 3.5.** *For any* $\angle A \geq \pi/2, \overline{L}_s/\overline{L}_m \geq \cos \pi/8$.

*Proof.* Suppose $\angle A \geq \pi/2$ (see Fig. 3.1(a)), then we have

$$\overline{L}_s = c\sin\alpha + c\cos\alpha + c\cos\alpha\frac{\sin\beta}{\sin(\pi/4-\beta)}, \quad \text{and}$$

$$\overline{L}_m \leq L_m(AC) \leq \frac{c\sin(\alpha+\beta-\pi/8)}{c\cos\pi/8} + \frac{\cos\alpha\sin\pi/4}{\sin(\pi/4-\beta)} \equiv L_m^*(AC).$$

Now define $f(\alpha,\beta) \equiv \overline{L}_s/L_m^*(AC)$. Note $f'(\alpha) \leq 0$. Thus

$$f(\alpha,\beta) \geq f(\frac{\pi}{2},\beta) = \frac{\cos\pi/8}{\sin(\beta+3\pi/8)} \geq \cos\frac{\pi}{8},$$

which implies $\overline{L}_s/\overline{L}_m \geq f(\alpha,\beta) \geq \cos\pi/8$. $\qquad\square$

**Lemma 3.6.** *For any* $\angle A \leq \pi/2, \ \overline{L}_s/\overline{L}_m \geq (\sqrt{6} - \sqrt{3} + 1)/2$.

*Proof.* First note that

$$\begin{aligned}
\overline{L}_s &= c(\sin\alpha + \cos\alpha) + c\cos\alpha\frac{\sin\beta}{\sin(\pi/4-\beta)} \\
&= a\frac{\sin\gamma}{\sin\pi/4} + a\frac{\sin(\pi/4-\gamma)}{\sin\pi/4} + a\frac{\sin(\pi/4-\gamma)}{\sin\pi/4}\frac{\sin(\pi/4-\beta)}{\sin\beta}.
\end{aligned}$$

When $AC$ is a legal orientation, we have

$$\overline{L}_m \leq L_m(AC) \leq \frac{c\sin(\alpha+\beta-\pi/8)}{\cos\pi/8} + \frac{c\cos\alpha\sin\pi/4}{\sin(\pi/4-\beta)} \equiv L_m^*(AC).$$

Now let $f_1(\alpha,\beta) \equiv s/L_m^*(AC)$. When $BC$ is a legal orientation, we have

$$\overline{L}_m \leq L_m(BC) \leq a + a\frac{\sin((\pi/4-\beta)+\gamma+\pi/8)}{\cos\pi/8}\frac{\sin(\pi/4-\gamma)}{\sin\beta} \equiv L_m^*(BC).$$

Let $f_2(\beta,\gamma) \equiv \overline{L}_s/L_m^*(BC)$. We then have $\overline{L}_s/\overline{L}_m \geq f_1(\alpha,\beta)$ and $\overline{L}_s/\overline{L}_m \geq f_2(\beta,\gamma)$. In the following we consider two cases.

Case 1. $\beta \leq \pi/12$. In this case we have $\gamma \geq \pi/6$. Since $f_2'(\beta) \leq 0$, we obtain

$$f_2(\beta,\gamma) \geq f_2(\frac{\pi}{12},\gamma) \geq f_2(\frac{\pi}{12},\frac{\pi}{6}) = \frac{\sqrt{6}-\sqrt{3}+1}{2},$$

where the equality holds if and only if $\angle A = \angle B = \angle C$ (see Fig. 3.2(a)).

Case 2. $\beta \geq \pi/12$. In this case, since $f_1'(\alpha) \geq 0$, we have

$$f_1(\alpha,\beta) \geq f(\frac{\pi}{4},\beta) \geq f(\frac{\pi}{4},\frac{\pi}{12}) = \frac{\sqrt{6}-\sqrt{3}+1}{2},$$

where the equality holds if and only if $\angle A = \angle B = \angle C$ (see Fig. 3.2(b)).
From the above discussion, we deduce $\overline{L}_s/\overline{L}_m \geq (\sqrt{6} - \sqrt{3} + 1)/2$. $\qquad\square$

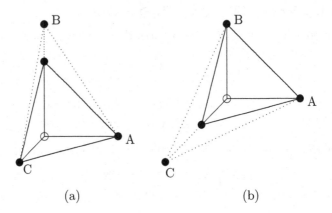

**Fig. 3.2.** Illustration for the proof of Lemma 3.6

**Theorem 3.7.** *For any three points in octilinear plane, $\bar{r}_4 = (\sqrt{6} - \sqrt{3} + 1)/2$.*

*Proof.* Note that, on the one hand, when $a = b = c = 1$ (see Fig.3.2), we have

$$\bar{L}_s = \sqrt{2} + \frac{\sqrt{3} - 1}{2}, \quad \bar{L}_m = 1 + \frac{\sqrt{3} + \sqrt{2} - 1}{2}, \quad \frac{\bar{L}_s}{\bar{L}_m} = \frac{\sqrt{6} - \sqrt{3} + 1}{2},$$

Thus $\bar{r}_4 \le (\sqrt{6} - \sqrt{3} + 1)/2$; On the other hand, by Lemmas 3.5-6, $\bar{r}_4 \ge (\sqrt{6} - \sqrt{3} + 1)/2$, we obtain $\bar{r}_4 = (\sqrt{6} - \sqrt{3} + 1)/2$ for three points.     □

### 3.3   General $\lambda$ Case

In Subsections 3.1-2 for cases of $\lambda = 2, 4$, we see that rational Steiner ratios achieve the minimum when three points in $P$ make an equilateral triangle. So in this subsection we also focus on this case for general $\lambda$. The ratios of three points of particular configuration could serve as an upper-bound on the ratio of three points in general case.

**Theorem 3.8.** *For any $\lambda$-geometry, $\bar{r}_\lambda \le \sqrt{3}/2$.*

*Proof.* Suppose $a = b = c = 1$. Using condition on angles of $\triangle ABC$ [14], we could compute the length of SMT. For this purpose, we consider the following three cases.

Case 1. $\lambda = 3m$. In this case we have $\bar{L}_s = 2$, $\bar{L}_m = \sqrt{3}$, $\bar{L}_s/\bar{L}_m = \sqrt{3}/2$.

Case 2. $\lambda = 3m + 1$. In this case we have

$$\bar{L}_s = \frac{1}{\sin m\pi/\lambda} + \frac{\sqrt{3}}{2} - \frac{1}{2} \frac{\sin(m+1)\pi/2\lambda}{\sin m\pi/\lambda} \text{ and } \bar{L}_m = 1 + \frac{\sin \pi/3\lambda + \sin 2\pi/3\lambda}{\sin \pi/\lambda}.$$

Note that

$$\frac{\bar{L}_s}{\bar{L}_m} \le \frac{\sqrt{3}}{2} \text{ if and only if } \frac{\bar{L}_s - \sqrt{3}}{\bar{L}_m - 2} \le \frac{\sqrt{3}}{2},$$

and

$$\frac{\overline{L}_s - \sqrt{3}}{\overline{L}_m - 2} = \frac{\sin\frac{\pi}{6\lambda}\cos\frac{\pi}{2\lambda}}{\sin\frac{m\pi}{\lambda}\sin\frac{\pi}{3\lambda}} \leq \frac{\cos\frac{\pi}{2\lambda}}{2\sin\frac{\pi}{4}\cos\frac{\pi}{6\lambda}} \leq \frac{1}{2\sin\frac{\pi}{4}} < \frac{\sqrt{3}}{2}.$$

Thus we have $\overline{L}_s/\overline{L}_m \leq \sqrt{3}/2$.

Case 3. $\lambda = 3m + 2$. In this case we have

$$\overline{L}_s = \frac{1}{\sin\frac{(m+1)\pi}{\lambda}} + \frac{\sqrt{3}}{2} - \frac{1}{2}\frac{\sin\frac{m\pi}{2\lambda}}{\sin\frac{(m+1)\pi}{\lambda}} \quad \text{and} \quad \overline{L}_m = 1 + \frac{\sin\frac{\pi}{3\lambda} + \sin\frac{2\pi}{3\lambda}}{\sin\frac{\pi}{\lambda}}.$$

Note that

$$\frac{\overline{L}_s}{\overline{L}_m} \leq \frac{\sqrt{3}}{2} \quad \text{if and only if} \quad \frac{\overline{L}_s - \sqrt{3}}{\overline{L}_m - 2} \leq \frac{\sqrt{3}}{2},$$

and

$$\frac{\overline{L}_s - \sqrt{3}}{\overline{L}_m - 2} = \frac{\sin\frac{\pi}{6\lambda}\cos\frac{\pi}{2\lambda}}{\sin\frac{(m+1)\pi}{\lambda}\sin\frac{\pi}{3\lambda}} \leq \frac{\cos\frac{\pi}{2\lambda}}{2\sin\frac{\pi}{3}\cos\frac{\pi}{6\lambda}} \leq \frac{1}{2\sin\frac{\pi}{3}} < \frac{\sqrt{3}}{2}.$$

Thus we have $\overline{L}_s/\overline{L}_m \leq \sqrt{3}/2$.

From the above discussion, we deduce that $\sqrt{3}/2$ is an upper-bound of $\overline{r}_\lambda$ for three points, which could also be an upper-bound on $\overline{r}_\lambda$ for arbitrary points. The theorem is then proved.     □

**Corollary 3.9.** *For any* $\lambda = 6m$, $\overline{r}_\lambda = \sqrt{3}/2$ .

*Proof.* Note that, on one hand, $r_\lambda = \sqrt{3}/2$ for $\lambda = 6m$ [11], and by Theorem 2.3, $\overline{r}_\lambda \geq r_\lambda \geq \sqrt{3}/2$. On the other hand, by Theorem 3.8, we obtain $\overline{r}_\lambda \leq \sqrt{3}/2$. Hence we have $\overline{r}_\lambda = \sqrt{3}/2$.     □

# 4   Second Rotational Steiner Ratio on 3 Points

In Section 3 we have studied the first rotational Steiner ratio $\overline{r}_\lambda$. In this section we study the second rotational Steiner ratio $\widehat{r}_\lambda$.

## 4.1   Rectilinear Case ($\lambda = 2$)

**Lemma 4.1.** *Let* $P$ *be a set of three points that make an equilateral triangle, then* $\widehat{r}_2(P) = (3 + \sqrt{3})/6$.

*Proof.* Suppose that the coordinate system is rotated in an angle of $\theta$ anti-clockwise away from $AB$, where $0 \leq \theta \leq \frac{\pi}{6}$. Let $\overline{L}_s(\theta)$ denote the length of SMT after rotation (see Fig.4.1(a)), and let $\overline{L}_m(\theta)$ denote the length of corresponding MST. Then we have $\overline{L}_s(\theta) = \sin\theta + \cos\theta + \sin(\pi/3 - \theta)$ and $\overline{L}_m(\theta) = \sqrt{2}(\sin(\pi/4 + \theta)) + \sqrt{2}(\sin(7\pi/12 + \theta))$.

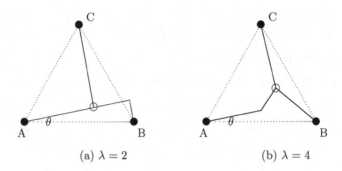

(a) $\lambda = 2$                    (b) $\lambda = 4$

**Fig. 4.1.** Illustration of constructing SMT under rotation

Note that

$$\frac{\overline{L}_s(\theta)}{\overline{L}_m(\theta)} = \frac{\sin\theta + \cos\theta + \sin(\frac{\pi}{3} - \theta)}{\sqrt{2}(\sin(\frac{\pi}{4} + \theta)) + \sqrt{2}(\sin(\frac{7\pi}{12} + \theta))} = \frac{3 + \sqrt{3}}{6},$$

which proves the lemma.                                                                    □

**Theorem 4.2.** *For three points in rectilinear plane,* $\hat{r}_2 = (3 + \sqrt{3})/6$.
*Proof.* Note that, one one hand by Lemma 4.1, we have $\hat{r}_2 \leq (3+\sqrt{3})/6$; On the other hand, by Theorem 2.3 and Theorem 3.4, we have $\hat{r}_2 \geq \overline{r}_2 = (3 + \sqrt{3})/6$. The theorem is then proved.                                                                    □

## 4.2    Octilinear Case ($\lambda = 4$)

**Lemma 4.3.** *Let $P$ be a set of three points that make an equilateral triangle, then* $\hat{r}_4(P) = (\sqrt{6} - \sqrt{3} + 1)/2$.

*Proof.* Suppose that the coordinate system is rotated in an angle of $\theta$ anticlockwise away from $AB$, where $0 \leq \theta \leq \pi/12$. Let $\overline{L}_s(\theta)$ denote the length of SMT after rotation (see Fig. 4.1(b)), and let $\overline{L}_m(\theta)$ denote the length of corresponding MST. We have

$$\overline{L}_s(\theta) = \cos(\frac{\pi}{3} - \theta) + \sin(\frac{\pi}{3} - \theta) + (\sqrt{2} - 2)[\cos\theta - \cos(\frac{\pi}{3} - \theta) - \sin\theta]$$
$$+ \sqrt{2}[\cos\theta - \cos(\frac{\pi}{3} - \theta)],$$

and

$$\overline{L}_m(\theta) = \frac{\cos(\theta - \frac{\pi}{8})}{\cos\frac{\pi}{8}} + \frac{\cos(\theta + \frac{\pi}{24})}{\cos\frac{\pi}{8}} = \frac{2\cos\frac{\pi}{12}\cos(\theta - \frac{\pi}{24})}{\cos\frac{\pi}{8}}.$$

Since $L_s(P, \theta)/L_m(P, \theta) = (\sqrt{6} - \sqrt{3} + 1)/2$, we have $\hat{r}_4(P) = (\sqrt{6} - \sqrt{3} + 1)/2$ for $P$ being a set of three points that make an equilateral triangle.                    □

**Theorem 4.4.** $\widehat{r}_4 = (\sqrt{6} - \sqrt{3} + 1)/2$ *for three points.*

*Proof.* Note that, on one hand, by Lemma 4.4, we have $\widehat{r}_4 \leq (\sqrt{6} - \sqrt{3} + 1)/2$; on the other hand, by Theorem 2.3 and Theorem 3.7, we have $\widehat{r}_4 \geq \overline{r}_4 = (\sqrt{6} - \sqrt{3} + 1)/2$. The theorem is then proved.                                      □

## 5    Conclusions

We introduced two rotational Steiner ratios in $\lambda$-geometry where the coordinate system is allowed to rotate in order to reduce the lengths of SMTs and MSTs. We obtain the exact values of ratios for three points of $\lambda = 2, 4$. We also give an upper-bound of these two ratios for some special cases and also the exact value of first rotational Steiner ratio for $\lambda = 6m$. We conjecture that the values of the rotational Steiner ratios we obtained for $\lambda = 2, 4$ are not only true for $|P| = 3$ but also for $P$ of any size, that is

$$\overline{r}_2 = \widehat{r}_2 = \frac{3 + \sqrt{3}}{6}, \qquad \overline{r}_4 = \widehat{r}_4 = \frac{\sqrt{6} - \sqrt{3} + 1}{2}.$$

Since the Steiner minimal tree problems in metric spaces have Polynomial Time Approximation Schemes (PTASs) [1], it is worth study whether a PTAS still exists when the orientations can be rotated.

## References

1. S. Arora, Polynomial-time approximation schemes for Euclidean TSP and other geometric problems, *Journal of the ACM*, 45(5)(1998), pp. 753-782.
2. M. Brazil, B. K. Nielsen, P. Winter, and M. Zachariasen, Rotationally Optimal Spanning and Steiner Trees in Uniform Orientation Metrics, *Computational Geometry: Theory and Applications*, 29(3)(2004), pp. 251-263.
3. M. Brazil, D. A. Thomas and J. F. Wang, Minimum Networks in Uniform Orientation Metrics, *SIAM Journal on Computing*, 30(2000), pp. 1579-1593.
4. D. Cieslik, *The Steiner Ratio*, Kluwer Academic Publishers, Boston, 2001.
5. D. Z. Du and F. K. Hwang, A Proof of the Gilbert-Pollak Conjecture on the Steiner Ratio, *Algorithmica*, 7(2-3)(1992), pp. 121-135.
6. M. R. Garey and D. S. Johnson, The Rectilinear Steiner Tree Problem is NP-Complete, *SIAM Journal of Applied Mathematics*, 32(4)(1977), pp. 826-834.
7. M. R. Garey and D. S. Johnson, *Computers and Intractability: a Guide to the Theory of NP-completeness*, Freeman, San Francisi, 1979.
8. M. Hanan, On Steiner's Problem with Rectilinear Distance, *SIAM J. Appl. Math.*, 14(2)(1966), pp. 255-265.
9. F. K. Hwang, On Steiner Minimal Trees with Rectilinear Distance, *SIAM Journal of Applied Mathematics*, 30(1)(1976), pp. 104-114.
10. F. K. Hwang, D. S. Richards and P. Winter, *The Steiner Tree Problem*, Springer Verlag, North Holland, 1994.
11. D. T. Lee and C. F. Shen, The Steiner Minimal Tree Problem in the $\lambda$-Geormetry Plane, in *Proc. ISAAC*, 1996, pp. 247-255.

12. D. T. Lee, C. F. Shen, and C.-L. Ding, On Steiner Tree Problem with 45° Routing, in *Proc. IEEE Int. Symp. on Circuit and Systems*, 1995, pp. 1680-1682.
13. S. L. Teig, The X-Architecture: Not Your Father's Diagonal Wiring, in *Proc. ACM International Workshop on System -Level Interconnect Prediction*, 2002, pp. 33-37.
14. P. Widmayer, Y. F. Wu, and C. K. Wong, On Some Distance Problem in Orientation Metrics, *Journal on Computing*, 16(1987), pp. 728-746.
15. X Initiative Homepage, www.xinitiative.org, March 2003.
16. Q. Zhu, H. Zhou, T. Jing, X.-L. Hong, and Y. Yang, Spanning Graph-Based Non-rectilinear Steiner Tree Algorithms, *IEEE Trans. on CAD*, 24(7)(2005), pp. 1066-1075.

# Two Classes of Simple MCD Graphs*

Yong-Bing Shi[1], Yin-Cai Tang[2], Hua Tang[1], Ling-Liu Gong[1], and Li Xu[1]

[1] Department of Applied Mathematics, Shanghai Teachers' University
Shanghai 200234, P.R. China
[2] Department of Statistics, East China Normal University
Shanghai 200062, P.R. China

**Abstract.** Let $S_n$ be the set of simple graphs on $n$ vertices in which no two cycles have the same length. A graph $G$ in $S_n$ is called a *simple maximum cycle-distributed graph* (*simple MCD graph*) if there exists no graph $G'$ in $S_n$ with $|E(G')| > |E(G)|$. A planar graph $G$ is called a *generalized polygon path* (*GPP*) if $G^*$ formed by the following method is a path: corresponding to each interior face $f$ of $\tilde{G}$ ($\tilde{G}$ is a plane graph of $G$) there is a vertex $f^*$ of $G^*$; two vertices $f^*$ and $g^*$ are adjacent in $G^*$ if and only if the intersection of the boundaries of the corresponding interior faces of $\tilde{G}$ is a simple path of $\tilde{G}$. In this paper, we prove that there exists a simple *MCD* graph on $n$ vertices such that it is a 2-connected graph being not a *GPP* if and only if $n \in \{10, 11, 14, 15, 16, 21, 22\}$. We also prove that, by discussing all the natural numbers except for 75 natural numbers, there are exactly 18 natural numbers, for each $n$ of which, there exists a simple *MCD* graph on $n$ vertices such that it is a 2-connected graph.

## 1 Introduction

Let $F_n(S_n)$ be the set of graphs(simple graphs) on $n$ vertices in which no two cycles have the same length. A graph $G$ in $F_n(S_n)$ is called a *MCD graph* (*simple MCD graph*) if there exists no graph $G'$ in $F_n(S_n)$ with $|E(G')| > |E(G)|$.

The number of edges of a *MCD* graph (simple *MCD* graph) on $n$ vertices is denoted by $f(n)(f^*(n))$. The question of determining $f(n)$ raised by Erdös is an unsolved problem (see [1, Problem 11, p. 247]). The related question is determining all *MCD* graphs.

In [2], we proved that, for every $n \geq 2$, there exists a *MCD* graph on $n$ vertices which contains one loop and one 2-cycle. Thus we may construct a *MCD* graph on $n$ vertices from a simple *MCD* graph on $n - 1$ vertices by adding one vertex and three edges, and hence $f(n) = f^*(n-1) + 3$. Therefore the questions of determining $f(n)$ and *MCD* graphs are transformed into that of determining $f^*(n)$ and simple *MCD* graphs.

A path $P$ in $G$ is called a *simple path* of $G$ if, for each interior vertex $v$ of $P$, $d_G(v) = 2$. A 2-connected planar graph $G$ is called a *generalized polygon tree*

* This work was supported by the Foundation of the Development of Science and Technology of Shanghai Higher Learning (04DB25) and Natural Science Foundation of China (10571057).

(*path*) if $G^*$ formed by the following method is a tree (path): corresponding to each interior face $f$ of $\tilde{G}$ ($\tilde{G}$ is a plane graph of $G$) there is a vertex $f^*$ of $G^*$; two vertices $f^*$ and $g^*$ are adjacent in $G^*$ if and only if the intersection of the boundaries of the corresponding interior faces of $\tilde{G}$ is a simple path of $\tilde{G}$. If $G$ is a generalized polygon tree (generalized polygon path), then $G$ is called a *GPT* (*GPP*). Xu in [3] has proved the following result:

**Proposition 1.** *A 2-connected graph $G$ is a GPT if and only if $G$ contains no subgraph homeomorphic to $K_4$.*

Shi in [4] has proved the following result.

**Theorem 1.** *There exists a simple MCD graph on $n$ vertices such that it is a 2-connected graph containing a subgraph homeomorphic to $K_4$ if and only if $n \in \{10, 11, 14, 15, 16, 21, 22\}$.*

In the third section, we first extend the theorem above and obtain the following result.

**Theorem 2.** *There exists a simple MCD graph on $n$ vertices such that it is a 2-connected graph being not a GPP if and only if $n \in \{10, 11, 14, 15, 16, 21, 22\}$.*

We introduce several sets, each of which consists of some natural numbers. Let

$$D_1 = \{n \mid n = (j+1)(j+2)/2 + 4, \, 5 \le j \le 21\},$$

$$D_2 = \{n \mid n = (j+1)(j+2)/2 + 5, \, 5 \le j \le 21\},$$

$$D_3 = \{n \mid n = (j+1)(j+2)/2 + 6, \, 6 \le j \le 21\},$$

$$D_4 = \{n \mid n = (j+1)(j+2)/2 + 7, \, 5 \le j \le 21\},$$

$$D_5 = \{n \mid n = (j+1)(j+2)/2 + 8, \, 7 \le j \le 21\},$$

$$M_1 = \{n \mid 3 \le n \le 27\}$$

$$M_2 = \{12, 17, 18, 23, 24, 25, 26\},$$

$$D = \bigcup_{i=1}^{5} D_i \quad \text{and} \quad M = M_1 - M_2.$$

Then $|D| = 75$, $|M| = 18$.

In the fourth section, we further extend Theorems 1 and 2 above and obtain the following results.

**Theorem 3.** *For each integer $n \notin M \bigcup D$, there does not exist a 2-connected simple MCD graph on $n$ vertices.*

**Theorem 4.** *For each integer $n \in M$, there exists a 2-connected simple MCD graph on $n$ vertices.*

The main results of this paper are Theorems 2, 3 and 4.

## 2   Definitions and Preliminary Results

Let $G$ be a 2-connected simple graph and let $C$ be any cycle in $G$. The set of vertices of $G$ is denoted by $V(G)$, and the set of edges of $G$ is denoted by $E(G)$. We assume that all vertices in $V(G) - V(C)$ and all edges in $E(G) - E(C)$ are drawn inside the bounded region of $C$.

Let $P_0 = C$. We form a path sequence $P_1, P_2, \ldots, P_k$ by the following procedure: for each $i = 1, 2, \ldots, k$, let $P_i$ be a path contained in $G - \cup_{j=0}^{i-1} E(P_j)$ with two and only two end vertices lying on $\cup_{j=0}^{i-1} P_j$, and $\cup_{j=0}^k E(P_j) = E(G)$. We call $(P_1, P_2, \ldots, P_k)$ a *path decomposition* of $G - E(C)$ (The path decomposition defined here is the same with the ear decomposition). Each $P_i (i = 1, 2, \ldots, k)$ is said to be an *inner path* of $G$ with respect to $C$.

Note that for a given 2-connected simple graph $G$ and a cycle $C$ of $G$, $G - E(C)$ may have many path decompositions. But for a give 2-connected simple graph $G$ and any cycle $C$ of $G$, the number of inner paths in any path decomposition of $G - E(C)$ is unique. This results in the following proposition.

**Proposition 2 (Shi [2]).** *Given a 2-connected simple graph $G$, let $C$ be any cycle of $G$. Then the number of inner paths in any path decomposition of $G - E(C)$ is equal to $|E(G)| - |V(G)|$.*

To avoid repetition, when $C$ is a given cycle of $G$, we usually abbreviate 'inner path of $G$ with respect to $C$' to 'inner path of $G$' or 'inner path'. In the coming discussion, all inner paths will be understood to be inner paths of a given path decomposition of $G - E(C)$.

If the end vertices of two inner paths lie on $C$, then clearly these two inner paths are internally disjoint.

Two inner paths $P_1$ and $P_2$ are said to *skew* if there are four distinct vertices $u$, $v$, $u'$ and $v'$ of $C$ such that $u$ and $v$ are the end vertices of $P_1$, $u'$ and $v'$ are the end vertices of $P_2$ and the four vertices appear in the cyclic order $u$, $u'$, $v$, $v'$ on $C$.

Two inner paths are said to be *parallel* if the end vertices of these two inner paths lie on $C$ and they are not skew.

A family of inner paths (at least three inner paths) is said to be parallel if any two inner paths of it are parallel.

Two inner paths $P$ and $R$ are said to be *independent* if there exists an inner path $Q$ such that $P$, $Q$ and $R$ are parallel, and $P$ and $R$ are separated by $Q$ (though they may have end vertices in common).

A family of parallel inner paths is said to be *dependent* if no two inner paths of it are independent.

A family of parallel inner paths (at least three inner paths) is said to be *independent* if no three inner paths of it are dependent.

These definitions are illustrated in Figure 1. There are five inner paths. The paths $S$ and $T$ are skew; the paths $P$, $Q$, $R$ and $S$ are parallel; the paths $P$ and $R$ are independent; the paths $Q$, $R$ and $T$ are dependent; the paths $P$, $Q$ and $T$ are independent.

Let $O^j$ denote the set of 2-connected simple graphs, each of which has exactly $j$ inner paths and let $m(G)$ denote the number of cycles in a graph $G$. Then we have

**Proposition 3 (Shi [2]).** *If $G \in O^j$, then $j = |E(G)| - |V(G)|$ and $m(G) \geq (j+1)(j+2)/2$.*

**Proposition 4 (Shi [4]).** *If $G \in O^j$ and $G$ contains a subgraph homeomorphic to $K_4$, then $m(G) \geq (j^2 + 5j)/2$.*

**Proposition 5 (Shi [5]).** *If $G$ is a 2-connected graph being not a GPP and $G \in O^j$, then $m(G) \geq (j^2 + 5j)/2 - 1$.*

Let $G$ be a Hamilton graph and let $C$ be a Hamilton cycle of $G$. Each edge in $E(G) - E(C)$ is an inner path of $G$ which is also called *bridge*. If two bridges of $G$ are skew each other, then they are said to be *a pair of skew bridges*.

Let $r \geq 3$ be an integer. A graph $G$ on $n$ vertices is said to be a *uniquely r-pancyclic graph (r-UPC graph)* if $G$ contains exactly one cycle of length $l$, for each integer $r \leq l \leq n$, and $G$ contains no cycle of length less than $r$. An *r-UPC* graph $G$ is said to be an *r-UPC[k] graph* if $G$ has exactly $k$ pair of skew bridges. In particular, a 3-UPC[0] graph is called an *outerplanar UPC-graph*.

The following theorems have been proved.

**Theorem 5 (Shi [6]).** *$G$ is a 3-UPC[1] graph if and only if $G \in \{G_{14}^{(i)} | i = 1, 2, 3\}$, where $G_{14}^{(i)}$ is a 3-UPC graph having exactly 14 vertices and three bridges.*

**Theorem 6 (Shi [7]).** *$G$ is an outerplanar UPC graph if and only if $G \in \{K_3, G_5, G_8^{(1)}, G_8^{(2)}\}$, where $G_5$ is a 3-UPC[0] graph having exactly 5 vertices and one bridge, and $G_8^{(i)} (i = 1, 2)$ is a 3-UPC[0] graph having exactly 8 vertices and two bridges.*

**Theorem 7 (Shi [4]).** *Let $G \in O^j$ and let $C$ be a Hamilton cycle of $G$. If $m(G) = (j^2 + 5j)/2$, then $G$ has at most one pair of skew bridges.*

**Theorem 8 (Shi [8]).** *$f^*(n) \geq n + k + \left[\left(\sqrt{8n - 24k^2 + 8k + 1} - 5\right)/2\right]$, where $k = \left[\left(\sqrt{21n - 5} + 11\right)/21\right]$.*

**Theorem 9 (Shi [4]).** *If there exists a 2-connected simple MCD graph on $n$ vertices, then $f^*(n) \leq n + \left[\left(\sqrt{8n - 15} - 3\right)/2\right]$.*

**Theorem 10 (Shi [2]).** *For each integer $n \geq 3$, $f(n) = f^*(n-1) + 3$.*

**Theorem 11 (Shi [2]).** *For each integer $3 \leq n \leq 17$,*

$$f(n) = n + \left[\left(\sqrt{8n - 23} + 1\right)/2\right]$$

**Theorem 12 (Sun et al. [9]).** *For each integer* $18 \leq n \leq 36$,

$$f(n) = \begin{cases} n + 1 + \left[\left(\sqrt{8n - 23} + 1\right)/2\right], & n = 30, \\ n + \left[\left(\sqrt{8n - 23} + 1\right)/2\right], & n \neq 30. \end{cases}$$

From Theorems 10, 11 and 12, we obtain immediately

**Theorem 13.** *For each integer* $2 \leq n \leq 35$,

$$f^*(n) = \begin{cases} n + 1 + \left[\left(\sqrt{8n - 15} - 3\right)/2\right], & n = 29, \\ n + \left[\left(\sqrt{8n - 15} - 3\right)/2\right], & n \neq 29. \end{cases}$$

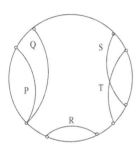

**Fig. 1.** Illustration for the definitions

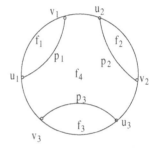

**Fig. 2.** Three dependent inner paths

## 3   2-Connected Non *GPP* Simple *MCD* Graphs

We first prove the following.

**Proposition 6.** *If $G$ is a 2-connected GPT being not a GPP, then there exists a cycle $C$ such that any path decomposition of $G - E(C)$ has three inner paths which are dependent.*

*Proof.* Let $G$ be a *GPT* being not a *GPP*. Then $G$ is a planar graph. Let $\tilde{G}$ be a plane graph of $G$, then there exist four inner faces $f_1$, $f_2$, $f_3$ and $f_4$ such that the boundary $C_4$ of $f_4$ and the boundary $C_i$ of $f_i (i = 1, 2, 3)$ intersect a simple path $P_i$. Let $P_i'$ denote the nontrivial component of $C_i - E(P_i)$ for $i = 1, 2, 3$. Replacing by paths $P_1'$, $P_2'$ and $P_3'$ the paths $P_1$, $P_2$ and $P_3$ contained in $C_4$, respectively, results in a cycle $C$. Clearly any path decomposition of $G - E(C)$ has three inner paths $P_1$, $P_2$, $P_3$ which are dependent (see Figure 2).     □

*Proof of Theorem 2.* The sufficiency follows from Theorem 1 immediately. We shall prove the necessity. By contradiction. Suppose that the theorem does not hold. Then for some positive integer $n \notin \{10, 11, 14, 15, 16, 21, 22\}$, there exists a simple *MCD* graph $G$ on $n$ vertices such that it is a 2-connected graph being not a *GPP*.

By Theorem 8, we have

$$f^*(n) \geq n + k + \left[ \left( \sqrt{8n - 24k^2 + 8k + 1} - 5 \right) / 2 \right],$$

where

$$k = \left[ \left( \sqrt{21n - 5} + 11 \right) / 21 \right]. \tag{1}$$

Clearly $G$ has at least $j$ inner paths, where

$$j = k + \left[ \left( \sqrt{8n - 24k^2 + 8k + 1} - 5 \right) / 2 \right]. \tag{2}$$

By Proposition 5, $m(G) \geq (j^2 + 5j)/2 - 1$. Thus

$$n \geq (j^2 + 5j)/2 + 1. \tag{3}$$

From (2), we have

$$n < \left( j^2 + 5j \right) / 2 + 5 + (7k - 2 - 2j)(k - 1)/2. \tag{4}$$

From (1), we have

$$21k^2 - 22k + 6 \leq n < 21k^2 + 20k + 5. \tag{5}$$

Combining (2) and (5) we have

$$j - 1 < 7k \leq j + 6. \tag{6}$$

It follows from (4) and the right hand side of (6) that

$$n < \left( j^2 + 5j \right) / 2 + 5 + (4 - j)(k - 1)/2. \tag{7}$$

If $k \geq 3$, then $j \geq 15$ follows from (6). In this case, we obtain from (7) that $n < \left( j^2 + 5j \right) / 2 - 6$. This contradicts (3).

If $k = 2$, then from (5) and (6) we find $46 \leq n < 129$ and $8 \leq j < 15$ respectively. In this case, we obtain from (7) that

$$n < \left( j^2 + 5j \right) / 2 + 3. \tag{8}$$

Combining (3) and (8), we have

$$n = \left( j^2 + 5j \right) / 2 + i, \ i = 1, 2. \tag{9}$$

If $k = 1$, then $5 \leq n < 46$ and $1 \leq j < 8$. We obtain from (7) that

$$n < \left( j^2 + 5j \right) / 2 + 5. \tag{10}$$

Combining (3) and (10), we have

$$n = \left( j^2 + 5j \right) / 2 + i, \ i = 1, 2, 3, 4. \tag{11}$$

Since $G$ is not a *GPP*, $j \geq 2$. Combining (9) and (11), we have only three cases:

*Case 1.* $n = \left(j^2 + 5j\right)/2 + i$ $(i = 3, 4$ and $10 \leq n \leq 45)$.

It follows that $n \in \{10, 11, 15, 16, 21, 22, 28, 29, 36, 37, 45\}$. Since $n \notin \{10, 11, 14, 15, 16, 21, 22\}$, $n \in \{28, 29, 36, 37, 45\}$. A similar proof to that of Main Theorem in [10] yields that there does not exist a simple *MCD* graph on 28 vertices such that it is a 2-connected graph being not a *GPP*.

For each $n \in \{29, 36, 37, 45\}$, we may form a simple graph $G_n$ with $n$ vertices and $n + 1 + \left[\left(\sqrt{8n - 15} - 3\right)/2\right]$ edges in which no two cycles have the same length (see Shi [4]). Thus $f^*(n) \geq n + 1 + \left[\left(\sqrt{8n - 15} - 3\right)/2\right]$. This contradicts Theorem 9.

*Case 2.* $n = \left(j^2 + 5j\right)/2 + 1$ and $8 \leq n < 129$.

By Proposition 5, $m(G) \geq \left(j^2 + 5j\right)/2 - 1$. Also $G$ is a simple *MCD* graph. It follows that $m(G) = (j^2 + 5j)/2 - 1$. Clearly $G$ is a *UPC* graph. From Proposition 4, $G$ contains no subgraph homeomorphic to $K_4$. Thus $G$ is an outerplanar *UPC* graph. It follows from Theorem 6 and $n \geq 8$ that $G \in \left\{G_8^{(1)}, G_8^{(2)}\right\}$. Clearly $G$ is a *GPP*, a contradiction.

*Case 3.* $n = \left(j^2 + 5j\right)/2 + 2$, and $9 \leq n < 129$.

Since $G$ is a simple *MCD* graph, $m(G) \leq \left(j^2 + 5j\right)/2$. We consider only two subcases.

*Subcase 3.1.* $m(G) = \left(j^2 + 5j\right)/2$. Clearly $G$ is a *UPC* graph. If $G$ contains no skew bridge, then $G$ is an outerplanar *UPC* graph. Thus, by Theorem 6, $G \in \left\{K_3, G_5, G_8^{(1)}, G_8^{(2)}\right\}$, i.e., $n \leq 8$, a contradiction. Therefore, $G$ contains skew bridges. By Theorem 7, $G$ has at most one pair of skew bridges. Thus $G$ is a 3-*UPC[1]* graph. Consequently, by Theorem 5, $G \in \left\{G_{14}^{(1)}, G_{14}^{(2)}, G_{14}^{(3)}\right\}$. i.e., $n = 14$, again a contradiction.

*Subcase 3.2.* $m(G) \leq \left(j^2 + 5j\right)/2 - 1$. By Proposition 5, $m(G) \geq \left(j^2 + 5j\right)/2 - 1$. Therefore, $m(G) = \left(j^2 + 5j\right)/2 - 1$ and $G$ contains no subgraph homeomorphic to $K_4$. Since $G$ is not *GPP* and $n \neq 14$, $j \geq 4$. Let $C$ be the longest cycle of $G$, then the length of $C$ is at most $\left(j^2 + 5j\right)/2 + 1$. Thus there is at most one vertex in $V(G) - V(C)$. If $V(G) - V(C) \neq \Phi$, then let $V(G) - V(C) = \{u\}$. It is easy to see that $d(u) = 2$ since $G$ is a simple graph and $G$ contains no subgraph homeomorphic to $K_4$. Thus the end vertices of all inner paths of $G$ lie on $C$. By Proposition 6, $G$ has three dependent inner paths, say $P_1, P_2, P_3$.

It is easily seen that there do not exist four dependent inner paths of $G$. Otherwise we may assume that $P, P_1, P_2, P_3$ are four dependent inner paths of $G$. Let $G^* = G - E(P)$. Then, by Proposition 5, $m(G^*) \geq \left[(j-1)^2 + 5(j-1)\right]/2 - 1$. Let $B$ be the set of all inner paths of $G$. Then $|B| = j$. If $B' \subseteq B$, then the number of cycles containing all inner paths in $B'$ but no inner paths in $B - B'$ is denoted by $c(B')$. If $B' = \{P_1, P_2, \ldots, P_k\}$, then $c(B')$ is also denoted by $c(P_1, P_2, \ldots, P_k)$. Now we count the number of cycles in $G$ containing P. Clearly

$c(P) = 2$; $c(P_1, P_2, P_3, P) = 1$; for each $P' \in B - \{P\}$, $c(P, P') = 1$; for each pair of distinct elements $i, j \in \{1, 2, 3\}$, $c(P, P_i, P_j) = 1$. Thus the number of cycles containing $P$ is at least $2 + 1 + (j - 1) + 3 = j + 5$. Consequently, $m(G) \geq m(G^*) + j + 5 \geq (j^2 + 5j)/2 + 2$, a contradiction.

Let $P_1 = (u_1, v_1)$, $P_2 = (u_2, v_2)$ and $P_3 = (u_3, v_3)$. (We allow the end vertices in common, see Figure 2). We denote by $C[u_i, v_i]$ the $(u_i, v_i)$-path which follows the clockwise orientation of $C$. If $P \notin \{P_1, P_2, P_3\}$ and the end vertices of $P$ lie on $C[u_i, v_i]$ for some $i \in \{1, 2, 3\}$, then $P$ and $P_i$ are said to be *companion*. Let $C^*$ be the cycle containing the dependent inner paths $P_1$, $P_2$ and $P_3$. We may assume that there are no inner paths in the interior of $C^*$. Because if there exists a inner path $P$ lying in the interior of $C^*$, then there must exist two inner paths $P_i$, $P_j \in \{P_1, P_2, P_3\}$ such that $P_i$, $P_j$ and $P$ are dependent. In this case let $C^*$ be the cycle containing $P_i$, $P_j$ and $P$. Continuing this procedure, we can obtain the cycle $C^*$ containing three dependent inner paths such that there is no inner path lying in the interior of $C^*$. Let $B_1 = B - \{P_1, P_2, P_3\}$. Then any inner path in $B_1$ companies to one of the three inner paths $P_1$, $P_2$ and $P_3$.

Clearly all inner paths company to only one of the three inner paths $P_1$, $P_2$ and $P_3$. Otherwise we may assume that there exist inner paths $P'$, $P'' \in B_1$ such that $P'$ companies to $P_1$ and $P''$ companies to $P_2$ (see Figure 3). Let $G^* = G - E(P')$. Then $m(G^*) \geq [(j - 1)^2 + 5(j - 1)]/2 - 1$. Clearly $c(P') = 2$; $c(P', P_3, P'') = c(P', P_3, P_2) = 1$; for any $\bar{P} \in B - \{P'\}$, $c(P', \bar{P}) = 1$. Thus the number of cycles containing $P'$ is at least $2+1+1+j-1=j+3$. Consequently $m(G) \geq m(G^*) + j + 3 \geq (j^2 + 5j)/2$, a contradiction.

We may assume that all inner paths in $B_1$ company to $P_1$. Clearly $B_1 \cup \{P_1\}$ is independent (see Figure 4), otherwise there exist three dependent inner paths in $B_1 \cup \{P_1\}$, and hence $G$ must has four dependent inner paths, a contradiction. We can prove that $G$ has two cycles which have the same length (See, A Class of Almost Uniquely Pancyclic Graphs, to appear in J. Sys. Sci. Math. Sci. (China) 4 (2006)), a contradiction.    □

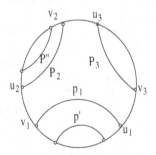

**Fig. 3.** The companies of $P_1$ and $P_2$

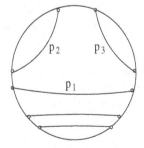

**Fig. 4.** $B_1 \cup \{P_1\}$ being independent

## 4    2-Connected Simple *MCD* Graphs

*Proof of Theorem 3.* By contradiction. Suppose that the theorem does not hold. Then for some positive integer $n \notin M \cup D$, there exists a 2-connected simple *MCD* graph on $n$ vertices. By Theorem 8, we have

$$f^*(n) \geq n + k + \left[ \frac{1}{2} \left( \sqrt{8n - 24k^2 + 8k + 1} - 5 \right) \right], \tag{12}$$

where $k = \left[ \frac{1}{21} \left( \sqrt{21n - 5} + 11 \right) \right]$, G has at least j inner paths, where

$$j = k + \left[ \frac{1}{2} \left( \sqrt{8n - 24k^2 + 8k + 1} - 5 \right). \right] \tag{13}$$

By Proposition 3, $m(G) \geq (j+1)(j+2)/2$. Thus

$$n \geq \frac{1}{2}(j+1)(j+2) + 2 \tag{14}$$

From (13), we have

$$n < \frac{1}{2} \left( j^2 + 5j \right) + 5 + \frac{1}{2}(7k - 2 - 2j)(k - 1). \tag{15}$$

From (12), we find that

$$21k^2 - 22k + 6 \leq n < 21k^2 + 20k + 5. \tag{16}$$

Combining (13) and (16), we have

$$j - 1 < 7k \leq j + 6. \tag{17}$$

It follows from (15) and the right-hand side of (17) that

$$n < \frac{1}{2} \left( j^2 + 5j \right) + 5 + \frac{1}{2}(4 - j)(k - 1). \tag{18}$$

If $k \geq 4$, then $j \geq 22$ follows from (17). In this case, we obtain from (18) that

$$n < \left( j^2 + 5j \right) / 2 + 5 + 3(4 - j)/2$$
$$= (j+1)(j+2)/2 + 10 - j/2$$
$$\leq (j+1)(j+2) - 1.$$

This contradicts (14).

If $k = 3$, then $15 \leq j \leq 21$ follows from (17). In this case, we obtain from (18) that

$$n < (j+1)(j+2)/2 + 8. \tag{19}$$

If $k = 2$, then $8 \leq j \leq 14$ follows from (17). In this case, we obtain from (18) that

$$n < (j+1)(j+2)/2 + 6 + j/2. \tag{20}$$

If $k = 1$, then $1 \leq j \leq 7$ follows from (17). In this case, we obtain from (18) that

$$n < (j+1)(j+2)/2 + 4 + j. \tag{21}$$

By hypothesis of $n$, $k \neq 0$. Since Fang in [11] has already obtained that

$$f^*(n) \geq n + 1 + \left[\frac{1}{2}\left(\sqrt{8n - 71} - 3\right)\right] \tag{22}$$

for $n \geq 30$. Also, by Theorem 9, we have

$$f^*(n) \leq n + \left[\frac{1}{2}\left(\sqrt{8n - 15} - 3\right)\right]. \tag{23}$$

Putting $(j+1)(j+2)/2 + 9 \leq n \leq (j+1)(j+2)/2 + 6 + j/2$ and $8 \leq j \leq 14$ in (22) and (23), respectively, we have $f^*(n) \geq n + 1 + j$ and $f^*(n) \leq n + j$, a contradiction.

Putting $(j+1)(j+2)/2 + 9 \leq n \leq (j+1)(j+2)/2 + 3 + j$ and $6 \leq j \leq 7$ in (22) and (23), respectively, we have $f^*(n) \geq n + 1 + j$ and $f^*(n) \leq n + j$, again a contradiction.

Therefore, by (19), (20), (21) and (14), we consider only the following integer values of n:

$$(j+1)(j+2)/2 + 2 \leq n \leq (j+1)(j+2)/2 + 3 + j, \quad \text{for } 1 \leq j \leq 5;$$

$$(j+1)(j+2)/2 + 2 \leq n \leq (j+1)(j+2)/2 + 8, \quad \text{for } 6 \leq j \leq 14;$$

$$(j+1)(j+2)/2 + 2 \leq n \leq (j+1)(j+2)/2 + 7, \quad \text{for } 15 \leq j \leq 21.$$

Also, by $n \notin M \cup D$, $n$ takes only the following integer values:

**(1)** $n = (j+1)(j+2)/2 + 2$ for $3 \leq j \leq 21$;
**(2)** $n = (j+1)(j+2)/2 + 3$ for $4 \leq j \leq 21$;
**(3)** $n = (j+1)(j+2)/2 + 8$ for $5 \leq j \leq 6$;

Now we consider three cases above.

*Case 1.* $n = (j+1)(j+2)/2 + 2$, for $3 \leq j \leq 21$.

By Proposition 3, $m(G) = (j+1)(j+2)/2$. Clearly the $j$ inner paths of $G$ are independent. Thus $G$ is an outerplanar *UPC* graph. It follows from Theorem 6 that $G \in \left\{K_3, G_5, G_8^{(1)}, G_8^{(2)}\right\}$, i.e., $n \in \{3, 5, 8\}$, a contradiction.

*Case 2.* $n = (j+1)(j+2)/2 + 3$, for $4 \leq j \leq 21$.

In this case, we can prove that there does not exist a 2-connected simple *MCD* graph (See, On Almost Uniquely Pancyclic Graphs, to appear in Advance in Mathematics (China), 5 (2006)).

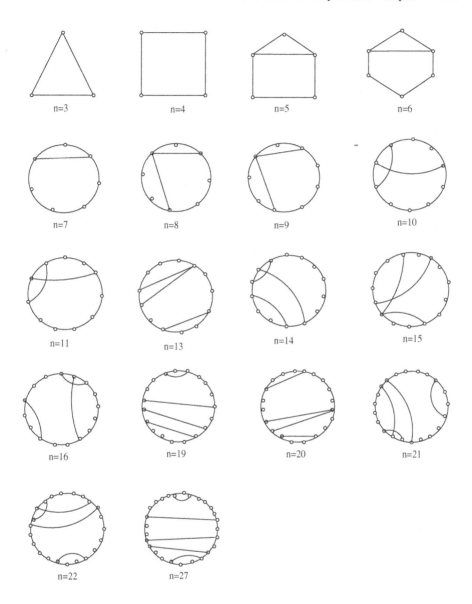

**Fig. 5.** 2-connected simple MCD graphs

*Case 3.* $n = (j+1)(j+2)/2 + 8$, for $5 \leq j \leq 6$.

Clearly $n \in \{29, 36\}$. Since Shi in [4] has formed a simple graph $G_n$ with $n$ vertices and $n + 1 + \left[\frac{1}{2}\left(\sqrt{8n-15} - 3\right)\right]$ edges in which no two cycles have the same length. Thus $f^*(n) \geq n + 1 + \left[\frac{1}{2}\left(\sqrt{8n-15} - 3\right)\right]$. This contradicts Theorem 9. This ends the proof of Theorem 3. $\qquad\square$

*Proof of Theorem 4.* For each $n \in M$, there exists a 2-connected simple graph with $n$ vertices and $n + \left[\frac{1}{2}\left(\sqrt{8n - 15} - 3\right)\right]$ edges in which no two cycles have the same length (see Figure 5). By Theorem 13, for each $n \in M$, $f^*(n) = n + \left[\frac{1}{2}\left(\sqrt{8n - 15} - 3\right)\right]$. Therefore each graph in Figure 5 is a simple $MCD$ graph. This ends the proof of the theorem.

# References

1. Bondy, J.A., Murty, U.S.R.: Graph Theory with Applications. Macmillan, New York (1976)
2. Shi, Y.: On Maximum Cycle-Distributed Graphs. Discrete Math. **71** (1988) 57–71
3. Xu, S: Classes of Chromatically Equivalent Graphs and Polygontrees. Discrete Math. **133** (1994) 267–278
4. Shi, Y.: On Simple $MCD$ Graphs Containing a Subgraph Homeomorphic to $K_4$. Discrete Math. **126** (1994) 325–338
5. Shi, Y.: The Number of Cycles in a 2-Connected Simple Graphs being not Generalized Polygon Paths. J. Shanghai Teachers' Univ. (Natural Science) **28** (3) (1999) 17–20 (Chinese)
6. Shi, Y: Futher Results on Uniquely Pancyclic Graphs (I), (II). J. Shanghai Teachers' Univ. (Natural Science) **15** (2) (1986) 21–27 **15** (4) (1986) 1–7 (Chinese)
7. Shi, Y.: Some Theorems of Uniquely Pancyclic Graphs. Discrete Math. **59** (1986) 167–180
8. Shi, Y.: The Number of Edges in a Maximum Cycle-Distributed Graph. Discrete Math. **104** (1992) 205–209
9. Sun, J., Shi, Y.: Determining the Number of Edges in a Maximum Cycle-Distributed Graph. J. Shanghai Teachers' Univ. (Natural Science) **23** (Special Issue on Mathematics) (1994) 30–39 (Chinese)
10. Shi, Y.: Simple $MCD$ Graphs on 28 Vertices. J. Shanghai Teachers' Univ. (Natural Science) **24** (2) (1995) 8–16
11. Fan, Y.: A New Lower Bound of the Number of Edges in a Simple MCD Graph. J. Shanghai Teachers' Univ. (Natural Science) **22** (4) (1993) 33–36 (Chinese)

# Core Stability of Flow Games[*]

Xiaoxun Sun and Qizhi Fang[**]

Department of Mathematics, Ocean University of China, Qingdao 266071, P.R. China

**Abstract.** In this paper, we study the problem of core stability for flow games, introduced by Kalai and Zemel (1982), which arises from the profit distribution problem related to the maximum flow in networks. Based on the characterization of dummy arc (i.e., the arc which satisfies that deleting it does not change the value of maximum flow in the network), we prove that the flow game defined on a simple network has the stable core if and only if there is no dummy arc in the network. We also show that the core largeness, the extendability and the exactness of flow games are equivalent conditions, which strictly imply the stability of the core.

## 1 Introduction

Originating with the pioneering work of Ford and Fulkerson [7], network flow models have been widely investigated and found its applications in various fields. In cooperative game theory, flow game was first discussed by Kalai and Zemel [9,10], which arose from the profit distribution problem related to the maximum flow in a network. There have been many results presented after the introduction of flow games, most of them focused on the core, the most important concept solution in cooperative game models. The characterization of the cores of flow games is due to Kalai and Zemel [9, 10] and Deng *etc.* [4]. They showed that flow games are totally balanced, and the profit allocations corresponding to minimum cuts in the network always belong to the core. On the other hand, it was proved that for the flow game, the problem of checking whether a given allocation belongs to the core is *co-NP*-complete [6].

Neumann and Morgenstern [13] proposed a solution concept called stable set, which turns out to be useful in the analysis of a lot of bargaining situations. But it is too difficult to survey its fundamental properties, because the stable set may not exist for all TU-games (Lucas [12]). Deng and Papadimitriou [5] pointed out that determining the existence of a stable set for a given cooperative game is not known to be computable, and it is still unsolved. When the game is convex, the core is a stable set. In general, however, the core and the stable set are different. Hence there is a question that people are interested in: when do the core and the stable set coincide, that is, when is the core stable?

The main purpose of this paper is to identify those flow games whose cores are stable. We first characterize a special kind of arcs, called dummy arcs, which

---

[*] Research is supported by NCET, NSFC (No.10371114, No.70571040/G0105).
[**] Corresponding author.

satisfy that deleting any of them does not change the value of maximum flow in the network. Based on this characterization, we prove that the flow game defined on a simple network (arc capacities are all equal to 1) has the stable core if and only if it contains no dummy arc. Furthermore, we consider the problems on the core largeness, the extendability and the exactness of flow games, which are the concepts closely related to core stability. We show that the three properties are equivalent for simple flow games, and also equivalent to a common condition that every $(s, t)$-cut contains a minimum $(s, t)$-cut in the simple network (where $s, t$ are the source and the sink of the network, respectively).

Several sufficient conditions for core stability have been discussed in the literature. Subconvexity of the game and largeness of the core were introduced by Sharkey [16] who showed that convexity implies subconvexity; subconvexity implies largeness of the core; which in turn implies stability of the core. In an unpublished paper, Kikuta and Shapley [11] investigated another condition, baptized to extendability of the game in Gellekom *etc.* [8], and proved that it is necessary for core largeness and still sufficient for core stability. However, as far as the core stability for concrete cooperative game model is concerned, few results have been obtained. The only exceptions are the work done by Solymosi and Raghavan [17], who studied the core stability of assignment games and Bietenhader and Okamoto [1], who studied the core stability of minimum coloring games defined on perfect graphs.

The organization of the paper is as follows. In Section 2 we introduce the basic definitions and some known results on flow games. In Section 3 we present the sufficient and necessary condition for the flow game defined on a simple network possessing the stable core. Finally, in Section 4 we discuss the equivalence of core largeness, extendability and exactness of flow games.

## 2    Definitions and Preliminaries

### 2.1    Cooperative Game

A cooperative (profit) game $\Gamma = (N, v)$ consists of a player set $N = \{1, 2, \ldots, n\}$ and a characteristic function $v\colon 2^N \to R$ with $v(\emptyset) = 0$. For each coalition $S \subseteq N$, $v(S)$ represents the profit achieved by the players in $S$ without participation of other players. The central problem in cooperative game is how to allocate the total profit $v(N)$ among the individual players in a 'fair' way. Any $x \in R^n$ with $\sum_{i \in N} x_i = v(N)$ is an allocation. Different requirements for fairness, stability and rationality lead to different sets of allocations which are generally referred to solution concepts. Among all solution concepts, the core is the most important, and it has been extensively studied for many game models.

A vector $x = (x_1, x_2, \ldots, x_n)$ is called an imputation if $\sum_{i \in N} x_i = v(N)$ and for any $i \in N\colon x_i \geq v(\{i\})$. Denote by $X(\Gamma)$ the set of imputations of $\Gamma$. The core of $\Gamma$ is defined by

$$C(\Gamma) = \{x \in R^n : x(N) = v(N) \text{ and } x(S) \geq v(S), \forall S \subseteq N\}. \tag{1}$$

Throughout this paper, we use the shorthand notation

$$x(S) = \sum_{i \in S} x_i$$

for each $x \in R^{|N|}$ and each $S \subseteq N$. The constraints imposed on $C(\Gamma)$ ensure that no coalition would have an incentive to split from the grand coalition $N$, and do better on its own. The game is called *ballanced*, if $C(\Gamma) \neq \emptyset$. For a subset $S \subseteq N$, we define the induced subgame $(S, v_S)$ on $S$: $v_S(T) = v(T)$ for each $T \subseteq S$. A cooperative game $\Gamma$ is called *totally balanced* if all its subgames are balanced, i.e., all its subgames have non-empty cores.

## 2.2  Core Stability

In their classical work in game theory, von Neumann and Morgenstern [13] proposed the first solution concept, the stable set, which is very useful for the analysis of a lot of bargaining situations. Suppose that $x$ and $y$ are imputations. We say that $x$ dominates $y$ if there is a coalition $S$ such that (a) $x(S) \leq v(S)$, and (b) $x_i > y_i$ for each $i \in S$. A set $\mathcal{F}$ of imputations is a *stable set* if there is no two imputations in $\mathcal{F}$ dominating each other and any imputation not in $\mathcal{F}$ is dominated by some imputation in $\mathcal{F}$. Formally, the core $C(\Gamma)$ of a game $\Gamma = (N, v)$ is *stable* if for every $y \in X(\Gamma) \setminus C(\Gamma)$, there exists a core allocation $x \in C(\Gamma)$ and a nonempty coalition $S \subset N$ such that $x(S) = v(S)$ and $x_i > y_i$ for each $i \in S$. That is, a game $\Gamma$ has the stable core if any imputation not in $C(\Gamma)$ is dominated by some core allocation.

There are three concepts related to the core stability. Let $\Gamma = (N, v)$ be a cooperative game with $|N| = n$. The core $C(\Gamma)$ is called *large* if for every $y \in R^n$ satisfying that $y(S) \geq v(S)$ for all $S \subseteq N$, there exists $x \in C(\Gamma)$ such that $x \leq y$. The game $\Gamma$ is called *extendable* if for every nonempty $S \subset N$ and every core element $y$ of the subgame $(S, v_S)$, there exists $x \in C(\Gamma)$ such that $x_i = y_i$ for all $i \in S$. The game $\Gamma$ is called *exact* if for every $S \subset N$, there exists $x \in C(\Gamma)$ such that $x(S) = v(S)$.

Kikuta and Shapley [11] showed that if a balanced game has the large core, then it is extendable; and if a balanced game is extendable, then it has the stable core. Sharkey [16] showed that if a totally balanced game has the large core then it is exact. Biswas, Pathasarathy, Potters and Voorneveld [2] pointed out that extendability implies exactness. For completeness of this paper, we summarize these results in the following theorem.

**Theorem 1.  [2, 11, 16]** *let $\Gamma = (N, v)$ be a totally balanced game. Then*

$$Largeness \Rightarrow Extendability \Rightarrow \begin{cases} Exactness, \\ Core \ Stability. \end{cases}$$

## 2.3  Flow Game

Flow games were first discussed by Kalai and Zemel [9, 10]. Consider a directed network $D = (V, E; \omega)$, where $V$ is the vertex set, $E$ is the arc set and $\omega \colon E \to R^+$

is the arc capacity function. Let $s$ and $t$ be two distinct vertices of $D$ which we denote the 'source' and the 'sink' of the network respectively. We assume that each player controls one arc in the network, i.e., we can identify the set of arcs with the set of players.

**Definition 1.** *Given a network $D = (V, E; \omega)$, the associated flow game $\Gamma = (E, v)$ is defined as*:

**(1)** *The player set is $E$;*
**(2)** *For each $S \subseteq E$, $v(S)$ is the value of a maximum flow from the 'source' $s$ to the 'sink' $t$ that uses the arcs owned by $S$ only.*

An $(s, t)$-path in $D$ is a simple path (it visits each node at most once) from $s$ to $t$. Let $\mathcal{P}$ be the set of $(s, t)$-paths on $D$, each regarded as a subset of edges. For any partition $S$ and $T$ of $V$ with $s \in S$ and $t \in T$, the arc set $C = \{e = (u, v) \in E : u \in S, v \in T\}$ is called an $(s, t)$-cut. The capacity of a cut is the sum of the capacities of its members. An $(s, t)$-cut is called *minimum* if it has least capacity among all $(s, t)$-cuts. For an $(s, t)$-cut $C$, we denote its indicator vector by $\mathbf{1}_C \in \{0, 1\}^{|E|}$, where $\mathbf{1}_C(e) = 1$ if $e \in C$, and 0 otherwise.

We call a network $D$ simple, if the arc capacities are all equal to 1. The flow games related to simple networks also include in the class of packing/covering games introduced in Deng *etc.* [4]. The following result on the characterization of the core of a flow game is due to Kalai and Zemel [10] and Deng *etc.* [4].

**Theorem 2.** **[4, 10]** *Let $\Gamma = (E, v)$ be the flow game defined on the network $D = (V, E; \omega)$. Then $\Gamma = (E, v)$ is totally balanced. In the case $D$ is a simple network, $C(\Gamma)$ is exactly the convex hull of the indicator vectors of the minimum cuts of $D$.*

However, Fang *etc.* [6] proved that for general flow games, the problem of testing whether an imputation belongs to the core is *co-$\mathcal{NP}$-complete*.

## 3   Core Stability of Flow Games

In this section we restrict our attention to flow games defined on simple networks. Let $D = (V, E)$ be a simple network ($|E| = n$) with source $s$ and sink $t$.

### 3.1   Dummy Arc in Simple Network

**Definition 2.** *Let $\Gamma = (E, v)$ be the flow game defined on $D = (V, E)$, and let $e \in E$ be contained in some $(s, t)$-path. The arc $e \in E$ is called a dummy arc (player) if $v(E \setminus \{e\}) = v(E)$. Furthermore, we distinguish the dummy arcs into two types*:

**(1)** *A dummy arc $e$ is called Type I, if there exists a maximum flow $f$ with $f(e) > 0$;*
**(2)** *A dummy arc $e$ is called Type II, if it holds that $f(e) = 0$ for any maximum flow $f$.*

For example, the network $D$ in Fig.1 is a simple network. It is easy to verify that $e_1$, $e_2$, $e_3$, $e_4$ are all dummy arcs. And $e_1$, $e_2$ and $e_3$ are the Type I dummy arcs, $e_4$ is a Type II dummy arc.

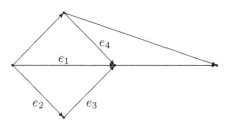

**Fig. 1.** Network $D$

We note that if an arc is not contained in any $(s, t)$-path, then it does not affect the values of all allocations in the flow game. Therefore, in the rest of the paper we assume that each arc is contained in some $(s, t)$-path.

We will give the descriptions of dummy arcs according to the core of the flow game. Followed from the the max-flow min-cut theorem [7] and Theorem 2, it is easy to show the correctness of the following results.

**Proposition 1.** *An arc $e \in E$ is not a dummy arc in the network $D$ if and only if there is a minimum $(s, t)$-cut $C$ of $D$ such that $e \in C$.*

**Proposition 2.** *An arc $e \in E$ is a dummy arc in the network $D$ if and only if $x(e) = 0$ for each core allocation $x \in C(\Gamma)$.*

In the rest of this subsection, we consider the algorithm to seek out the dummy arcs of Type I and II. The problem is stated as:

---

**Problem**: Seeking Out Dummy Arc

---

**Instance**: A simple network $D = (V, E)$
**Question**: Find all dummy arcs of Type I and Type II, respectively

---

We show that this problem can be solved in polynomial time as follows. For each $e \in E$, we compute the maximum flow in the subnetwork $D \backslash \{e\}$ and identify the dummy arc set $\{e \in E \colon v(E \backslash \{e\}) = v(E)\}$. The value of a maximum flow in a simple network is equal to the maximum number of arc-disjoint $(s, t)$-paths, and this can be computed in time $O(|E|^{3/2})$ (see, e.g., [14]).

To seek out the dummy arcs of Type I and II, one can do as follows: We first find a maximum flow $f$ in the network $D$ and construct the residual network $D_f = (V, E_f)$ from $D$ and $f$, where

$$E_f = \{(u, v) \colon e = (u, v) \in E, f(e) < 1\} \cup \{(v, u) \colon e = (u, v) \in E, f(e) > 0\}.$$

Then for each dummy arc $e = (u, v)$, $e$ is Type I if there is a $(v, u)$-path in $D_f$, and Type II otherwise. Since finding a shortest path tree rooted at a given vertex in the residual network $D_f$ can be done in time $O(|E|)$ (see, e.g., [14]), it follows that the set of Type I and II dummy arcs can be identified in time $O(|E||V|)$.

## 3.2   Condition on Core Stability

In this subsection, we present a sufficient and necessary condition for a flow game possessing the stable core. We first state our main result.

**Theorem 3.** *Let $D = (V, E)$ be a simple network. Then the corresponding flow game $\Gamma = (E, v)$ has the stable core if and only if $D$ contains no dummy arc.*

To prove this theorem, we need several lemmas.

**Lemma 1.** *The core $C(\Gamma)$ is stable if and only if for any $y \in X(\Gamma) \backslash C(\Gamma)$, there exists a core element $z$ and an $(s, t)$-path $P$ such that $z(P) = 1$ and $z(e) > y(e)$ for each $e \in P$.*

*Proof.* Sufficiency is directly by the definition of core stability. Suppose that $C(\Gamma)$ is stable. Then for any $y \in X(\Gamma) \backslash C(\Gamma)$, there exists a core element $z$ and a coalition $S$ such that $z(S) = v(S)$ and $z(e) > y(e)$ for each $e \in S$. Obviously, $v(S) > 0$ and there are $v(S)$ arc-disjoint $(s, t)$-paths in the induced network $D[S]$, namely $P_1, P_2, \ldots, P_{v(S)}$. Denote $S' = P_1 \cup P_2 \cup \cdots \cup P_{v(S)}$. Then we have $z(S') = v(S) = v(S')$, and consequently, there must be a path $P_k$ such that $z(P_k) \leq 1$ ($1 \leq k \leq v(S)$). Also $z \in C(\Gamma)$ implies that $z(P_k) = 1$. Since $z(e) > y(e)$ for each $e \in S$ and $P_k \subseteq S' \subseteq S$, we have $z(e) > y(e)$ for each $e \in P_k$, as desired.    □

**Lemma 2.** *Let $P$ be an $(s, t)$-path in $D$ containing no dummy arc. Then $|P \cap C| = 1$ for each minimum cut $C$ in $D$.*

*Proof.* Suppose that $C^*$ is a minimum cut with $|C^* \cap P| \geq 2$. Since $P$ contains no dummy arc, there is a maximum flow $f$ such that $f(e) = 1$ for all $e \in P$. It follows that $v(E \setminus P) = v(E) - 1$. But by our assumption on $C^*$ and $P$, $E \setminus P$ contains at most $|C^*| - 2 = v(E) - 2$ arcs in $C^*$, which contradicts to the fact that $C^*$ is a minimum cut of $D$.    □

To describe the next result, let us give some notations. For a simple network $D = (V, E)$ and $e_0 \in E$, let $\overline{D} = D \setminus e_0$ and $\overline{\Gamma}$ be the flow game defined on $\overline{D}$. For an $|E|$-dimensional vector $x = \{x(e) : e \in E\}$, let $x_{-e_0}$ be the restriction of $x$ on $E \setminus \{e_0\}$. That is, $x_{-e_0}$ is an $(|E| - 1)$-dimensional vector satisfying that

$$\forall e \in E \setminus \{e_0\}, \ x_{-e_0}(e) = x(e).$$

Accordingly, we define two new sets $X^*(\overline{\Gamma})$ and $C^*(\overline{\Gamma})$ by

$$X^*(\overline{\Gamma}) = \left\{ x \in R^{|E|} : x(e_0) = 0 \text{ and } x_{-e_0} \in X(\overline{\Gamma}) \right\};$$

$$C^*(\overline{\Gamma}) = \left\{ x \in R^{|E|} : x(e_0) = 0 \text{ and } x_{-e_0} \in C(\overline{\Gamma}) \right\}.$$

Obviously, there is a one-to-one correspondence between $X^*(\overline{\Gamma})$ and $X(\overline{\Gamma})$, also between $C^*(\overline{\Gamma})$ and $C(\overline{\Gamma})$.

**Lemma 3.** *Let $\Gamma$ and $\overline{\Gamma}$ be the flow games corresponding to network $D$ and $\overline{D} = D \setminus e_0$, respectively. If $e_0$ is a dummy arc in $D$, then we have*

**(1)** $X^*(\overline{\Gamma}) \subseteq X(\Gamma)$ *and* $C(\Gamma) \subseteq C^*(\overline{\Gamma})$;
**(2)** $C(\Gamma)$ *is stable implies that* $C(\overline{\Gamma})$ *is stable.*

*Proof.* The correctness of (1) can be verified easily. (2) Suppose that $C(\Gamma)$ is stable. According to our definition given above and Lemma 1, it suffices to show that for a given $y \in X^*(\overline{\Gamma}) \setminus C^*(\overline{\Gamma})$, there exists $z \in C^*(\overline{\Gamma})$ and an $(s, t)$-path $P \subseteq E \setminus \{e_0\}$ such that $z(P) = 1$ and $z(e) > y(e)$ for each $e \in P$. By the result of (1), we have $y \in X(\Gamma) \setminus C(\Gamma)$. Also by Lemma 1, there exists a vector $z \in C(\Gamma) \subseteq C^*(\overline{\Gamma})$ and an $(s, t)$-path $P \subseteq E$ such that $z(P) = 1$ and $z(e) > y(e) \geq 0$, $\forall e \in P$. According to Proposition 2, $P$ contains no dummy arc of $D$, which implies that $P$ is also an $(s, t)$-path in $\overline{D}$, as desired. □

*Proof of Theorem 3.* Sufficiency. Suppose that $D$ contains no dummy arc. Given $y \in X(\Gamma) \setminus C(\Gamma)$, then there is an $(s, t)$-path $P^* = \{e_{i_1}, e_{i_2}, \ldots, e_{i_k}\}$ with $y(P^*) < 1$ and $y(e_{i_j}) \geq 0$ for $j = 1, 2, \ldots, k$. Since $P^*$ contains no dummy arc, each arc $e_{i_j}$ corresponds to a minimum $(s, t)$-cut of $D$, denoted by $C_j$, such that $e_{i_j} \in C_j$, $\forall j = 1, 2, \ldots, k$. Followed from Lemma 2, $C_j \neq C_l$ for all $j \neq l$ $(j, l = 1, 2, \ldots, k)$.

Denote the indicator vectors of $C_1, C_2, \ldots, C_k$ by $\mathbf{x}^1, \mathbf{x}^2, \ldots, \mathbf{x}^k$, respectively. Construct a core allocation $z \in C(\Gamma)$ as follows:

$$z = \lambda_1 \mathbf{x}^1 + \lambda_2 \mathbf{x}^2 + \cdots + \lambda_k \mathbf{x}^k,$$

$$\lambda_j = y(e_{i_j}) + \frac{1 - y(P^*)}{k}. \quad j = 1, 2, \ldots, k \tag{2}$$

Obviously $\lambda_j > 0$ and $\sum_{j=1}^{k} \lambda_j = 1$. By Theorem 2 we have that $z \in C(\Gamma)$. Moreover,

$$z(P^*) = \sum_{j=1}^{k} \left[ y(e_{i_j}) + \frac{1 - y(P^*)}{k} \right] \mathbf{x}^j(P^*)$$

$$= y(P^*) + k \cdot \frac{1 - y(P^*)}{k} = 1,$$

and $z(e_{i_j}) > y(e_{i_j})$ for each $j = 1, 2, \ldots, k$. Hence, followed from Lemma 1, $C(\Gamma)$ is stable.

Necessity. Denote the number of dummy arcs in $D$ by $d_0$. We prove the necessity by the induction on $d_0$. We first show that the statement is true for $d_0 = 1$ and $d_0 = 2$ by distinguishing in three cases.

*Case 1.* If $d_0 = 1$ and let $e_0$ be the unique dummy arc of $D$, then $e_0$ must be a type II dummy arc. Let $\overline{D} = D \setminus e_0$ and $\overline{\Gamma}$ be the corresponding flow game. We claim that $C(\Gamma)$ is a proper subset of $C^*(\overline{\Gamma})$.

By Lemma 3, $C(\Gamma) \subseteq C^*(\overline{\Gamma})$. On the other hand, there must be a vector $z \in C^*(\overline{\Gamma}) \setminus C(\Gamma)$. In fact, given an $(s, t)$-path $P_0$ containing $e_0$, denoted by $P_0 = \{s, v_1, \ldots, v_j, v_{j+1}, \ldots, v_l, t\}$ and $e_0 = (v_j, v_{j+1})$. Consider an $(s, v_{j+1})$ path $P_1 \subset E \setminus \{e_0\}$ and a $(v_j, t)$-path $P_2 \subset E \setminus \{e_0\}$. In the network $\overline{D}$, we duplicate all the arcs in $P_0 \setminus \{e_0\}$ to get a new network $D'$ and let $C'$ be a minimum $(s, t)$-cut of $D'$. It is easy to see that the values of the maximum flows in $D$, $\overline{D}$ and $D'$ are all equal, and $C'$ is also a minimum $(s, t)$-cut of $\overline{D}$. Because each arc $e \in P_0 \setminus \{e_0\}$ is a dummy arc in $D'$, so $P_0 \cap C' = \emptyset$ and $C'$ is not a cut of $D$. It follows that the indicator vector $\mathbf{1}_{C'}$ of $C'$ satisfies that $\mathbf{1}_{C'} \in C^*(\overline{\Gamma}) \setminus C(\Gamma)$.

Choose a vector $y \in C^*(\overline{\Gamma}) \setminus C(\Gamma)$, then $y \in X(\Gamma) \setminus C(\Gamma)$. If $P$ is an $(s, t)$-path not containing $e_0$, then $y(P) \geq 1$ because $P$ is also an $(s, t)$-path in $\overline{D}$. Hence, any core allocation of $\Gamma$ can not dominate $y$ via $P$. If $P'$ is an $(s, t)$-path containing $e_0$, then by Proposition 2, $x(e_0) = 0$ for any core allocation $x \in C(\Gamma)$. It implies that $x$ also does not dominate $y$ via $P'$. Followed from Lemma 1, $C(\Gamma)$ is not stable.

*Case 2.* If $d_0 = 2$ and $e_0$ and $e_0'$ are both dummy arcs of Type II, then by Lemma 3 (2) and Case 1, $C(\Gamma)$ is also not stable.

*Case 3.* If $d_0 = 2$ and let $e_0$ and $e_0'$ are two dummy arcs. If one of $e_0$ and $e_0'$ is a Type I dummy arc, then followed by the definition of Type I dummy arc, the other dummy arc must be Type I either. Therein, dropping any one of them makes the other arc no longer being a dummy arc. Without loss of generality, consider the network $\overline{D} = D \setminus e_0$ and the associated flow game $\overline{\Gamma}$. It is easy to see that the minimum $(s, t)$-cut in $\overline{D}$ containing the arc $e_0'$ is not a $(s, t)$-cut in $D$, implying that $C(\Gamma) \subset C^*(\overline{\Gamma})$. Similar argument as that in Case 1 yields that $C(\Gamma)$ is not stable.

Let us assume that the necessity holds for all networks $D$ with $d_0 = k \geq 2$. Consider a network $D$ with $d_0 = k+1$. Let $e_0$ be a dummy arc in $D$. If $\overline{D} = D \setminus e_0$ contains no dummy arcs, then similar argument as that in Case 3 shows that $C(\Gamma)$ is not stable. If the network $\overline{D} = D \setminus e_0$ also contains dummy arc and the number of dummy arcs is no more than $k$. By applying the induction hypothesis on $\overline{D}$ and the result of Lemma 3, we conclude that $C(\Gamma)$ is not stable.  $\square$

## 4  Exactness, Extendability and Core Largeness

In this section, we prove that for the flow game defined on a simple network, the properties of exactness, extendability and core largeness are equivalent. Since flow games are totally balanced, this equivalence strictly implies the stability of the core of a flow game. Let $D = (V, E)$ be a simple network. An $(s, t)$-cut is called *minimal* in $D$ if none of its proper subsets is also an $(s, t)$-cut; an $(s, t)$-cut is called *minimum* if it has least cardinality among all $(s, t)$-cuts.

**Theorem 4.** *Let $D = (V, E)$ be a simple network with 'source' $s$ and 'sink' $t$, and $\Gamma = (E, v)$ be the corresponding flow game. Then the following statements are equivalent:*

(a) *The flow game $\Gamma$ is exact;*
(b) *The flow game $\Gamma$ is extendable;*
(c) *The core $C(\Gamma)$ is large;*
(d) *Every $(s, t)$-cut contains a minimum $(s, t)$-cut.*

Since flow games are totally balanced (Theorem 2), the implication "$(c) \Rightarrow (b) \Rightarrow (a)$" is true followed from Theorem 1. It remains to prove "$(a) \Rightarrow (d)$" and "$(d) \Rightarrow (c)$".

*Proof of "$(a) \Rightarrow (d)$".* Suppose that the flow game $\Gamma = (E, v)$ is exact. Let $S$ be an $(s, t)$-cut of $D$ and $\bar{S} = E \setminus S$. Then by the definition of the exactness, there exists $x \in C(\Gamma)$ such that $x(\bar{S}) = v(\bar{S}) = 0$, which implies that $x(S) = x(E) - x(\bar{S}) = v(E) - x(\bar{S}) = v(E)$. Denoting by $\mathcal{C}$ the set of minimum $(s, t)$-cuts of $D$. Followed from Theorem 2, $x$ can be expressed as

$$x = \sum_{C \in \mathcal{C}} \lambda_C \mathbf{1}_C, \tag{3}$$

where $\lambda_C \geq 0$ for every $C \in \mathcal{C}$ and $\sum_{C \in \mathcal{C}} \lambda_C = 1$. Then we have

$$v(E) = x(S) = \sum_{C \in \mathcal{C}} \lambda_C \mathbf{1}_C(S) = \sum_{C \in \mathcal{C}} \lambda_C |S \cap C| \leq \sum_{C \in \mathcal{C}} \lambda_C |C| = |C|. \tag{4}$$

Since $C$ is a minimum $(s, t)$-cut of $D$, $v(E) = |C|$. Hence, the equality holds throughout the expressions, meaning that $S \cap C = C$, for any $C \in \mathcal{C}$ with $\lambda_C > 0$. That is, $S$ contains at least a minimum $(s, t)$-cut of $D$. □

To show "$(d) \Rightarrow (c)$", we need some more facts. The first one is due to van Gellekom, Potters and Reijinierse [8]. For a profit game $\Gamma = (N, v)$ with $|N| = n$, the set of upper vectors is defined as:

$$L(\Gamma) = \{y \in R^n : y(S) \geq v(S), \forall S \subseteq N\}. \tag{5}$$

**Lemma 4.** [8] *Let $\Gamma = (N, v)$ be a balanced profit game. Then $\Gamma = (N, v)$ has the large core if and only if $y(N) \leq v(N)$ for all extreme points $y$ of $L(\Gamma)$.*

In order to characterize the extreme points of $L(\Gamma)$, We discuss a similar polyhedron for flow games. Let $\Gamma = (E, v)$ be a flow game corresponding to a simple network $D = (V, E)$ with $|E| = n$. Recall that $\mathcal{P}$ is the set of $(s, t)$-paths in $D$. Define

$$L'(\Gamma) = \{y \in R^n : y(P) \geq 1, \forall P \in \mathcal{P}; y(e) \geq 0, \forall e \in E\} \tag{6}$$

**Lemma 5.** *Let $\Gamma = (E, v)$ be the flow game defined on a simple network. Then it holds that $L(\Gamma) = L'(\Gamma)$.*

*Proof.* It is easy to see that $L(\Gamma) \subseteq L'(\Gamma)$. To show the other direction of inclusion $L(\Gamma) \supseteq L'(\Gamma)$, let $y \in L'(\Gamma)$. We have to check that $y(S) \geq v(S)$ for every $S \subseteq N$. Assume that $v(S) = k$ and let us construct a sub-network $G[S]$.

Then the maximum flow in this sub-network yields $k$ arc disjoint paths, namely, $P_1, P_2, \ldots, P_k$. Since $y \in L'(\Gamma)$, we have $y(P_i) \geq 1$ for each $i \in \{1, 2, \ldots k\}$. Therefore,

$$y(S) \geq \sum_{i=1}^{k} y(P_i) \geq k = v(S) \tag{7}$$

That is, $y \in L(\Gamma)$. This completes the proof.    □

Lemma 5 means that $L(\Gamma)$ has the same extreme points as $L'(\Gamma)$, so in order to give the proof of Theorem 4, it suffices to investigate the extreme points of $L'(\Gamma)$ for the flow game $\Gamma$.

**Lemma 6.** *Let $D = (V, E)$ be a simple flow network and $\Gamma = (E, v)$ be the corresponding flow game. Then each extreme point of $L'(\Gamma)$ is the indicator vector of a minimal $(s, t)$-cut of $D$.*

*Proof.* Let $y^*$ be an extreme point of $L'(\Gamma)$. Based on the theory of linear programming, there exists a non-negative integer function $c \colon E \to Z^+$, such that $y^*$ is the unique optimal solution of the following linear program:

$$\min \sum_{e \in E} c(e) y(e)$$
$$\text{s.t.} \begin{cases} y(P) \geq 1, & \forall P \in \mathcal{P}, \\ y(e) \geq 0, & \forall e \in E. \end{cases} \tag{8}$$

The dual program of (8) is:

$$\max \sum_{P \in \mathcal{P}} x(P)$$
$$\text{s.t.} \begin{cases} \sum_{e \in P \in \mathcal{P}} x(P) \leq c(e), & \forall e \in E, \\ x(P) \geq 0, & \forall P \in \mathcal{P}. \end{cases} \tag{9}$$

It is easy to see that the dual program (9) is the formulation of the max-flow problem in the network $D$ under the arc capacity function $c$. Because of the integrity of $c$, Ford-Fulkerson's theorem of max-flow and min-cut implies that the linear program (9), as well as the linear program (8), has integer optimal solution. Since $y^*$ is the unique optimal solution of (8), $y^*$ is integral. Furthermore, it is easy to see that $y^*$ must be a $\{0, 1\}$-vector.

Let $C^* = \{e \in E \colon y^*(e) = 1\}$. Since $y^*$ satisfies $y^*(P) \geq 1$ for any $P \in \mathcal{P}$, $C^*$ is an $(s, t)$-cut of $D$. On the other hand, assume that $C^*$ is not a minimal $(s, t)$-cut and the proper subset $C'$ of $C^*$ is an $(s, t)$-cut of $D$. Then it is obvious that the indicator vector $y'$ of $C'$ is also an optimal solution of linear program (8), which contradicts to the fact that $y^*$ is the unique optimal solution of (8). Therefore, $C^*$ is a minimal $(s, t)$-cut of $D$.    □

Combined with Lemma 5 and 6, we are able to show "$(d) \Rightarrow (c)$".

*Proof of* "$(d) \Rightarrow (c)$". Let $D$ be a simple network such that every $(s, t)$-cut contains a minimum $(s, t)$-cut of $D$. Choose an extreme point of $L(\Gamma)$. Followed from Lemma 6, this extreme point is the characteristic vector of some minimal $(s, t)$-cut of $D$, namely $C$. By our assumption, $C$ is certainly a minimum $(s, t)$-cut $C$ of $D$. Therefore, we have $\mathbf{1}_C(N) = |C| = v(E)$. According to Lemma 4, the core of the flow game is large. Thus, we complete the proof of Theorem 4. $\square$

# References

1. Bietenhader, T., Okamoto, Y.: Core Stability of Minimum Coloring Games. Proceedings of 30th International Workshop on Graph-Theoretic Concept in Computer Science. Lecture Notes in Computer Science **3353** (2004) 389–401
2. Biswas, A.K., Parthasarathy, T., Potters, J.A.M., Voorneveld, M.: Large Cores and Exactness. Games Econom. Behav. **28** (1999) 1–12
3. Bondareva, O.N.: Some Applications of Linear Programming Methods to the Theory of Cooperative Games(in Russian). Problemy Kibernetiki **10** (1963) 119–139
4. Deng, X., Ibaraki, T., Nagamochi, H.: Algorithmic Aspects of the Core of Combinatorial Optimization Games. Math. Oper. Res. **24** (1999) 751–766
5. Deng, X., Papadimitriou, C.H.: On the Complexity of Cooperative Solution Concepts. Math. Oper. Res. **19** (1994) 257–266
6. Fang, Q., Zhu, S., Cai, M., Deng, X.: Membership for Core of LP Games and Other Games. Lecture Notes in Computer Science **2108** (2001) 247–256 (COCOON 2001)
7. L.R. Ford, Jr.; Fulkerson, D.R.: Flows in Networks, Princeton University Press, Princeton, New Jersey (1962)
8. van Gellekom, J.R.G., Potters, J.A.M., Reijnierse, J.H.: Prosperity Properties of TU-Games. Internat. J. Game Theory **28** (1999) 211–227
9. Kalai, E., Zemel, E.: Totally Balanced Games and Games of Flow. Math. Oper. Res. **7** (1982) 476–478
10. Kalai, E., Zemel, E.: Generalized Network Problems Yielding Totally Balanced Games. Operations Research **30** (1982) 498–1008
11. Kikuta, K., Shapley, L.S.: Core Stability in $n$-Person Games. Manuscript (1986)
12. Lucas, W.F.: A Game with no Solution. Bull. Amer. Math. Soc. **74** (1968) 237–239
13. von Neumann, J., Morgenstern, O.: Theory of Games and Economic Behaviour. Princeton Univeristy Press, Princeton (1944)
14. Schrijver, A.: Combinatorial Optimization: Polyhedra and Efficiency. Springer-Verlag, Berlin Heidelberg (2003)
15. Shapley, L.S.: On Balanced Sets and Cores. Naval Research Logistics Quarterly **14** (1967) 453–460
16. Sharkey, W.W.: Cooperative Games with Large Cores. Internat. J. Game Theory **11** (1982) 175–182
17. Solymosi, T., Raghavan, T.E.S: Assignment Games with Stable Cores. Internat. J. Game Theory **30** (2001) 177–185

# The (Adjacent) Vertex-Distinguishing Total Coloring of the Mycielski Graphs and the Cartesian Product Graphs*

Yanli Sun[1] and Lei Sun[2]

[1] College of Information and Engineering, Taishan Medical University, Taian 271016, P.R. China
sunyanli1999@126.com
[2] Department of Mathematics, Shandong Normal University, Jinan 250014, P.R. China
leisun@beelink.com

**Abstract.** The graph obtained by the famous Mycielski's construction is called the Mycielski graph. This paper focuses on the relation between the basic graphs and two classes of constructed graphs on the (adjacent) vertex-distinguishing total coloring. And some sufficient conditions with which the Mycielski graphs and the Cartesian product graphs satisfy the adjacent vertex-distinguishing total coloring conjecture are given.

**AMS Subject Classifications:** 05C15.

## 1 Introduction

The coloring problem of of graph is one of the primary fields of the graph theory. With the development of the field, some scholars presented a few coloring problem with different restrictions. The vertex-distinguishing total coloring and the adjacent vertex-distinguishing total coloring were presented by Zhang [9].

All graphs concerned in this paper are simple, finite and undirected. We denote the set of vertices and edges of graph $G$ by $V(G)$ and $E(G)$, respectively. Let $\Delta(G)$, $\delta(G)$ respectively denote the maximum degree and the minimum degree of $G$.

The Mycielski's construction [6] and graph product [7,8] are two important topics in graph theory.

The other terminologies we refer to references [1,2,3,4,5].

**Definition 1.** *The Mycielski graphs are obtained by the Mycielski's construction. Given a graph $G$ with $V(G) = \{v_1, v_2, \ldots, v_n\}$, the Mycielski graph of $G$ is denoted by $M(G)$ with $V(M(G)) = V(G) \cup \{v'_1, v'_2, \ldots, v'_n, \omega\}$ and $E(M(G)) = E(G) \cup \{v_i v'_j \mid v_i v_j \in E(G),\ i, j = 1, 2, \ldots, n\} \cup \{v'_i \omega \mid i = 1, 2, \ldots, n\}$.*

---

* Supported by the National Natural Science Council(10471078).

J. Akiyama et al. (Eds.): CJCDGCGT 2005, LNCS 4381, pp. 200–205, 2007.
© Springer-Verlag Berlin Heidelberg 2007

**Definition 2.** *For graphs $G_1 = (V_1, E_1)$ and $G_2 = (V_2, E_2)$, the Cartesian product $G_1 \times G_2$ has vertex set $V_1 \times V_2$, and in which $(v_1, u_1)$ is adjacent to $(v_2, u_2)$ if and only if either $v_1v_2 \in E_1$ and $u_1 = u_2 \in V_2$ or $v_1 = v_2 \in V_1$ and $u_1u_2 \in E_2$. Clearly, $G_1 \times G_2 \cong G_2 \times G_1$ and $\Delta(G_1 \times G_2) = \Delta(G_1) + \Delta(G_2)$.*

**Definition 3.** *Suppose $G(V(G), E(G))$ is a graph, $k$ is a positive integer and $S$ is a $k$-element set. Let $f$ be a map from $V(G) \cup E(G)$ to $S$. If*

1) *for any $uv, vw \in E(G)$, $u \neq w$, there is $f(uv) \neq f(vw)$;*
2) *for any $uv \in E(G)$, there are $f(u) \neq f(v)$, $f(u) \neq f(uv)$, $f(uv) \neq f(v)$;*
3) *for any $u, v \in V(G)$, there is $C(u) \neq C(v)$; where $C(u) = \{f(u)\} \cup \{f(uv)| \\ uv \in E(G)\}$ is called the color set of vertex $u$,*

*then $f$ is called a vertex-distinguishing $k$-total coloring of graph $G$ (briefly noted as $k$-VDTC) and $\chi_{vt}(G) = \min\{k \mid G \text{ has } k\text{-VDTC}\}$ is called vertex-distinguishing total chromatic number of graph $G$.*

*If condition 3) is changed to*

**3')** *for any $uv \in E(G)$, there is $C(u) \neq C(v)$,*

*then $f$ is called a adjacent vertex-distinguishing $k$-total coloring of graph $G$ (in brief, noted as $k$-AVDTC) and $\chi_{at}(G) = \min\{k \mid G \text{ has } k\text{-AVDTC}\}$ is called the adjacent vertex-distinguishing total chromatic number of graph $G$.*

It is trivial that for any graph $G$, $\chi_{vt}(G) \geq \chi_{at}(G) \geq \Delta(G) + 1$. Furthermore, if $G$ has two adjacent vertices of maximum degree, then $\chi_{at}(G) \geq \Delta(G) + 2$. Zhang proposed the following conjectures in [9]:

*Conjecture 1. For any simple graph $G$, $\chi_{at}(G) \leq \Delta(G) + 3$.*

*Conjecture 2. For any simple graph $G$, $k_T(G) \leq \chi_{vt}(G) \leq k_T(G) + 1$, where $k_T(G) = \min\left\{l \mid C_l^{d+1} \geq n_d, \delta \leq d \leq \Delta, n_d \text{ is the number of the vertices with degree } d \text{ in graph } G\right\}$.*

The conjectures have been verified for a few classes of special graphs and $\chi_{at}(G)$, $\chi_{vt}(G)$ are given for these graphs, such as cycle, complete graph, complete bipartite graph, star, fan, wheel, tree and their joint graph [9,10,11]. The Mycielski graphs of these special graphs are only studied on the adjacent vertex-distinguishing total coloring. In this paper, we will study the Mycielski graph and the Cartesian product graphs of any graphs on the adjacent vertex-distinguishing total coloring and the vertex-distinguishing total coloring.

## 2   Main Results

**Theorem 1.** *Suppose graph $G$ has $n$ vertices with $\delta(G) \geq 2$. If $\chi_{vt}(G) \geq n$, then $\chi_{vt}(M(G)) \leq \chi_{vt}(G) + \Delta(G)$; if $\chi_{vt}(G) < n$, then $\chi_{vt}(M(G)) \leq n + \Delta(G)$.*

*Proof.* $1^0$ $\chi_{vt}(G) \geq n$.

Let $f'$ be a $\chi_{vt}(G)$-VDTC with $\chi_{vt}(G)$-element set $S = \{1, 2, \ldots, \chi_{vt}(G)\}$. We construct a $\chi_{vt}(G) + \Delta(G)$-VDTC $f$ of $M(G)$:

**1)** for any $x \in V(G) \cup E(G)$, let $f(x) = f'(x)$.

**2)** for edges between $V(G)$ and $V' = \{v'_1, v'_2, \ldots, v'_n\}$, we color with $\Delta(G)$-element set $S'(S' \cap S = \emptyset)$. In fact, for bipartite graph $G(V(G), V')$ with edge-set $\{v_i v'_j \mid v_i v_j \in E(G), i, j = 1, 2, \ldots, n\}$, the number of edge-coloring $\chi'(G(V(G), V')) = \Delta(G(V(G), V')) = \Delta(G)$, thus there is a proper $\Delta(G)$-edge coloring.

**3)** let $f(v'_i) = f'(v_i)$ $i = 1, 2, \ldots, n$.

**4)** suppose $\{f(v'_i) \mid i = 1, 2, \ldots, n\} = \{1, 2, \ldots, l\} \subseteq S(l \le \chi_{vt}(G))$. We partition the vertices of $V'$ by their colors, let $W_j$ denote the vertex subset with color $j$ in $V'$, and $E_j = \{v'_i \omega \mid v'_i \in W_j\}$, $n_j = |W_j| = |E_j|$, $j = 1, 2, \ldots, l$. Clearly, $n_1 + n_2 + \cdots + n_l = n$, we color the edges in $E_1, E_2, \ldots, E_l$ successively:

the edges in $E_1$ are colored with $\{2, 3, \ldots, n_1 + 1\}$; the edges in $E_2$ are colored with $\{n_1 + 2, n_1 + 3, \cdots, n_1 + n_2 + 1\}; \cdots$; the edges in $E_{l-1}$ are colored with $\{n_1 + n_2 + \cdots + n_{l-2} + 2, \ldots, n_1 + n_2 + \cdots + n_{l-1} + 1\}$; if $n_l = 1$, then we color the edge in $E_l$ with color 1, otherwise we color the edges with $\{n_1 + n_2 + \cdots + n_{l-1} + 2, \ldots, n_1 + n_2 + \cdots + n_l, 1\}$.

**5)** let $f(\omega) \in S'$.

Now we verify $f$ is proper: for any vertex $u$, $C'(u) = \{f'(u)\} \cup \{f'(uv) \mid uv \in E(G)\}$, $C(u) = \{f(u)\} \cup \{f(uv) \mid uv \in E(M(G))\}$.

**1)** For any $v_i, v_j \in V(G)$, $C'(v_i) \neq C'(v_j)$ and there are only some more $S'$'s elements in $C(v_i)$ than in $C'(v_i)$, thus $C(v_i) \neq C(v_j)$.

**2)** $\delta(G) \ge 2$, so there are at least three $S$'s elements and two $S'$'s elements in $C(v_i)$, there are just two $S$'s elements and at least two $S'$'s elements in $C(v'_j)$, and $C(\omega)$ just has one $S'$'s element, thus $C(v_i) \neq C(v'_j) \neq C(\omega)$, $C(v_i) \neq C(\omega)$, $i, j = 1, 2, \ldots, n$.

**3)** For any $v'_i, v'_j \in V'$, we only prove the colors which don't belong to $S'$ in the color-set of vertices are distinguishing, then the color-set of vertices are distinguishing too. If $v'_i, v'_j \in V'$ are all belong to $W_t$, then $f(v'_i) = f(v'_j)$ but $f(v'_i \omega) \neq f(v'_j \omega)$, so it is easy to see they are distinguishing. Otherwise suppose they are belong to $W_t, W_r (t < r)$ respectively:

   **a)** If $1 = t < r = l$ and $n_1 \ge l - 1$, then the color-set of the vertex whose incident edge is colored $l$ may be same to the color-set of the vertex whose incident edge is colored 1. Since $1 = t < r = l$, $l \neq 1$, and it is trivial for $l = 2$, when $l \ge 3$, we change the color of the edge colored $l$ with the edge colored $n_1 + 2$.

   **b)** Otherwise, $f(v'_i) \neq f(v'_j) \neq f(v'_i \omega)$, that is to say the colors which belong to $S$ in the color-set of vertices are distinguishing, then $C(v'_i) \neq C(v'_j)$.

$2^0$ $\chi_{vt}(G) < n$, the proof is same to $1^0$ except we construct a $(n + \Delta(G))$-VDTC $f$ of $M(G)$ and $S' \cap \{1, 2, \ldots, n\} = \emptyset$. $\qquad\square$

**Theorem 2.** *Suppose graph $G$ has $n$ vertices with $\delta(G) \ge 2$ and $k$ is a positive integer. If $\chi_{vt}(G) \le \Delta(G) + k$ and $\Delta(G) + k \ge n$, then $\chi_{vt}(M(G)) \le 2\Delta(G) + k \le \Delta(M(G)) + k$.*

*Proof.* Let $f'$ be a $(\Delta(G)+k)$-VDTC with color-set $S = \{1, 2, \ldots, \Delta(G)+k\}$. We construct a $(2\Delta(G)+k)$-VDTC $f$ of $M(G)$, and the methods of construction and verification are same to Theorem 1's $1^0$. Clearly $\Delta(M(G)) = \max\{2\Delta(G), n\}$, thus $\chi_{vt}(M(G)) \leq 2\Delta(G) + k \leq \Delta(M(G)) + k$. □

**Corollary 1.** *Suppose $G$ is a graph with $n$ vertices and $k$ is a positive integers. If $\chi_{at}(G) \leq \Delta(G) + k$ and $\Delta(G) + k \geq n$, then $\chi_{at}(M(G)) \leq 2\Delta(G) + k \leq \Delta(M(G)) + k$.*

**Corollary 2.** *Suppose graph $G$ satisfies the conditions of Corollary 1. If $\chi_{at} \leq \Delta(G) + 3$, then $\chi_{at}(M(G)) \leq \Delta(M(G)) + 3$. Furthermore, if $k = 1$, then $\chi_{vt}(M(G)) = \Delta(M(G)) + 1$ and $\chi_{at}(M(G)) = \Delta(M(G)) + 1$.*

We focus on the Cartesian product graph in the following.

**Theorem 3.** *Suppose $G$, $H$ are two graphs, and $k > 1$ is a positive integer. If $\Delta(G) + 2 \leq \chi_{at}(G) \leq \Delta(G) + k$, $\chi'(H) = \Delta(H)$ and $\chi(H) \leq \Delta(G) + k$, then $\chi_{at}(G \times H) \leq \Delta(G \times H) + k$.*

*Proof.* Suppose $\chi(H) = m$, $V(H) = \bigcup_{i=0}^{m-1} U_i$, where $U_i$ are the vertex-subset partitioned with the color.

$V(G \times H) = \bigcup_{i=0}^{m-1}\{(x, v) \mid x \in V(G), v \in U_i\}$, $E(G \times H) = (\bigcup_{i=0}^{m-1}\{(x, v)(y, v) \mid xy \in E(G), v \in U_i\}) \cup \{(x, v)(x, u) \mid x \in V(G), vu \in E(H)\}$.

Suppose mapping $f_1\colon V(G) \cup E(G) \longrightarrow \{0, 1, \ldots, \Delta(G)+k-1\}$ is a $(\Delta(G)+k)$-AVDTC of $G$, and mapping $f_2\colon E(H) \longrightarrow \{1, 2, \ldots, \Delta(H)\}$ is a $\Delta(H)$-edge coloring of $H$. $C_1(x)$, $C_2(v)$ respectively denote the color-sets of $x \in V(G)$ and $v \in V(H)$ induced by $f_1$ and $f_2$, i.e., $C_1(x) = \{f_1(x)\} \cup \{f_1(xy) \mid xy \in E(G)\}$ and $C_2(v) = \{f_2(vu) \mid vu \in E(H)\}$. "$\oplus$" denotes the addition under mod $(\Delta(G)+k)$. With $f_1$, $f_2$, we construct a $(\Delta(G \times H)+k)$-AVDTC $f\colon V(G \times H) \cup E(G \times H) \longrightarrow \{0, 1, \ldots, \Delta(G) + \Delta(H) + k - 1\}$:

1) $f((x, v)) = f_1(x) \oplus i$, for any $x \in V(G)$, $v \in U_i$, $i = 0, 1, \ldots, m - 1$.

2) $f((x, v)(y, u)) = \begin{cases} f_1(xy) \oplus i, & xy \in E(G), v = u \in U_i, \\ & i = 0, 1, \ldots, m - 1, \\ \Delta(G) + k - 1 + f_2(vu), & vu \in E(H), x = y \in V(G). \end{cases}$

Because $\Delta(G) + k \geq \chi(H) = m$, thus $t, t \oplus 1, \ldots, t \oplus (m - 1)$ are different one another for any $0 \leq t \leq \Delta(G) + k - 1$, so it is easy to see $f$ is a proper $(\Delta(G \times H) + k)$-total coloring of $G \times H$.

Now we verify the adjacent vertices are distinguishing. If $(x, v)(y, u) \in E(G \times H)$, then either $xy \in E(G)$ and $v = u \in V(H)$ or $vu \in E(H)$ and $x = y \in V(G)$:

(1) $xy \in E(G)$, $v = u \in V(H)$.

Suppose $v = u \in U_i$, then $C((x, v)) = \{i \oplus t \mid t \in C_1(x)\} \cup \{t + \Delta(G) + k - 1 \mid t \in C_2(v)\}$ and $C((y, u)) = \{i \oplus t \mid t \in C_1(y)\} \cup \{t + \Delta(G) + k - 1 \mid t \in C_2(u)\}$. Because $C_1(x) \neq C_1(y)$ and $v = u$, thus $C((x, v)) \neq C((y, u))$.

(2) $vu \in E(H)$, $x = y \in V(G)$.

Suppose $v \in U_i$, $u \in U_j$ $(i < j)$, then $C((x, v)) = \{i \oplus t \mid t \in C_1(x)\} \cup \{t + \Delta(G)+k-1 \mid t \in C_2(v)\}$ and $C((y, u)) = \{j \oplus t \mid t \in C_1(y)\} \cup \{t+\Delta(G)+k-1 \mid t \in C_2(u)\}$. Because $C_1(x) = C_1(y)$, $i < j$ and $|C_1(x)| = |C_1(y)| \leq \Delta(G) + 1 < \Delta(G) + k$, thus $\{i \oplus t \mid t \in C_1(x)\} \neq \{j \oplus t \mid t \in C_1(y)\}$, so $C((x, v)) \neq C((y, u))$. $\square$

**Theorem 4.** *If* $\chi_{at}(G) \leq \Delta(G) + 3$, $\chi'(H) = \Delta(H)$ *and* $\chi(H) \leq \chi_{at}(G)$, *then* $\chi_{at}(G \times H) \leq \Delta(G \times H) + 3$.

*Proof.* (1) When $\chi_{at}(G) = \Delta(G) + 2, \Delta(G) + 3$, it can be obtained by Theorem 3.

(2) When $\chi_{at}(G) = \Delta(G)+1$, the proof is similar to Theorem 3 except for a little change: let $f_1 \colon V(G) \cup E(G) \longrightarrow \{0, 1, \ldots, \Delta(G)+1\}$ be a $(\Delta(G)+2)$-AVDTC of $G$; "$\oplus$" denotes the addition under $\mathrm{mod}\,(\Delta(G)+2)$. $\square$

The following result can be obtained by extending Theorem 4.

**Theorem 5.** *Suppose* $G$, $H$ *are two graphs, and* $k > 1$ *is a positive integer. If* $\Delta(G) + 2 \leq \chi_{at}(G) \leq \Delta(G) + k$ *and* $\chi(H) \leq \Delta(G) + k$, *then* $\chi_{at}(G \times H) \leq \Delta(G \times H) + k + 1$.

*Proof.* The proof is similar to Theorem 3 except for a little change: Let $f_2 \colon E(H) \longrightarrow \{1, 2, \ldots, \Delta(H) + 1\}$ be a $(\Delta(H) + 1)$-edge coloring of $H$. $\square$

**Corollary 3.** *Suppose* $G$, $H$ *are two graphs, and* $k > 1$ *is a positive integer. If* $\chi_{at}(G) \leq \Delta(G) + 2$ *and* $\chi(H) \leq \Delta(G)$, *then* $\chi_{at}(G \times H) \leq \Delta(G \times H) + 3$.

*Proof.* (1) When $\chi_{at}(G) = \Delta(G) + 2$, $\chi_{at}(G \times H) \leq \Delta(G \times H) + 3$ is obtained by Theorem 5.

(2) When $\chi_{at}(G) = \Delta(G)+1$, the proof is similar to Theorem 3 except for a little change. let $f_1 \colon V(G) \cup E(G) \longrightarrow \{0, 1, \ldots, \Delta(G)+1\}$ be a $(\Delta(G)+2)$-AVDTC of $G$ and $f_2 \colon E(H) \longrightarrow \{1, 2, \ldots, \Delta(H) + 1\}$ be a $(\Delta(H) + 1)$-edge coloring of $H$; "$\oplus$" denotes the addition under $\mathrm{mod}\,(\Delta(G)+2)$. $\square$

## References

1. Bollobas, B.: Modern Graph Theory. Springer-Verlag New York Inc(1998)
2. Chartrand, G., Lesniak-Foster, L.: Graph and Digraphs. Ind. Edition, Wadsworth Brooks/Cole, Monterey CA (1986)
3. Reinhard, D.: Graph Theory. Springer Verlag New York, Inc (1997)
4. Hansen, P., Marcotte, O.: Graph Coloring and Application. AMS providence, Rhode Island USA (1999)
5. Bondy, J.A., Murty, U.S.R.: Graph Theory with Applications. North-Holland (1976)
6. Mycielski, J.: Sur Le Coloring des Graphes. Colloq. Math. **3** (1955) 161–162
7. Čižek, N., Klavžar, S.: On the Chromatic Number of the Lexicographic Product and the Cartesian Sum of Graphs. Discrete Math. **134** (1994) 17–24

8. Klavžar, S.: Coloring Graph Products—A Survey. Diserete Math. **155** (1996) 135–145
9. Zhang, Z.F., Li, J.W., Chen, J.W. et al: On The Vertex-Distinguish Total Coloring of Graphs. Science Report of Northwest Normal University, **168** (2003)
10. Zhang, Z.F., Liu, L.Z., Wang, J.F.: Adjacent Strong Edge Coloring of Graphs. Applied Math. Letters **15** (2002) 623–626
11. Zhang, Z.F., Li, J.W., Chen, X.E. et al: $D(\beta)$-Vertex-distinguishing Total Coloring of Graphs. Science in China (to appear)

# Three Classes of Bipartite Integral Graphs*

Ligong Wang** and Hao Sun

Department of Applied Mathematics, School of Science, Northwestern Polytechnical
University, Xi'an, Shaanxi 710072, P.R. China
ligongwangnpu@yahoo.com.cn

**Abstract.** A graph $G$ is called integral if all zeros of its characteristic polynomial $P(G, x)$ are integers. In this paper, the bipartite graphs $K_{p,q(t)}$, $K_{p(s),q(t)}$ and $K_{p,q} \equiv K_{q,r}$ are defined. We shall derive their characteristic polynomials from matrix theory. We also obtain their sufficient and necessary conditions for the three classes of graphs to be integral. These results generalize some results of Balińska et al. The discovery of these integral graphs is a new contribution to the search of integral graphs.

**Keywords:** Integral graph, characteristic polynomial, graph spectrum.

**AMS Subject Classifications:** 05C50.

## 1 Introduction

We use $G$ to denote a simple graph with vertex set $V(G) = \{v_1, v_2, \ldots, v_n\}$ and edge set $E(G)$. The adjacency matrix $A = A(G) = [a_{ij}]$ of $G$ is an $n \times n$ symmetric matrix of 0's and 1's with $a_{ij} = 1$ if and only if $v_i$ and $v_j$ are joined by an edge. The characteristic polynomial of $G$ is the polynomial $P(G) = P(G, x) = det(xI_n - A)$, where and in the sequel $I_n$ always denotes the $n \times n$ identity matrix. The spectrum of $A(G)$ is also called the *spectrum* of $G$.

The notion of integral graphs was first introduced by Harary and Schwenk in 1974 [12]. A graph $G$ is called *integral* if all zeros of its characteristic polynomial $P(G, x)$ are integers. In general, the problem of characterizing integral graphs seems to be very difficult. Thus, it makes sense to restrict our investigations to some interesting families of graphs, for instance, cubic graphs [5, 21], complete $r$-partite graphs [20, 24], graphs with maximum degree 4 [3, 4], 4-regular integral graphs [10], integral graphs which belong to the class $\overline{\alpha K_{a,b}}$ or $\overline{\alpha K_a \cup \beta K_{b,b}}$ [18, 19], etc. Trees present another important family of graphs for which the problem has been considered in [1, 6, 7, 12, 13, 14, 15, 16, 17, 22, 25, 26, 27, 28]. Some graph operations, which when applied on integral graphs produce again integral graphs, are described in [12] or [9]. Other results on integral graphs can be found in [1, 9, 16]. For all other facts or terminology on graph spectra, see [9].

A complete bipartite graph $K_{p,q}$ is a graph with vertex classes $V_1$ and $V_2$ if $V = V_1 \cup V_2$, $V_1 \cap V_2 = \emptyset$, where $V_i$ are nonempty disjoint sets, $|V_1| = p$ and

---

* Supported by NSFC (No.70571065) and DPOP in NPU.
** Corresponding author.

J. Akiyama et al. (Eds.): CJCDGCGT 2005, LNCS 4381, pp. 206–215, 2007.
© Springer-Verlag Berlin Heidelberg 2007

$|V_2| = q$, such that two vertices in $V$ are adjacent if and only if they belong to different classes. The graph $K_{p,q(t)}$ on $p + q(t + 1)$ vertices is obtained by attaching t new endpoints to each vertex in the vertex set $V_2$ of the graph $K_{p,q}$. The graph $K_{p(s),q(t)}$ on $p(s + 1) + q(t + 1)$ vertices is obtained by attaching $s$ and $t$ new endpoints to each vertex in the vertex sets $V_1$ and $V_2$ of the graph $K_{p,q}$, respectively. The graph $K_{p,q} \equiv K_{q,r}$ on $p + 2q + r$ vertices is obtained by adding the edges $\{v_i w_i | i = 1, 2, \ldots, q\}$ from two disjoint graphs $K_{p,q}$ with vertex classes $V_1 = \{u_i | i = 1, 2, \ldots, p\}$, $V_2 = \{v_i | i = 1, 2, \ldots, q\}$ and $K_{q,r}$ with vertex classes $U_1 = \{w_i | i = 1, 2, \ldots, q\}$, $U_2 = \{z_i | i = 1, 2, \ldots, r\}$, respectively. Let the tree $T(q, t)$ of diameter 4 be formed by joining the centers of $q$ copies of $K_{1,t}$ to a new vertex $v$. Let the tree $K_{1,s} \bullet T(q, t)$ of diameter 4 be obtained by identifying the center $w$ of $K_{1,s}$ and the center $v$ of $T(q, t)$. Let the tree $T[s, t]$ of diameter 3 be formed by joining the centers of $K_{1,s}$ and $K_{1,t}$ with a new edge. The graph $G^t$ be obtained by attaching $t$ new endpoints to each vertex of the graph $G$. In [2], Balińska et al proved the graph $K_{n+2,n(1)}$ is integral. In [6, 22, 25, 26, 27], the authors gave some sufficient or necessary conditions for the graphs $K_{1(s),1(t)} = T[s, t]$, $K_{p(t),q(t)} = K_{p,q}^t$, $K_{1(s),q(t)} = K_{1,s} \bullet T(q, t)$ and $K_{p(s),1(t)} = K_{1,t} \bullet T(p, s)$ to be integral. In [23], we proved the graph $K_{n,n+1} \equiv K_{n+1,n}$ is integral. In this paper, we shall derive characteristic polynomials of the graphs $K_{p,q(t)}$, $K_{p(s),q(t)}$ and $K_{p,q} \equiv K_{q,r}$ from matrix theory. We also obtain their sufficient and necessary conditions for the three classes of graphs to be integral. These results generalize some results of [2, 6, 22, 23, 25, 26, 27]. The discovery of these integral graphs is a new contribution to the search of integral graphs.

## 2   The Characteristic Polynomials of Three Classes of Graphs

In this section, we shall determine the characteristic polynomials of the graphs $K_{p,q(t)}$, $K_{p(s),q(t)}$ and $K_{p,q} \equiv K_{q,r}$ from matrix theory.
    Next we shall give some notations on matrices.

**(1)** $R$ denotes the set of real numbers.
**(2)** $R^{m \times n}$ denotes the set of $m \times n$ matrices whose entries are $R$.
**(3)** $A^T$ denotes the transpose of the matrix $A$.
**(4)** $J_{m \times n}$ and $0_{m \times n}$ denote the $m \times n$ matrix with all entries equal to 1 and the $m \times n$ matrix with all entries equal to 0, respectively.
**(5)** Denote $J_n = J_{n \times n}$ and $0_n = 0_{n \times n}$.

**Theorem 1.** $P\left(K_{p,q(t)}, x\right) = x^{q(t-1)+p} \left(x^2 - t\right)^{q-1} \left(x^2 - pq - t\right)$.

*Proof.* By properly ordering the vertices of the graph $K_{p,q(t)}$, the adjacency matrix $A = A(K_{p,q(t)})$ of $K_{p,q(t)}$ can be written as the $(p+q(t+1)) \times (p+q(t+1))$ matrix such that

$$A = A\left(K_{p,\,q(t)}\right) = \begin{bmatrix} 0_t & 0_t & \dots 0_t & 0_{t\times p} & B_1 \\ 0_t & 0_t & \dots 0_t & 0_{t\times p} & B_2 \\ \dots & \dots & \ddots \dots & \dots & \dots \\ 0_t & 0_t & \dots 0_t & 0_{t\times p} & B_q \\ 0_{p\times t} & 0_{p\times t} & \dots 0_{p\times t} & 0_p & J_{p\times q} \\ B_1^T & B_2^T & \dots B_q^T & J_{q\times p} & 0_q \end{bmatrix},$$

where $B_k = \left[a_{ij}^{(k)}\right] = \begin{cases} 1, \text{ if } j = k, \\ 0, \text{ otherwise.} \end{cases}$, $B_k \in R^{t\times q}$, for $k = 1, 2, \dots, q$. Then we have

$$P\left(K_{p,\,q(t)},\,x\right) = \left|xI_{p+q(t+1)} - A\left(K_{p,\,q(t)}\right)\right|$$

$$= \begin{vmatrix} xI_t & 0_t & \dots & 0_t & 0_{t\times p} & -B_1 \\ 0_t & xI_t & \dots & 0_{t\times t} & 0_{t\times p} & -B_2 \\ \dots & \dots & \ddots & \dots & \dots & \dots \\ 0_t & 0_t & \dots & xI_t & 0_{t\times p} & -B_q \\ 0_{p\times t} & 0_{p\times t} & \cdots & 0_{p\times t} & xI_p & -J_{p\times q} \\ -B_1^T & -B_2^T & \dots & -B_q^T & -J_{q\times p} & xI_q \end{vmatrix}.$$

By careful calculation, we can obtain that the characteristic polynomial of $K_{p,\,q(t)}$ is

$$P\left(K_{p,\,q(t)},\,x\right) = x^{q(t-1)+p}\left(x^2 - t\right)^{q-1}\left(x^2 - pq - t\right).$$

Thus this theorem is proved.    □

**Theorem 2.** $P\left(K_{p(s),\,q(t)},\,x\right) = x^{p(s-1)+q(t-1)}\left(x^2 - s\right)^{p-1}\left(x^2 - t\right)^{q-1}\left[x^4 - (pq + s + t)x^2 + st\right]$.

*Proof.* By properly ordering the vertices of the graph $K_{p(s),\,q(t)}$, the adjacency matrix $A = A\left(K_{p(s),\,q(t)}\right)$ of $K_{p(s),\,q(t)}$ can be written as the $(p(s+1) + q(t+1)) \times (p(s+1) + q(t+1))$ matrix such that

$$A = A(K_{p(s),\,q(t)})$$

$$= \begin{bmatrix} 0_s & 0_s & \dots 0_s & 0_{s\times t} & 0_{s\times t} & \dots 0_{s\times t} & B_1 & 0_{s\times q} \\ 0_s & 0_s & \dots 0_s & 0_{s\times t} & 0_{s\times t} & \dots 0_{s\times t} & B_2 & 0_{s\times q} \\ \dots & \dots & \dots\dots & \dots & \dots & \dots\dots & \dots & \dots \\ 0_s & 0_s & \dots 0_s & 0_{s\times t} & 0_{s\times t} & \dots 0_{s\times t} & B_p & 0_{s\times q} \\ 0_{t\times s} & 0_{t\times s} & \dots 0_{t\times s} & 0_t & 0_t & \dots 0_t & 0_{t\times p} & C_1 \\ 0_{t\times s} & 0_{t\times s} & \dots 0_{t\times s} & 0_t & 0_t & \dots 0_t & 0_{t\times p} & C_2 \\ \dots & \dots & \ddots\dots & \dots & \dots & \ddots\dots & \dots & \dots \\ 0_{t\times s} & 0_{t\times s} & \dots 0_{t\times s} & 0_t & 0_t & \dots 0_t & 0_{t\times p} & C_q \\ B_1^T & B_2^T & \dots B_p^T & 0_{p\times t} & 0_{p\times t} & \dots 0_{p\times t} & 0_p & J_{p\times q} \\ 0_{q\times s} & 0_{q\times s} & \dots 0_{q\times s} & C_1^T & C_2^T & \dots C_q^T & J_{q\times p} & 0_q \end{bmatrix},$$

where

$$B_k = \left[a_{ij}^{(k)}\right] = \begin{cases} 1, & \text{if } j = k, \\ 0, & \text{otherwise.} \end{cases}, \ B_k \in R^{s \times p}, \text{ for } k = 1, 2, \ldots, p,$$

and

$$C_m = \left[b_{ij}^{(m)}\right] = \begin{cases} 1, & \text{if } j = m, \\ 0, & \text{otherwise.} \end{cases}, \ C_m \in R^{t \times q}, \text{ for } m = 1, 2, \ldots, q.$$

Then we have

$$P\left(K_{p(s),q(t)}, x\right) = \left| x I_{p(s+1)+q(t+1)} - A\left(K_{p(s),q(t)}\right) \right|$$

$$= \begin{vmatrix} xI_s & 0_s & \ldots 0_s & 0_{s \times t} & 0_{s \times t} & \ldots 0_{s \times t} & -B_1 & 0_{s \times q} \\ 0_s & xI_s & \ldots 0_s & 0_{s \times t} & 0_{s \times t} & \ldots 0_{s \times t} & -B_2 & 0_{s \times q} \\ \ldots & \ldots & \ldots & \ldots & \ldots & \ldots & \ldots & \ldots \\ 0_s & 0_s & \ldots xI_s & 0_{s \times t} & 0_{s \times t} & \ldots 0_{s \times t} & -B_p & 0_{s \times q} \\ 0_{t \times s} & 0_{t \times s} & \ldots 0_{t \times s} & xI_t & 0_t & \ldots 0_t & 0_{t \times p} & -C_1 \\ 0_{t \times s} & 0_{t \times s} & \ldots 0_{t \times s} & 0_t & xI_t & \ldots 0_t & 0_{t \times p} & -C_2 \\ \ldots & \ldots & \ddots & \ldots & \ldots & \ddots & \ldots & \ldots \\ 0_{t \times s} & 0_{t \times s} & \ldots 0_{t \times s} & 0_t & 0_t & \ldots xI_t & 0_{t \times p} & -C_q \\ -B_1^T & -B_2^T & \ldots -B_p^T & 0_{p \times t} & 0_{p \times t} & \ldots 0_{p \times t} & xI_p & -J_{p \times q} \\ 0_{q \times s} & 0_{q \times s} & \ldots 0_{q \times s} & -C_1^T & -C_2^T & \ldots -C_q^T & -J_{q \times p} & xI_q \end{vmatrix}.$$

By careful calculation, we can obtain that the characteristic polynomial of $K_{p(s),q(t)}$ is

$$P\left(K_{p(s),q(t)}, x\right) = x^{p(s-1)+q(t-1)} \left(x^2 - s\right)^{p-1} \left(x^2 - t\right)^{q-1}$$

$$\left[x^4 - (pq + s + t)x^2 + st\right].$$

Thus this theorem is proved.                                                      □

**Theorem 3**

$$P\left(K_{p,q} \equiv K_{q,r}, x\right) = x^{p+r-2}(x+1)^{q-1}(x-1)^{q-1}$$

$$\left[x^4 - (pq + qr + 1)x^2 + pq^2r\right].$$

*Proof* By properly ordering the vertices of the graph $K_{p,q} \equiv K_{q,r}$, the adjacency matrix $A = A\left(K_{p,q} \equiv K_{q,r}\right)$ of $K_{p,q} \equiv K_{q,r}$ can be written as the $(p+2q+r) \times (p+2q+r)$ matrix such that

$$A = A\left(K_{p,q} \equiv K_{q,r}\right) = \begin{bmatrix} 0_p & J_{p \times q} & 0_{p \times r} & 0_{p \times q} \\ J_{q \times p} & 0_q & 0_{q \times r} & I_q \\ 0_{r \times p} & 0_{r \times q} & 0_r & J_{r \times q} \\ 0_{q \times p} & I_q & J_{q \times r} & 0_q \end{bmatrix}.$$

Then we have

$$P\left(K_{p,q} \equiv K_{q,r},\, x\right) = |xI_{p+2q+r} - A\left(K_{p,q} \equiv K_{q,r}\right)|$$

$$= \begin{bmatrix} xI_p & -J_{p\times q} & 0_{p\times r} & 0_{p\times q} \\ -J_{q\times p} & xI_q & 0_{q\times r} & -I_q \\ 0_{r\times p} & 0_{r\times q} & xI_r & -J_{r\times q} \\ 0_{q\times p} & -I_q & -J_{q\times r} & xI_q \end{bmatrix}.$$

By careful calculation, we can obtain that the characteristic polynomial of $K_{p,q} \equiv K_{q,r}$ is

$$P\left(K_{p,q} \equiv K_{q,r},\, x\right) = x^{p+r-2}(x+1)^{q-1}(x-1)^{q-1}$$

$$\left[x^4 - (pq + qr + 1)x^2 + pq^2r\right].$$

Thus this theorem is proved.    □

## 3    Integral Graphs

In this section, we shall give some sufficient and necessary conditions for the three classes of graphs $K_{p,q(t)}$, $K_{p(s),q(t)}$ and $K_{p,q} \equiv K_{q,r}$ to be integral.

**Theorem 4.** The graph $K_{p,q(t)}$ is integral if and only if (i) $t$ and $pq + t$ are perfect squares, or (ii) $q = 1$ and $pq + t$ is a perfect square.

*Proof.* By Theorem 1, we know that the graph $K_{p,q(t)}$ is integral if and only if the equation

$$x^{q(t-1)+p}(x^2 - t)^{q-1}\left(x^2 - pq - t\right) = 0$$

has only integral roots.

Hence it is easy to check the validity of this theorem.    □

**Corollary 1.** (1) ([15]) If $q = 1$, then $K_{p,q(t)} = K_{1,p+t}$ is integral if and only if $p + t$ is a perfect square.
(2) If $t = k^2$, $pq = m^2 - k^2 > 0$, then $K_{p,q(t)}$ is integral, where $k$ and $m$ are positive integers.
(3) ([2] or [11]) If $p = n + 2$, $q = n$, and $t = 1$, then $K_{p,q(t)}$ is integral, where $n$ is a positive integer.

*Proof.* It is easy to check the validity by Theorem 1 or Theorem 4.    □

**Theorem 5.** The graph $K_{p(s),q(t)}$ is integral if and only if there exist natural numbers $a$ and $b$ such that $x^4 - (pq + s + t)x^2 + st$ can be factorized as $(x^2 - a^2)(x^2 - b^2)$, and one of the following three conditions holds:

(i) $s$ and $t$ are perfect squares,
(ii) $p = q = 1$,
(iii) $p = 1$, $t$ is a perfect square (or $q = 1$, $s$ is a perfect square).

*Proof.* From Theorem 2, it is easy to check the validity by using a method similar to that used in Theorem 4. □

**Corollary 2.** ([6]) *If $p = q = 1$, $1 \leq s \leq t$, then the graph $K_{p(s), q(t)} = T[s, t]$ is an integral tree of diameter 3 if and only if (i) $s = t = k(k+1)$, where $k$ is a positive integer, or (ii) when $1 \leq s < t$, let $(s, t) = d$, and $d$ is a positive integer but not a perfect square, let $s = d(\frac{y_k - y_l}{2})^2$, $t = d(\frac{y_k + y_l}{2})^2$, $k > l > 0$, where $y_k, y_l$ are odd or even, and $y_k, y_l \in \{y_n \mid y_0 = 0, y_1 = b_1, y_{n+2} = 2a_1 y_{n+1} - y_n, n \geq 0\}$, and $a_1 + b_1\sqrt{d}$ is the fundamental solution of the Pell equation*

$$x^2 - dy^2 = 1. \tag{1}$$

**Corollary 3.** *If at least one of positive integers $p$ and $q$ is non-one, then the graph $K_{p(s), q(t)}$ is integral if and only if $s = k^2$, $t = m^2$, $pq = a^2 + b^2 - k^2 - m^2$, $a^2b^2 = k^2m^2$, where $a$, $b$, $k$ and $m$ are positive integers.*

*Proof.* By Theorem 5, it is not difficult to prove this corollary. □

**Corollary 4. (1)** *If $s = t$, then the graph $K_{p(s), q(t)} = K_{p, q}^t$ is integral if and only if $s = t = ab$, $pq = (a - b)^2 > 0$, and $ab$ is a perfect square, where $a$ and $b$ are positive integers.*

**(2)** *([26]) If $s = t$, $pq$ is a perfect square, and there exist positive integers $t$, $t_1$ and $y$ such that $t = t_1^2 = y(y + \sqrt{pq})$, then the graph $K_{p(s), q(t)} = K_{p, q}^t$ is integral.*

**(3)** *([22, 25, 27]) If $p = 1$, then the tree $K_{p(s), q(t)} = K_{1, s} \bullet T(q, t)$ of diameter 4 is integral if and only if $t = k^2$, $s = \frac{a^2b^2}{k^2} \geq 1$, $q = a^2 + b^2 - k^2 - \frac{a^2b^2}{k^2} \geq 1$, where $a$, $b$ and $k$ are positive integers.*

**(4)** *([22, 25, 27]) If $q = 1$, then the tree $K_{p(s), q(t)} = K_{1, t} \bullet T(p, s)$ of diameter 4 is integral if and only if $s = k^2$, $t = \frac{a^2b^2}{k^2} \geq 1$, $p = a^2 + b^2 - k^2 - \frac{a^2b^2}{k^2} \geq 1$, where $a$, $b$ and $k$ are positive integers.*

*Proof.* (1) When $s = t$, by Theorem 5, then the graph $K_{p(s), q(t)} = K_{p, q}^t$ is integral if and only if $a^2b^2 = st$, $a^2 + b^2 = pq + s + t$, and $s = t = ab$ is a perfect square, where $a$ and $b$ are positive integers. Hence we deduce that $s = t = ab$, $ab$ is a perfect square, and $pq = (a - b)^2 \geq 1$. Thus (1) of this corollary is proved.

(2)-(4) By Theorem 5, we are easy to prove them. □

*Example 1.* Let $p, q, s, t$ be those positive integers in Table 1, then the graph $K_{p(s), q(t)}$ is integral.

*Proof.* It is easy to check the validity by Theorem 2 or Theorem 5. □

**Theorem 6.** *The graph $K_{p, q} \equiv K_{q, r}$ is integral if and only if there exist natural numbers $a$ and $b$ such that $x^4 - (pq + qr + 1)x^2 + pq^2r$ can be factorized as $(x^2 - a^2)(x^2 - b^2)$, where $p$, $q$ and $r$ are positive integers.*

*Proof.* From Theorem 3, it is easy to check the validity by using a method similar to that used in Theorem 4. □

**Table 1.** Integral graphs $K_{p(s),\,q(t)}$

| $p$ | $q$ | $s$ | $t$ | $p$ | $q$ | $s$ | $t$ | $p$ | $q$ | $s$ | $t$ | $p$ | $q$ | $s$ | $t$ |
|---|---|---|---|---|---|---|---|---|---|---|---|---|---|---|---|
| 1 | 9 | 4 | 4 | 3 | 3 | 4 | 4 | 9 | 1 | 4 | 4 | 1 | 24 | 4 | 9 |
| 1 | 24 | 9 | 4 | 2 | 12 | 4 | 9 | 2 | 12 | 9 | 4 | 3 | 8 | 4 | 9 |
| 3 | 8 | 9 | 4 | 4 | 6 | 4 | 9 | 4 | 6 | 9 | 4 | 6 | 4 | 4 | 9 |
| 6 | 4 | 9 | 4 | 8 | 3 | 4 | 9 | 8 | 3 | 9 | 4 | 12 | 2 | 4 | 9 |
| 12 | 2 | 9 | 4 | 24 | 1 | 4 | 9 | 24 | 1 | 9 | 4 | 1 | 1 | 2 | 2 |
| 1 | 1 | 6 | 6 | 1 | 1 | 12 | 12 | 1 | 1 | 20 | 20 | 1 | 1 | 30 | 30 |
| 1 | 1 | 42 | 42 | 1 | 1 | 56 | 56 | 1 | 1 | 50 | 98 | 1 | 1 | 98 | 50 |
| 1 | 1 | 147 | 192 | 1 | 1 | 192 | 147 | 1 | 1 | 486 | 726 | 1 | 1 | 726 | 486 |

**Corollary 5.** *For the graph $K_{p,q} \equiv K_{q,r}$, we have*

**(1)** *If $p = r$, then the graph $K_{p,q} \equiv K_{q,p}$ is integral if and only if $pq = k(k+1)$, where $p$, $q$ and $k$ are positive integers.*

**(2)** *If $p < r$, let $(p, r) = d_1$, and $d = d_1 q$, then we have the following results.*

   **(i)** *If $d$ is a perfect square, then the graph $K_{p,q} \equiv K_{q,r}$ is not an integral graph.*

   **(ii)** *If $d$ is a positive integer that is not a perfect square, then all the positive integral solutions for the graph $K_{p,q} \equiv K_{q,r}$ (where $1 \le p < r$) being an integral graph are given via*

$$pq = d\left(\frac{y_k - y_l}{2}\right)^2, \; rq = d\left(\frac{y_k + y_l}{2}\right)^2, \; k > l > 0,$$

   *where $y_k$, $y_l$ are odd or even, and $y_k$, $y_l \in \{y_n \,|\, y_0 = 0,\, y_1 = b_1,\, y_{n+2} = 2a_1 y_{n+1} - y_n,\, n \ge 0\}$, and $a_1 + b_1\sqrt{d}$ is the fundamental solution of the equation (1).*

*Proof.* (1) When $p = r$, we are easy to check the validity by Theorem 3 or Theorem 6.

(2) When $p < r$, by Theorem 6, we know that the necessary and sufficient condition for $K_{p,q} \equiv K_{q,r}$ being an integral graph is that there are positive integers $a$ and $b$ such that

$$\begin{cases} a^2 + b^2 = pq + qr + 1 \\ a^2 b^2 = pq^2 r \end{cases} \tag{2}$$

Let $(p, r) = d_1$, and $1 \le p < r$, by (2), we have

$$p = d_1 p_1^2, \; r = d_1 r_1^2, \; ab = d_1 q p_1 r_1, \tag{3}$$

where $p_1$ and $r_1$ are positive integers, and $(p_1, r_1) = 1$. By (2) and (3), we get that

$$(a + b)^2 - d(p_1 + r_1)^2 = 1, \tag{4}$$

where $d = d_1 q$.

Clearly, if $d$ is a perfect square, then the diophantine equation (4) has no integral solutions.

Let $d$ be a positive integer that is not a perfect square, then the equation (4) is a Pell equation. Let $\varepsilon = a_1 + b_1 \sqrt{d}$ be the fundamental solution of the equation (1). By (3), we deduce that

$$a + b = \frac{\varepsilon^k + \bar{\varepsilon}^k}{2}, \quad p_1 + r_1 = \frac{\varepsilon^k - \bar{\varepsilon}^k}{2\sqrt{d}}, \quad k > 0, \tag{5}$$

where $\bar{\varepsilon} = a_1 - b_1 \sqrt{d}$ and $\varepsilon \bar{\varepsilon} = 1$.

By $a$, $p_1$ of (5) and $ab = dp_1 r_1 = d_1 q p_1 r_1$ of (3), we get that

$$\left( 2b - \frac{\varepsilon^k + \bar{\varepsilon}^k}{2} \right)^2 - d \left( 2r_1 - \frac{\varepsilon^k - \bar{\varepsilon}^k}{2\sqrt{d}} \right)^2 = 1.$$

Thus, we have that

$$2b = \frac{\varepsilon^k + \bar{\varepsilon}^k}{2} + \frac{\varepsilon^l + \bar{\varepsilon}^l}{2}, \quad 2r_1 = \frac{\varepsilon^k - \bar{\varepsilon}^k}{2\sqrt{d}} + \frac{\varepsilon^l - \bar{\varepsilon}^l}{2\sqrt{d}}, \quad l > 0.$$

Hence, we have that

$$p_1 = \left( \frac{\varepsilon^k - \bar{\varepsilon}^k}{2\sqrt{d}} - \frac{\varepsilon^l - \bar{\varepsilon}^l}{2\sqrt{d}} \right) / 2, \quad r_1 = \left( \frac{\varepsilon^k - \bar{\varepsilon}^k}{2\sqrt{d}} + \frac{\varepsilon^l - \bar{\varepsilon}^l}{2\sqrt{d}} \right) / 2, \quad k > l > 0.$$

Let

$$y_n = \frac{\varepsilon^n - \bar{\varepsilon}^n}{2\sqrt{d}}, \quad n = 0, 1, 2, \ldots$$

Then we get that the Pell sequence (see [6] or [8])

$$y_0 = 0, \quad y_1 = b_1, \quad y_{n+2} = 2a_1 y_{n+1} - y_n, \quad (n \geq 0).$$

Hence, all the positive integral solutions for the graph $K_{p,q} \equiv K_{q,r}$ (where $1 \leq p < r$) being an integral graph are given via

$$pq = d \left( \frac{y_k - y_l}{2} \right)^2, \quad rq = d \left( \frac{y_k + y_l}{2} \right)^2, \quad k > l > 0.$$

The proof is now complete.    $\square$

**Corollary 6.** *For the graph $K_{p,q} \equiv K_{q,r}$, let $p$, $q$, $r$, $n$ be positive integers, we have.*

**(1) ([23])** *If $p = r = n$, $q = n + 1$, then the graph $K_{n,n+1} \equiv K_{n+1,n}$ is a $(n + 1)$-regular integral graph.*

**(2)** *If $p = r$, and $K_{p,q} \equiv K_{q,p}$ is integral, then the graph $K_{q,p} \equiv K_{p,q}$ is integral, too.*

*Proof.* It is easy to check the validity by Theorem 3 or Theorem 6.    □

*Example 2.* Let $p$, $q$, $r$, $n$ be those positive integers in Table 2, then the graph $K_{p,q} \equiv K_{q,r}$ is integral.

**Table 2.** Integral graphs $K_{p,q} \equiv K_{q,r}$

| $p$ | $q$ | $r$ | $p$ | $q$ | $r$ | $p$ | $q$ | $r$ | $p$ | $q$ | $r$ |
|---|---|---|---|---|---|---|---|---|---|---|---|
| 1 | 2 | 1 | 2 | 1 | 2 | 1 | 6 | 1 | 2 | 3 | 2 |
| 3 | 2 | 3 | 6 | 1 | 6 | $5^2$ | 2 | $7^2$ | $7^2$ | 2 | $5^2$ |
| $2 \cdot 5^2$ | 1 | $2 \cdot 7^2$ | $2 \cdot 7^2$ | 1 | $2 \cdot 5^2$ | $9^2$ | 6 | $11^2$ | $2 \cdot 9^2$ | 3 | $2 \cdot 11^2$ |
| $3 \cdot 9^2$ | 2 | $3 \cdot 11^2$ | $6 \cdot 9^2$ | 1 | $6 \cdot 11^2$ | / | / | / | / | / | / |

*Proof.* It is easy to check the validity by Theorem 3 or Theorem 6.    □

## Acknowledgements

The authors wish to thank the referees for their valuable comments and suggestions.

## References

1. Balińska, K.T., Cvetković, D., Radosavljević, Z., Simić, S.K., Stevanović, D.: A Survey on Integral Graphs. Univ. Beograd, Publ. Elektrotehn. Fak. Ser. Mat. **13** (2002) 42–65
2. Balińska, K.T., Kupczyk, M., Simić, S.K., Zwierzýnski, K.T.: On Generating all Integral Graphs on 11 Vertices. Tech. Rep. 469. Computer Science Center. Techn. Univ. of Poznán (1999/2000)
3. Balińska, K.T., Simić, S.K.: The Nonregular, Bipartite, Integral Graphs with Maximum Degree 4. Part I: Basic Properties. Discrete Math. **236** (1-3) (2001) 13–24
4. Balińska, K.T., Simić, S.K., Zwierzýnski, K.T.: Which Non-Regular Bipartite Integral Graphs with Maximum Degree Four Do not Have ±1 as Eigenvalues? Discrete Math. **286** (1-2) (2004) 15–24
5. Bussemaker, F.C., Cvetković, D.: There are Exactly 13 Connected Cubic Integral Graphs. Univ. Beograd. Publ. Elektrotehn. Fak. Ser. Mat. Fiz. **544** (1976) 43–48
6. Cao, Z.F.: On the Integral Trees of Diameter $R$ when $3 \leq R \leq 6$. J. Heilongjiang University **2** (1-3) (1988) p. 95
7. Cao, Z.F.: Some New Classes of Integral Trees with Diameter 5 or 6. J. Systems Sci. Math. Sci. **11** (1) (1991) 20–26
8. Cao, Z.F.: Introductory Diophantine Equations. Haerbin: Haerbin Polytechnical University Press (1989)
9. Cvetković, D., Doob, M., Sachs, H.: Spectra of Graphs—Theory and Application. New York, Francisco. London: Academic Press (1980)

10. Cvetković, D., Simić, S.K., Stevanović, D.: 4-Regular Integral Graphs. Univ. Beograde. Publ. Elektrotehn. Fak. Ser. Mat. **9** (1998) 89–102
11. Hansen, P., Mélot, H.: Computers and Discovery in Algebraic Graph Theory. Linear Algebra Appl. **356** (2002) 211–230
12. Harary, F., Schwenk, A.J.: Which Graphs Have Integral Spectra? In: Bari, R., Harary, F. (eds.): Graphs and Combinatorics. Lecture Notes in Mathematics, Vol. 406. Springer-Verlag, Berlin (1974) 45–51
13. Híc, P., Nedela, R.: Balanced Integral Trees. Math. Slovaca **48** (5) (1998) 429–445
14. Híc, P., Pokorný, M.: On Integral Balanced Rooted Trees of Diameter 10. Acta Univ. M. Belii Math **10** (2003) 9–15
15. Li, X.L., Lin, G.N.: On Trees with Integer Eigenvalues. Kexue Tongbao (Chinese Science Bulletin) **32** (11) (1987) 813–816 (in Chinese), or On Integral Trees Problems. Kexue Tongbao (English Ed.) **33** (10) (1988) 802–806 (in English)
16. Li, X.L., Wang, L.G.: Integral Trees—A Survey. Chinese Journal of Engineering Math. **17** (2000) 91–93
17. Liu, R.Y.: Integral Trees of Diameter 5. J. Systems Sci. Math. Sci. **8** (4) (1988) 357–360
18. Lepović, M.: On Integral Graphs which Belong to the Class $\overline{\alpha K_{a,b}}$. Graphs Combin. **19** (4) (2003) 527–532
19. Lepović, M.: On Integral Graphs which Belong to the Class $\overline{\alpha K_a \cup \beta K_{b,b}}$. Discrete Math. **285** (1-3) (2004) 183–190
20. Roitman, M.: An Infinite Family of Integral Graphs. Discrete Math. **52** (2-3) (1984) 313–315
21. Schwenk, A.J.: Exactly Thirteen Connected Cubic Graphs Have Integral Spectra. In: Alavi Y., Lick D.R. (eds.): Lecture Notes in Mathematics Series A, Vol. 642. Springer, Berlin (1978) 516–533
22. Wang, L.G., Li, X.L.: Some New Classes of Integral Trees with Diameters 4 and 6. Australasian J. Combinatorics **21** (2000) 237–243
23. Wang, L.G., Li, X.L., Hoede, C.: Two Classes of Integral Regular Graphs. Ars Combinatoria **76** (2005) 303–319
24. Wang, L.G., Li, X.L., Hoede, C.: Integral Complete $r$-Partite Graphs. Discrete Math. **283** (1-3) (2004) 231–241
25. Wang, L.G., Li, X.L., Yao, X.J.: Integral Trees with Diameters 4, 6 and 8. Australasian J. Combinatorics **25** (2002) 29–44
26. Wang, L.G., Li, X.L., Zhang, S.G.: Construction of Integral Graphs. Appl. Math. J. Chinese Univ. Ser. B **15** (3) (2000) 239–246
27. Wang, L.G., Li, X.L., Zhang, S.G.: Families of Integral Trees with Diameters 4, 6 and 8. Discrete Appl. Math. **136** (2) (2004) 349–362
28. Watanabe, M., Schwenk, A.J.: Integral Startlike Trees. J. Austral. Math. Soc. Ser. A **28** (1979) 120–128

# Reconfirmation of Two Results
# on Disjoint Empty Convex Polygons*

Liping Wu[1] and Ren Ding[2]

[1] College of Physical Science and Technology, Hebei University,
Baoding 071002, China
wlp123wlp@eyou.com
[2] College of Mathematics, Hebei Normal University, Shijiazhuang
050016, China
rending@hebtu.edu.cn

**Abstract.** For $k \geq 3$, let $m(k, k+1)$ be the smallest integer such that any set of $m(k, k+1)$ points in the plane, no three collinear, contains two different subsets $Q_1$ and $Q_2$, such that $CH(Q_1)$ is an empty convex $k-$gon, $CH(Q_2)$ is an empty convex $(k+1)-$gon, and $CH(Q_1) \cap CH(Q_2) = \emptyset$, where $CH$ stands for the convex hull. In this paper, we revisit the case of $k = 3$ and $k = 4$, and provide new proofs.

## 1 Introduction

In 1979, Erdős [1] asked the following combinatorial geometry problem: for $k \geq 3$, find the smallest integer $n(k)$ such that any set of $n(k)$ points in the plane, no three collinear, contains the vertex set of a convex $k-$gon, whose interior contains no point of the set. We call such a convex $k-$gon an empty convex $k-$gon or a $k$-hole. Klein [2] found $n(4) = 5$, Harborth [3] found $n(5) = 10$. Horton [4] showed that $n(k)$ dose not exist for $k \geq 7$. K.Hosono and M.Urabe [7]considered a related problem: for $k \geq 3$, find the smallest integer $m(k, k+1)$ such that any set of $m(k, k+1)$ points in the plane, no three collinear, contains two different subsets $Q_1$ and $Q_2$, such that $CH(Q_1)$ is a $k-$hole, $CH(Q_2)$ is a $(k+1)-$hole, and $CH(Q_1) \cap CH(Q_2) = \emptyset$, where $CH$ stands for the convex hull.

In this paper, we revisit the case of $k = 3$ and $k = 4$ and provide new proofs for the following two results obtained by [6] and [7].

**Theorem 1.** *[6]* $m(3, 4) = 7$.

**Theorem 2.** *[7]* $m(4, 5) \leq 14$.

## 2 Definitions and Notations

A point set in which no three points collinear is called to be in general position. In this paper, we consider only such point set. The following definitions and

---

* This research was supported by National Natural Science Foundation of China 10571042, NSF of Hebei A2005000144 and the Science Foundation of Hebei Normal University.

J. Akiyama et al. (Eds.): CJCDGCGT 2005, LNCS 4381, pp. 216–220, 2007.

notations are from [5]. For a given point set $P$, a convex region $R$ is called to be empty, denoted by $R \cong \emptyset$, if its interior contains no point of $P$. Consider the angular domain in the plane determined by the points $a, b$ and $c$, not on a line, such that $a$ is the apex and both $b$ and $c$ are on the boundary of the angular domain and that $\angle bac$ is acute. We call the interior region of this angular domain a convex cone, denoted by $C(a; b, c)$. If $C(a; b, c)$ contains some points of a given point set $P$, then we call the point $q \in P \cap C(a; b, c)$ the attack point, denoted by $A(a; b, c)$, if $C(a; b, q) \cong \emptyset$.

Moreover, for a given point set $P$, we denote the subset of $P$ on the boundary of $CH(P)$ by $V(P) = \{v_1, v_2, \ldots, v_t\}$ with the order anti-clockwise. We use the notation $\overline{ab}$ to refer to the line segment between $a$ and $b$, and $ab$ to refer to the extended straight line associated with two points $a$ and $b$.

## 3  Proofs

**Proof of Theorem 1**: Let $P$ be a set of 7 points in general position in the plane. The interior points of $P$ are the points of $P$ that are not on the boundary of $CH(P)$. Let $Q$ be a subset of $P$ which consists of the interior points of $P$.

**Case 1.** $|V(P)| \geq 4$. If $CH(P)$ is a convex 7-gon or a convex 6-gon, the Theorem is trivial. If $CH(P)$ is a convex 5-gon or a convex 4-gon, then there exists an extended straight line $l$ associated with an edge of $CH(Q)$ such that $l$ separates an $i$−hole ($i \geq 4$) from the remaining points of $P$, so we can find a 3-hole and a 4-hole which are disjoint.

For brevity, we use $(*)$ to stand for the statement:

$(*)$ *There does not exist the extended straight line associated with an edge of $CH(Q)$ which separates an $i$−hole ($i \geq 4$) from the remaining points of $P$.*

**Case 2.** $|V(P)| = 3$. Now we only need to prove the results by assuming $(*)$. Let $p_i = A(v_i; v_{i+1}, v_{i+2})$, where we identify indices modulo 3. Obviously, the convex cone $C(v_i; v_{i+1}, p_i)$ is empty. If $C(v_{i+1}; v_i, p_i)$ is not empty, let $y_i = A(p_i; v_i, v_{i+2})$, then the line segment $\overline{p_i y_i}$ is an edge of $CH(Q)$ and the extended straight line $p_i y_i$ separates the empty convex quadrilateral $v_i v_{i+1} p_i y_i$ disjoint from the remaining points of $P$, that is contrary to $(*)$. So we may assume $C(v_{i+1}; v_i, p_i) \cong \emptyset$.

Next we will show $p_i \neq p_j$ for any pair of indices $i \neq j(i, j = 1, 2, 3)$. If not, there must exist some $i$ such that $p_i = p_{i+1}$, then $C(v_i; v_{i+1}, p_i) \cup C(v_{i+1}; v_{i+2}, p_i) \cong \emptyset$ holds. Let $z = A(p_i; v_i, v_{i+2})$, then the line segment $\overline{p_i z}$ is an edge of $CH(P)$ and the extended straight line $p_i z$ separates the empty convex quadrilateral $v_i v_{i+1} p_i z$ disjoint from the other points of $P$, that is contrary to $(*)$. Thus we may assume

$(**)$ $C(v_i; v_{i+1}, p_i) \cup C(v_{i+1}; v_i, p_i) \cong \emptyset$, and $p_i \neq p_j$ for any pair of indices $i \neq j$ ($i, j = 1, 2, 3$).

Let $Q = \{p_1, p_2, p_3, u\}$, then we have two subcases.

**Subcase 2.1.** $|V(Q)| = 3$. Then $u \in \triangle p_1 p_2 p_3$, by $(*)$ $v_1 p_1 u p_3$ is a 4-hole, $v_3 p_2 v_2$ is a 3-hole (see Fig.1).

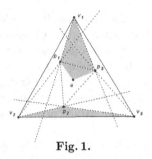

**Fig. 1.**

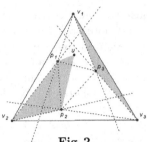

**Fig. 2.**

**Subcase 2.2.** $|V(Q)| = 4$. Then there must exist a triangle, say $\triangle v_1 p_1 p_3$ (see Fig.2) such that $u \in \triangle v_1 p_1 p_3$, then $v_2 p_2 u p_1$ is a 4-hole, $v_1 p_3 v_3$ is a 3-hole.

Hence $m(3,4) \leq 7$. $m(3,4) \geq 7$ is obvious, so $m(3,4) = 7$.    □

To prove Theorem 2, we need the following lemma which is obviously true by the fact that $n(5) = 10$.

**Lemma 1.** *Let* $|P| = 14$. *If* $(*)$ *is not true, then there exist a 4-hole and a 5-hole in* $P$, *and the two holes are disjoint.*

**Proof of Theorem 2:** Let $p_i = A(v_i; v_{i+1}, v_{i+2})(i = 1, 2, 3, \ldots, m)$. As usual, we only need to prove by assuming $(*)$. By the reasoning similar to that in Theorem 1, we may assume

$(***)$ $C(v_i; v_{i+1}, p_i) \cup C(v_{i+1}; v_i, p_i) \cong \emptyset$, *and* $p_i \neq p_j$ *for any pair of indices* $i \neq j$ $(i, j = 1, 2, \ldots, t)$.

**Case 1:** $|V(P)| \geq 5$. Consider the domain determined by two lines $v_{i+1} p_i$, $v_{i-1} p_{i-1}$ and $\overline{p_i p_{i-1}}$ for every $i$, and denote the interior of the domain by $T_i$. Denote the quadrilateral corresponding to $T_i$ by $Q_i = p_{i-1} p_i v_{i+1} v_{i-1}$. Since $|V(P)| \geq 5$, there exists some $T_i \cong \emptyset$. If $Q_i$ is not empty, then $v_i p_i v p_{i-1}$ is a 4-hole, where $v$ is the nearest point of $\overline{p_i p_{i-1}}$ in $P$. Again by $n(5) = 10$, there exists a 5-hole in the remaining points (see Fig.3). If $Q_i \cong \emptyset$, then $v_{i-1} p_{i-1} p_i v_{i+1} A(v_{i-1}; v_{i+1}, p_{i+1})$ is 5-hole, and $P \backslash \{v_i, v_{i+1}, p_i, p_{i-1}, v_{i-1}, A(v_{i-1}; v_{i+1}, p_{i+1})\}$ is a set of 8 points, then there exists a 4-hole by $n(4) = 5$ (see Fig.4).

**Case 2:** $|V(P)| = 4$. Then by $(*)$ and $(***)$ there exists a triangle domain defined by Case 1, say $T_1$, containing 0 or 1 point of $P$. If 0 is the case, we are done by the proof for case 1. Now suppose $T_1$ contains only one point $q$ of $P$.

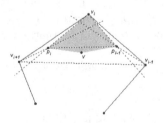

**Fig. 3.**

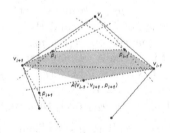

**Fig. 4.**

**Subcase 2.1:** $C(v_2; p_1, q)$ is not empty. Let $q_1 = A(v_2; p_1, q)$, then $v_1 p_1 q_1 A(q_1; v_1, q)$ is a 4-hole separated from the other 10 points (see Fig.5).

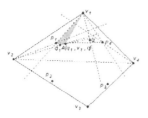

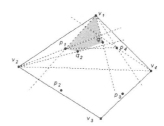

**Fig. 5.**                           **Fig. 6.**

**Subcase 2.2:** $C(v_2; p_1, q) \cong \emptyset$.

**Subsubcase 2.2.1:** $C(v_2; q, p_4)$ is not empty. Let $q_2 = A(v_2; q, p_4)$. If $q_2 \in C(v_1; p_1, q)$, then $v_1 p_1 q_2 q$ is a 4-hole separated from the other 10 points (see Fig.6). If $q_2 \in C(v_1; q, p_4)$ and $C(q; v_2, q_2) \cong \emptyset$, then $p_1 v_2 q_2 q$ is a 4-hole separated from the other 10 points (see Fig.7); if $q_2 \in C(v_1; q, p_4)$ but $C(q; v_2, q_2)$ is not empty, then $q_2 q p_1 v_2 A(q_2; v_2, p_2)$ is a 5-hole, $P \backslash \{v_1, p_1, q, q_2, v_2, A(q_2; v_2, p_2)\}$ contains a 4-hole (see Fig.8).

**Subsubcase 2.2.2:** $C(v_2; q, p_4) \cong \emptyset$. Then $p_4 q p_1 v_2 A(p_4; v_2, p_2)$ is a 5-hole, the 8 point set $P \backslash \{v_1, p_1, q, p_4, v_2, A(p_4; v_2, p_2)\}$ contains a 4-hole (see Fig.9).

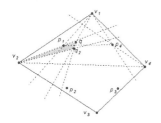

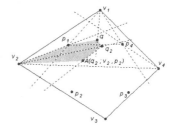

**Fig. 7.**                           **Fig. 8.**

**Case 3:** $|V(P)| = 3$. By assuming $(*)$ and $(***)$, there exists a triangular domain, say $T_1$, containing at most 2 points of $P$. If $T_1$ contains at most one point, we are done by the same argument as for case 2. Now consider the case of $T_1$ containing 2 points of $P$. Let $p = A(v_2; p_1, p_3)$, $q = A(v_3; p_3, p_1)$. If $p$ is not in $T_1$, then we have a 4-hole $v_1 p_1 p A(p; v_1, p_3)$ separated from the remaining 10 points and we are done. Thus we can assume that both $p$ and $q$ are in $T_1$ by symmetry. If $p = q$, either a 4-hole $v_1 p_1 r p$ or a 4-hole $v_1 p_3 r p$ can be separated from the remaining points, where $r$ is the remaining point of $P$ in $T_1$. So we suppose $p \neq q$. If $\{p, q, p_1, p_3\}$ is not in convex position, without loss of generality, let

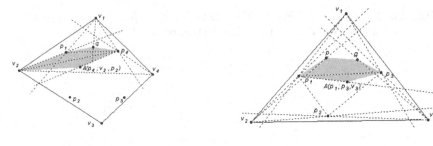

Fig. 9.                              Fig. 10.

$q \in \triangle p_1 p p_3$, then either a 4-hole $v_1 p q p_1$ or $v_1 p q p_3$ can be separated from the remaining points. Thus we assume that $\{p, q, p_1, p_3\}$ is in convex position, then $p_3 q p p_1 A(p_1; p_3, v_3)$ is a 5-hole, $P \backslash \{v_1, p_1, p, q, p_3, A(p_1; p_3, v_3)\}$ contains 8 points, and so contains a 4-hole (see Fig.10).                              $\square$

# References

1. P.Erdős, On some problems of elementary and combinatorial geometry, *Ann. Mat Pura. Appl.* 103(4)(1975)99-108.
2. P.Erdős and G.Szekeres, A combinatorial problem in geometry, *Compositio Math.* 2(1935)463-470.
3. H. Harborth, Konvexe Fünfecke in ebenen Punktmengen, *Elem. Math.* 33(1978) 116-118.
4. J.D.Horton, Sets with no empty convex 7-gon, *Canad. Math.Bull.* 26(1983)482-484.
5. Kiyoshi Hosono and Masatsugu Urabe, On the number of disjoint convex quadrilaterals for a planar point set, *Computational Geometry* 20(2001)97-104.
6. Masatsugu Urabe, On a partition into convex polygons, *Discrete Applied Mathematics* 64(1996)179-191.
7. Kiyoshi Hosono and Masatsugu Urabe, On the Minimum Size of a Point Set Containing Two Non-intersecting Empty Convex Polygons, *JCDCG 2004, LNCS 3742*, 117 - 122, 2005.

# The Binding Number of a Digraph[*]

Genjiu Xu[1], Xueliang Li[2], and Shenggui Zhang[1,3]

[1] Department of Applied Mathematics, Northwestern Polytechnical University,
Xi'an, Shaanxi 710072, P.R. China
xugenjiu@nwpu.edu.cn
[2] Center for Combinatorics and LPMC, Nankai University,
Tianjin 300071, P.R. China
x.li@eyou.com
[3] Department of Logistics, The Hong Kong Polytechnic University,
Hung Hom, Kowloon, Hong Kong, P.R. China
shengguizhang@nwpu.edu.cn

**Abstract.** Caccetta-Häggkvist's Conjecture discusses the relation between the girth $g(D)$ of a digraph $D$ and the minimum outdegree $\delta^+(D)$ of $D$. The special case when $g(D) = 3$ has lately attracted wide attention. For an undirected graph $G$, the binding number $bind(G) \geq \frac{3}{2}$ is a sufficient condition for $G$ to have a triangle (cycle with length 3). In this paper we generalize the concept of binding numbers to digraphs and give some corresponding results. In particular, the value range of binding numbers is given, and the existence of digraphs with a given binding number is confirmed. By using the binding number of a digraph we give a condition that guarantees the existence of a directed triangle in the digraph. The relationship between binding number and connectivity is also discussed.

**Keywords:** Binding number; girth; directed graph; Caccetta-Häggkvist Conjecture.

## 1 Introduction

Throughout the paper we consider only simple digraphs without loops and parallel arcs. Terminology and notation not defined here can be found in [4].

Let $D = (V(D), A(D))$ be a digraph. For a vertex $v \in V(D)$, by $N_D^+(v)$ and $N_D^-(v)$ we denote the set of outneighbors and the set of inneighbors of $v$ respectively, i.e., $N_D^+(v) = \{u \in V(D) \mid vu \in A(D)\}$ and $N_D^-(v) = \{u \in V(D) \mid uv \in A(D)\}$. For a subset $S$ of $V(D)$, we use $N_D^+(S)$ and $N_D^-(S)$ for $\cup_{v \in S} N_D^+(v)$ and $\cup_{v \in S} N_D^-(v)$, respectively. The *outdegree* and *indegree* of $v$, denoted by $d_D^+(v)$ and $d_D^-(v)$ respectively, are defined as $d_D^+(v) = |N_D^+(v)|$ and $d_D^-(v) = |N_D^-(v)|$. We set $\delta^+(D) = \min\{d_D^+(v) \mid v \in V(D)\}$, $\delta^-(D) = \min\{d_D^-(v) \mid v \in V(D)\}$ and $\delta(D) = \min\{\delta^+(D), \delta^-(D)\}$. When no confusion occurs, we use $N^+(v)$, $N^-(v)$, $N^+(S)$, $N^-(S)$, $d^+(v)$, $d^-(v)$, $\delta^+$, $\delta^-$ and $\delta$ for

---

[*] Supported by NSFC grant No. 10101021.

J. Akiyama et al. (Eds.): CJCDGCGT 2005, LNCS 4381, pp. 221–227, 2007.
© Springer-Verlag Berlin Heidelberg 2007

$N_D^+(v)$, $N_D^-(v)$, $N_D^+(S)$, $N_D^-(S)$, $d_D^+(v)$, $d_D^-(v)$, $\delta^+(D)$, $\delta^-(D)$ and $\delta(D)$, respectively. $D$ is called $d$-regular if $d^+(v) = d^-(v) = d$ for every vertex $v \in V(D)$. The girth of $D$, denoted by $g(D)$, is the length of a shortest directed cycle of $D$. In the following, we always assume $|V(D)| = n$ and $\delta(D) = \delta$.

In 1970, Behzad et al. [1] proposed the following

*Conjecture 1.* [1] Let $D$ be a $d$-regular digraph. Then $g(D) \le \lceil \frac{n}{d} \rceil$.

Caccetta and Häggkvist made the following more general conjecture in [7].

*Conjecture 2.* (Caccetta-Häggkvist Conjecture [7]) Let $D$ be a digraph with $\delta^+(D) \ge d$. Then $g(D) \le \lceil \frac{n}{d} \rceil$.

These conjectures have received much attention in recent years. Conjecture 1 has been proved for $d = 2$ by Behzad [2] and for $d = 3$ by Bermond [3]. Conjecture 2 has been proved for $d = 2$ by Caccetta and Häggkvist [7], for $d = 3$ by Hamidoune [11] and for $d = 4, 5$ by Hoàng and Reed [12]. In [8], Chvátal and Szemerédi established the bound $\frac{n}{d} + 2500$ for arbitrary values of $d$. Nishimura [14] reduced the additive constant from 2500 to 304. Some further developments on these conjectures can be found in [13, 16, 17].

An interesting special case of Conjecture 2 is: Any directed graph with $n$ vertices and minimum outdegree at least $n/3$ has a directed triangle (a directed cycle of length 3). Several papers have been devoted to this special case, see Bondy [5], Shen [15], Graaf et. al [10], Goodman [9], and Broersma and Li [6]. However, this conjecture has not been resolved to date.

The corresponding problem on the existence of triangles in undirected graphs has been studied by considering a graph invariant called the binding number of a graph. For a simple graph $G$ without loops and multiple edges, its *binding number* is defined as

$$bind(G) = \min_{\substack{\emptyset \ne S \subseteq V(G) \\ N(S) \ne V(G)}} \left\{ \frac{|N(S)|}{|S|} \right\},$$

where $N(S) = \{u \mid uv \in E(G), v \in S\}$. This parameter was introduced by Woodall [19] in 1973. He conjectured that: If a graph $G$ has $bind(G) \ge 3/2$, then it contains a triangle. This conjecture was confirmed by Shi in 1984.

**Theorem 1.** [18] *Let $G$ be a simple graph with $bind(G) \ge \frac{3}{2}$. Then $G$ contains a triangle.*

In this paper, motivated by the concept of binding number of graphs, we introduce the binding number of a digraph in Section 2. Some basic results on this parameter are also obtained in this section. In Section 3, we show that a digraph with binding number at least $\frac{\sqrt{5}+1}{2}$ has a directed triangle. Two conjectures on the girth of a digraph in terms of the binding number are also posed.

## 2    The Binding Number of a Digraph

**Definition 1.** *The binding number $bind(D)$ of a digraph $D$ is defined as*

$$bind(D) = \min_{\substack{\emptyset \neq S \subseteq V(D) \\ N(S) \neq V(D)}} \left\{ \frac{|N(S)|}{|S|} \right\},$$

$$where\ N(S) = \begin{cases} N^+(S), & if\ |N^+(S)| < |N^-(S)|; \\ N^+(S)\ or\ N^-(S), & if\ |N^+(S)| = |N^-(S)|; \\ N^-(S), & if\ |N^+(S)| > |N^-(S)|. \end{cases}$$

By the definition, for any digraph $D$, the binding number $bind(D)$ is a rational number, and for any arc $uv \notin A(D)$, we have $bind(D) \leq bind(D + uv)$. Let $v$ be the vertex of $D$ with $\delta = d^+(v)$ or $d^-(v)$. Set $S = \{v\}$. Then we have $bind(D) \leq \frac{|N(S)|}{|S|} = \delta$. The following theorem provides an alternative definition of the binding number.

**Theorem 2.** *Let $D$ be a digraph. Then $bind(D) = \min\limits_{\substack{\emptyset \neq S \subseteq V(D) \\ N(S) \neq V(D)}} \left\{ \frac{|V(D)|-|S|}{|V(D)|-|N(S)|} \right\}.$*

*Proof.* Firstly, we prove that $bind(D) \leq \frac{|V(D)|-|S|}{|V(D)|-|N(S)|}$ for any nonempty set $S \subseteq V(D)$ with $N(S) \neq V(D)$.

Given any nonempty set $S \subseteq V(D)$ with $N(S) \neq V(D)$. If $|N(S)| = |N^+(S)| \leq |N^-(S)|$, then let $T = V(D) \setminus N^+(S)$. If $s \in S$, then $s \notin N^-(T)$, so we know $N^-(T) \subseteq V(D) \setminus S$. By the definition of the binding number, we have $bind(D) \leq \frac{|N^-(T)|}{|T|} \leq \frac{|V(D) \setminus S|}{|T|} = \frac{|V(D)|-|S|}{|V(D)|-|N^+(S)|}$.

If $|N(S)| = |N^-(S)| \leq |N^+(S)|$, then we can show similarly that $bind(D) \leq \frac{|V(D)|-|S|}{|V(D)|-|N^-(S)|}$. Hence $bind(D) \leq \frac{|V(D)|-|S|}{|V(D)|-|N(S)|}$.

Next, we show that there exists a nonempty set $S_0 \subseteq V(D)$ such that $bind(D) = \frac{|V(D)|-|S_0|}{|V(D)|-|N(S_0)|}$. By the definition of the binding number, let the nonempty set $T_0 \subseteq V(D)$ be the vertex set satisfying $bind(D) = \frac{|N(T_0)|}{|T_0|}$. Without loss of generality, we may assume that $N(T_0) = N^+(T_0)$. Let $S_0 = V(D) \setminus N^+(T_0)$. Then $T_0 \cap N^-(S_0) = \emptyset$.

**(1)** If $bind(D) = 0$, then $N^+(T_0) = \emptyset$ and $S_0 = V(D)$. Moreover, $N(S_0) \neq V(D)$ since $T_0 \not\subseteq N^-(S_0)$. Therefore $\frac{|V(D)|-|S_0|}{|V(D)|-|N(S_0)|} = 0 = bind(D)$.

**(2)** If $bind(D) > 0$, we have $\delta \geq 1$. And if $t \notin T_0$, then it should be that $N^+(t) \not\subseteq N^+(T_0)$ and $N^+(t) \neq \emptyset$. Otherwise, let $T_1 = \{t\} \cup T_0$, we have $bind(D) = \frac{|N(T_0)|}{|T_0|} > \frac{|N(T_1)|}{|T_1|}$, which is a contradiction to the definition of $bind(D)$. Hence $N^+(t) \cap S_0 \neq \emptyset$, that is, $t \in N^-(S_0)$. So we have $T_0 \cup N^-(S_0) = V(D)$. We also have $T_0 = V(D) \setminus N^-(S_0)$ since $T_0 \cap N^-(S_0) = \emptyset$. Therefore $bind(D) = \frac{|N^+(T_0)|}{|T_0|} = \frac{|V(D) \setminus S_0|}{|V(D) \setminus N^-(S_0)|} = \frac{|V(D)|-|S_0|}{|V(D)|-|N(S_0)|}$.

This completes the proof.                                    □

The next two corollaries follow from Theorem 2 immediately.

**Corollary 1.** *Let $D$ be a digraph. Then $bind(D) \leq \frac{n-1}{n-\delta}$.*

*Proof.* Assume that $d^+(v) = \delta$, and let $S = \{v\}$. Then $|N(S)| = \delta$. By Theorem 2, we have $bind(D) \leq \frac{n-|S|}{n-|N(S)|} = \frac{n-1}{n-\delta}$.                □

**Corollary 2.** *Let $D$ be a digraph. Then $0 \leq bind(D) < 2$.*

*Proof.* It is trivial that $bind(D) \geq 0$. Since $D$ has no loops and parallel arcs, it is trivial that $\delta \leq \lfloor \frac{n-1}{2} \rfloor$ for the digraph $D$. By Corollary 1, we have $bind(D) \leq \frac{n-1}{n-\delta}$. It follows immediately that $bind(D) < 2$.

Hence $0 \leq bind(D) < 2$.                                    □

From Corollary 2, we see that the binding number of a digraph is a rational number in $[0, 2)$. A natural question is: For a given rational number $r \in [0, 2)$, does there exist a digraph with binding number $r$? The answer is positive.

**Theorem 3.** *For every rational number $r \in [0, 2)$, there exits a digraph $D$ with $bind(D) = r$.*

*Proof.* Suppose that $r = \frac{p}{q}$, where $p$ is a nonnegative integer and $q$ is a positive integer. We distinguish two cases.

*Case 1.* $0 \leq r \leq 1$.

Let $K(p, q, p)$ be a complete tripartite graph with three parts of vertex set $S_1$, $S_2$, $S_3$, and $|S_1| = |S_3| = p$, $|S_2| = q$. We construct a digraph $D$ by orienting $K(p, q, p)$ such that the arc set $A(D) = (S_1, S_2) \cup (S_2, S_3) \cup (S_3, S_1)$, where $(S_i, S_j)$ denotes the set of all possible arcs from $S_i$ to $S_j$.

It is easy to verify that the binding number of the digraph $D = \overrightarrow{K(p, q, p)}$ satisfies $bind(D) = \frac{p}{q}$.

*Case 2.* $1 < r < 2$.

Assume that $d = p - q + 1$. We construct a circulant digraph $D = C_S(N)$ such that $n = p+1$, and symbol set $S = \{1, 2, \ldots, d\}$. Here, a *circulant digraph* $C_S(N)$ is defined as the digraph with vertices the elements of $N = \{0, 1, 2, \ldots, n-1\}$ and arcs all pairs of the form $(i, i + s \pmod{n})$ with $i \in N$ and $s \in S$. By the definition of the binding number, for any nonempty vertex set $T \subseteq V(D)$ such that $N(T) \neq V(D)$, it is easy to check that if $T = \{0, 1, 2 \ldots, q-1\}$, then $N(T) = N^+(T) = \{1, 2, \ldots, q, q+1, \ldots, q+d-1\}$ and $bind(D) = \frac{|N(T)|}{|T|} = \frac{p}{q}$.
                                                              □

**Theorem 4.** *Let $D$ be a digraph.*

(1) *If $D$ is not strongly connected, then $bind(D) \leq 1$;*
(2) *If $D$ is $k$-connected, but not $(k + 1)$-connected, then $bind(D) \leq \frac{n+k}{n-k}$.*

*Proof.* (1) Since $D$ is not strongly connected, there exists a vertex set $S \subset V(D)$ such that $(S, \bar{S}) = \emptyset$ or $(\bar{S}, S) = \emptyset$, where $(S, \bar{S})$ means the set of arcs from $S$ to $\bar{S}$ in $D$. Without loss of generality, let $(S, \bar{S}) = \emptyset$. By the definition of $bind(D)$, we have $bind(D) \leq \frac{|N^+(S)|}{|S|} \leq \frac{|S|}{|S|} = 1$.

(2) Since $D$ is $k$−connected but not $(k+1)$-connected, there exists a vertex set $X \subset V(D)$ with $|X| = k$ such that $D \setminus X$ is not strongly connected. Let $T$ be the smallest strongly connected component of $D \setminus X$. Clearly $|T| \leq \frac{1}{2}(n - \kappa)$. Let $S = V(D) - (X \cup T)$. Then we have $|S| \geq \frac{1}{2}(n - \kappa)$. Considering $N(S)$, we distinguish three cases of $D$ as shown in Fig. 1.

Note that, in Fig. 1 an arrow from set $A$ to $B$ indicates that three exist arcs from some vertices in $A$ to some vertices in $B$. Note that $X$ is a minimum vertex cutset to destroy the strongly connectedness of $D$. So, for any vertex $x$ in $X$, there are an arc coming to $x$ as well as an arc going out from $x$. Moreover, there are at least $|X|$ vertices in $T$ belonging to $N^-(S)$ in Case 1 and belonging to $N^+(S)$ in Case 2. So it is easy to see that $|N(S)| \leq |X| + |S|$.
    Hence $bind(D) \leq \frac{|N(S)|}{|S|} \leq \frac{|X|+|S|}{|S|} \leq \frac{n+k}{n-k}$.     □

The following result gives some properties of the binding numbers for some special digraphs.

**Theorem 5.** *Let $D$ be a digraph.*

(1) *If $\delta = 0$, then $bind(D) = 0$;*
(2) *If $D$ is a directed cycle, then $bind(D) = 1$;*
(3) *If $D$ is a bipartite digraph with partitions $(A, B)$, then $bind(D) \leq \min\left\{ \frac{|A|}{|B|}, \frac{|B|}{|A|} \right\} \leq 1$;*
(4) *If $D$ is a tournament, then $bind(D) = 0$ or $bind(D) \geq \frac{1}{2}$.*

*Proof.* (1) and (2) are obvious.

(3) It is clear that $N(S) \subseteq B$ if $S = A$. Similarly, $N(S) \subseteq A$ if $S = B$. Then we immediately have $bind(D) \leq \min\left\{ \frac{|N(S)|}{|S|} \right\} \leq \min\left\{ \frac{|A|}{|B|}, \frac{|B|}{|A|} \right\} \leq 1$.

(4) If $\delta = 0$, we know $bind(D) = 0$ by conclusion (1). Otherwise, since $D$ is a tournament, for any pair of vertices $u, v \in V(D)$, it should be that $uv \in A(D)$ or $vu \in A(D)$. So at most one of $d^+(u) = 0$ and $d^+(v) = 0$ is true, the same to $d^-(u) = 0$ and $d^-(v) = 0$. For any $S \subseteq V(D)$, the deduced subdigraph by $S$ is also tournament, so $|N(S)| = \min\{|N^+(S)|, |N^-(S)|\} \geq |S| - 1$. Let $S_0$ be the nonempty vertex set such that $bind(D) = \frac{|N(S_0)|}{|S_0|}$. If $|S_0| = 1$, then $bind(D) = 0$ or $bind(D) \geq 1 > \frac{1}{2}$. And if $|S_0| \geq 2$, then $bind(D) = \frac{|N(S_0)|}{|S_0|} \geq \frac{|S_0|-1}{|S_0|} \geq \frac{1}{2}$.
    Hence $bind(D) = 0$ or $bind(D) \geq \frac{1}{2}$.     □

## 3    The Binding Number and Girth

As we pointed out earlier, Shi proved that a graph with binding number at least $3/2$ has a triangle. For digraphs, we can prove the following result.

**Theorem 6.** *Let $D$ be a digraph with $bind(D) \geq \frac{\sqrt{5}+1}{2}$. Then $g(D) = 3$.*

*Proof.* We prove it by contradiction. Assume that there is a digraph $D$ with $bind(D) \geq \frac{1+\sqrt{5}}{2}$ and $D$ does not contain a directed cycle of length 3.

Let $\delta = d^+(v)$. Then $(N^+(v), N^-(v)) = \emptyset$ since $D$ has no directed triangles. Denote $S = N^+(v)$. Then $N^+(S) \cap N^-(v) = \emptyset$, and $|N^-(v)| \geq \delta$. Therefore $bind(D) \leq \frac{|N^+(S)|}{|S|} \leq \frac{|V(D)|-|N^-(v)|}{|S|} \leq \frac{n-\delta}{\delta}$, hence $\delta \leq \frac{n}{bind(D)+1}$.

By Corollary 1, we have $bind(D) \leq \frac{n-1}{n-\delta}$, so $\delta \geq \frac{(bind(D)-1)n+1}{bind(D)}$. Therefore $\frac{(bind(D)-1)n+1}{bind(D)} \leq \delta \leq \frac{n}{bind(D)+1}$, it follows that $n(bind(D))^2-(n-1)bind(D)-n+1 \leq 0$. So we have $bind(D) \leq \frac{n-1+\sqrt{5n^2-6n+1}}{2n} < \frac{1+\sqrt{5}}{2}$, which is a contradiction to the hypothesis. This completes the proof. $\qquad\square$

Similar to the undirected graph case, we have the following conjectures.

*Conjecture 3.* Let $D$ be digraph with $bind(D) \geq \frac{3}{2}$. Then $g(D) = 3$.

The following more general conjecture may also hold.

*Conjecture 4.* Let $D$ be a digraph with $bind(D) \geq \frac{k}{k-1}$ $(k \geq 3)$. Then $g(D) \leq k$.

In fact, if $bind(D) \geq \frac{k}{k-1}$ $(k \geq 3)$, then it follows from Corollary 1 that $\delta \geq \frac{n+k-1}{k} > \frac{n}{k}$. Therefore, Caccetta-Häggkvist Conjecture is stronger than Conjecture 4. And, Conjecture 3 corresponds to Caccetta-Häggkvist Conjecture in the case of $\delta \geq \frac{n}{3}$.

## References

1. Behzad, M., Chartrand, G., Wall, C.: On Minimal Regular Digraphs with Given Girth. Fund. Math. **69** (1970) 227–231
2. Behzad, M.: Minimally 2-Regular Digraphs with Given Girth. J. Math. Soc. Japan **25** (1973) 1–6
3. Bermond, J.C.: 1-Graphs Réguliers de Girth Donné. Cahiers Centre Etudes Rech. Oper. Bruxelles **17** (1975) 123–135
4. Bondy, J.A., Murty, U.S.R. (eds.): Graph Theory with Applications. Macmillan London and Elsevier, New York (1976)
5. Bondy, J.A.: Counting Subgraphs: A New Approach to the Caccetta-Häggkvist Conjecture. Discrete Math. **165, 166** (1997) 71–80
6. Broerssma, H.J., Li, X.: Some Approaches to a Conjecture on Short Cycles in Digraphs. Discrete Appl. Math. **120** (2002) 45–53
7. Caccetta, L., Häggkvist, R.: On Minimal Digraphs with Given Girth. Congr. Numer. **21** (1978) 181–187

8. Chvátal, V., Szemerédi, E.: Short Cycles in Directed Graphs. J. Combin. Theory Ser. B **35** (1983) 323–327
9. Goodman, A.W.: Triangles in a Complete Chromatic Graph with Three Colors. Discrete Math. **5** (1985) 225–235
10. de Graaf, M., Schrijver, A.: Directed Triangles in Directed Graphs. Discrete Math. **110** (1992) 279–282
11. Hamidoune, Y.O.: A Note on Minimal Directed Graphs with Given Girth. J. Combin. Theory Ser. B **43** (1987) 343–348
12. Hoàng, C.T., Reed, B.: A Note on Short Cycle in Digraphs. Discrete Math. **66** (1987) 103–107
13. Li, Q., Brualdi, R.A.: On Minimal Directed Graphs with Girth 4. J. Czechoslovak Math. **33** (1983) 439–447
14. Nishimura, T.: Short Cycles in Digraphs. Discrete Math. **72** (1988) 295–298
15. Shen, J.: Directed Triangles in Digraphs. J. Combin. Theory Ser. B **73** (1998) 405–407
16. Shen, J.: On the Girth of Digraphs. Discrete Math. **211** (2000) 167–181
17. Shen, J.: On the Caccetta-Häggkvist Conjecture. Graphs and Combinatorics **18** (2002) 645–654
18. Shi, R.: The Binding Number of a Graph and its Triangle. Acta Mathematicae Applicatae Sinica **2** (1) (1985) 79–86
19. Woodall, D.R.: The Binding Number of a Graph and its Anderson Number. J. Combin. Theory Ser. B **15** (1973) 225–255

# The Kauffman Bracket Polynomial of Links and Universal Signed Plane Graph

Weiling Yang[1] and Fuji Zhang[2]

School of Mathematical Sciences, Xiamen, University,
Xiamen, Fujian 361005, P.R. China

**Abstract.** In this paper we show that a list of chain polynomials of 6 graphs are sufficient to give all the chain polynomials of signed plane graphs with cyclomatic number $1 - 5$ by special parametrization. Using this result the Kauffman bracket polynomials of 801 knots and 1424 links with crossing number no more than 11 can be obtained in a unify way.

**Keywords:** Link diagram, Signed plane graph, Chain polynomial, Kauffman bracket polynomial.

## 1 Introduction

It is well known that a link diagram corresponds to a signed plane graph and the Kauffman brackets polynomial of the link can be obtained by computing the generalized Tutte Polynomial of its corresponding signed plane graph [1]. Recently in [2] Jin and one of the present authors extended the concept of chain polynomial to signed graph and used it to deal with the Tutte polynomials of signed graphs. Furthermore they define an equivalence relation on the set of link diagrams according to the homeomorphic types of their corresponding signed plane graphs. For each equivalence class we only need to compute the chain polynomial of a unique signed plane graph (say universal graph with respect to the equivalence relation). Then the Kauffman brackets polynomial of any member of the class can be obtained easily from the chain polynomial of a universal graph by a special parametrization. [2] showed that when we deal with prime links the universal graphs(neglecting their sign) can be determined as all the plane graphs without cut vertex and the vertex with degree two. [2] also lists all the universal graphs (neglecting their sign) with cyclomatic number less than 5. The number of universal graph of cyclomatic number of 1, 2, 3 and 4 are 1, 2, 4 and 8 respectively. Examples are provided to compute the Kauffman bracket polynomials of sets of links in equivalence classes. If we go forward, we find that when cyclomatic number is 5, there are 111 2-connected plane graphs without the vertex with degree two which are also universal graphs (neglecting their sign). The 111 graphs had been found in [3] by computer search and proved in [4]. Clearly it is a tedious task to compute the chain polynomial of these 111 graphs and use it to compute Kauffman bracket polynomials of knots and links.

In this paper we attempt to reduce the universal graphs. Firstly, we reduce the set of universal graphs to be the set of 3-edge-connected signed plane graphs

J. Akiyama et al. (Eds.): CJCDGCGT 2005, LNCS 4381, pp. 228–244, 2007.
© Springer-Verlag Berlin Heidelberg 2007

without cut vertex. We list the universal graphs with cyclomatic number of 1, 2, 3, 4 and 5, that are 1, 1, 3, 10 and 56 graphs respectively (see Fig. 7). To our surprise we find that the 71 graphs can be obtained by contracting some edges of 6 graphs. So we can reduce the universal graphs with cyclomatic number of 1, 2, 3, 4 and 5 to these 1, 1, 1, 1, and 2 graphs respectively. Then we compute the chain polynomial of these 6 graphs by MAPLE. The Kauffman bracket polynomial of a prime link with crossing number no more than 11 can easily be computed by the chain polynomial of a signed plane graph in the set of reduced universal graph by a special parametrization. One example is provided to illustrate our approach.

## 2    Preliminaries

A graph $G$ is an ordered triple $(V(G), E(G), \psi_G)$ consisting of a non-empty set $V(G)$ of vertices, a set $E(G)$, disjoint from $V(G)$, of edges, and an incidence function $\psi_G$ that associates with each edge of $G$ an unordered pair of (not necessarily distinct) vertices of $G$. Throughout this paper, we use $Ch[G]$ to denote the chain polynomial of $G$ and use $n(G)$ to denote the cyclomatic number of $G$.

A link $L$ of $n$ components is a subset of $\mathbb{R}^3 \subset \mathbb{R}^3 \cup \{\infty\} = \mathbb{S}^3$, consisting of $n$ disjoint piecewise linear simple closed curves. A knot is a connected link. Although links live in $\mathbb{R}^3$, we usually represent them by link diagrams: the regular projections of links into $\mathbb{R}^2$ with overpassing curves specified.

*Example 1.* The (right-handed) trefoil knot and the Hopf link(See Fig. 1).

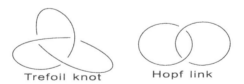

Trefoil knot          Hopf link

**Fig. 1.** Link diagrams

A signed graph is a graph whose edges are each labeled with a sign $(+1$ or $-1)$. There is a correspondence between link diagrams and signed plane graphs via medial construction. Now we give a brief account of it (for details, see [1]).

Given a link diagram $D$, we first shade its faces checker-boardly. Then associate an edge-colored multigraph $G[D]$ as follows: for each shaded face $F$, take a vertex $v_F$ in $F$, and for each crossing at which $F_1$ and $F_2$ meet, take an edge $v_{F_1}v_{F_2}$. Furthermore, give each edge $v_{F_1}v_{F_2}$ a sign according to the type of the crossing, as shown in Fig. 2. We call the graph obtained a signed plane graph. Note that every plane diagram has precisely two shadings.

Conversely, every signed plane graph $G$ can construct the corresponding link diagram $D$. First we construct the medial graph of $M(G)$. $M(G)$ of a connected non-trivial plane graph $G$ is a 4-regular plane graph obtained by inserting a vertex on every edge of $G$, and joining two new vertices by an edge lying in a face of $G$ if the vertices are on adjacent edges of the face; if $G$ is trivial (that is, it is an isolated vertex), its medial graph is a simple closed curve surrounding the vertex (strictly, it is not a graph); if $G$ is not connected, its medial graph $M(G)$ is the disjoint union of the medial graphs of its connected components. Then we turn $M(G)$ into a link diagram $D = D(G)$. Convert each vertex of $M(G)$ into a crossing according to the sign of the edge on which the vertex is inserted(see Fig. 2). For the details, see [6].

**Fig. 2.** Sign

**Example 2.** The knot $9_{49}$ and its signed plane graphs. See Fig. 3.

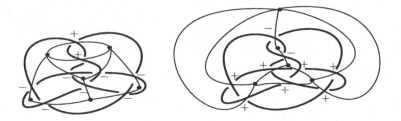

**Fig. 3.** $9_{49}$ and its signed plane graphs

**Definition 1 ([1, 2]).** *The Kauffman bracket polynomial $\langle D \rangle \in \mathbb{Z}\left[\mathbb{A}, \mathbb{A}^{-1}\right]$ of a link diagram $D$ is defined by the following two properties:*

(1) *The Kauffman bracket polynomial of a diagram consisting of $k$ disjoint simple closed curves in the plane is $d^{k-1}$, where $d = -A^2 - A^{-2}$.*
(2) *$\langle D \rangle = A\langle D_1 \rangle + A^{-1}\langle D_2 \rangle$, where $D$, $D_1$ and $D_2$ are identical diagrams except that a crossing of $D$ is nullified in two different ways to form $D_1$ and $D_2$, respectively (see Fig. 4).*

**Definition 2.** *The chain polynomial $Ch[M]$ of a graph $M$ is defined as*

$$Ch[M] = \sum_Y F[E(M) - Y](1 - w)\varepsilon(E(M) - Y) \tag{1}$$

*where the summation is over all subsets of $E(M)$; $F[Y]$ (respectively $F[E(M) - Y]$ is the flow polynomial (see [6, P. 347]) of $\langle Y \rangle$ (respectively $\langle E(M) - Y \rangle$), the*

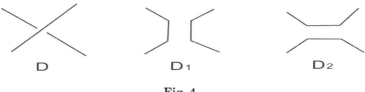

**Fig. 4.**

spanning subgraph of $M$ only containing all edges in $Y$ (respectively $E(M) - Y$); $\varepsilon(Y)$ (respectively $\varepsilon(E(M) - Y)$ is the product of labels of the edges (we usually identify the edge with its label) in $Y$ (respectively $(E(M) - Y)$).

The reduction $R(G)$ of a graph $G$ is the graph obtained by suppressing the vertices of degree 2 of $G$ successively until suppression is no longer possible. Obviously there is a one-to-one correspondence between the chains of $G$ and the edges of $R$.

**Definition 3 ([2]).** A signed graph $G$ is special if it satisfies the property: for any chain, all edges in this chain have the same sign.

As pointed out in [2], the signed graph of a prime link is special.

**Theorem 1 ([2]).** Let $D$ be a link diagram. Let $G$ be a special signed plane graph of $D$. Let $R$ be the reduction of $G$. In $Ch[R]$, if we replace $\omega$ by $-A^4 - 1 - A^{-4}$, and replace $a$ by $(-A^{-4})^{n_a}$ for every chain $a$, then

$$\langle D \rangle = \frac{A^m}{(-A^2 - A^{-2})^{q-p+1}} Ch[R], \qquad (2)$$

where $m$ is the sum of all signs of $G$, and $p$ and $q$ are the numbers of vertices and edges of graph $R$, respectively.

## 3 Main Results

[2] introduces an equivalence relation for knots and links as pointed out in the introduction, and shows that any prime knot of link correspond to a plane graph without cut-vertices. Then for each equivalence class, the Kauffman brackets polynomial of any member of the class can be obtained easily from the chain polynomial of a universal graph– signed plane graph without cut vertex and vertex of degree 2, by a special parametrization. Now we further reduce the set of universal graph in two steps based on the following theorems.

**Theorem 2.** Let $M$ be a signed plane graph without cut-vertices corresponding to link diagram $L$. Then there exist a 3-edge-connected signed plane graph $M'$ without cut-vertices whose corresponding link is $L$.

*Proof.* If $M$ has a 2-edge cut $\{a, b\}$, then its corresponding link $L$ and its corresponding special signed plane graph have the following local configuration:

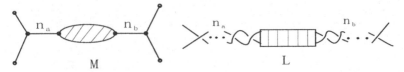

**Fig. 5.** Signed plane graph $M$ and link $L$

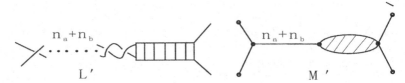

**Fig. 6.** Diagram $L'$ and corresponding signed plane graph $M'$ of $L'$

Writhing $L$ $-n_b$ times, we get diagram $L'$ of the same link. Showing in Fig. 6.

But in its corresponding signed plane graph $M'$, the 2-edge cut is eliminated. Note that in this procedure, for the corresponding signed plane graph, no

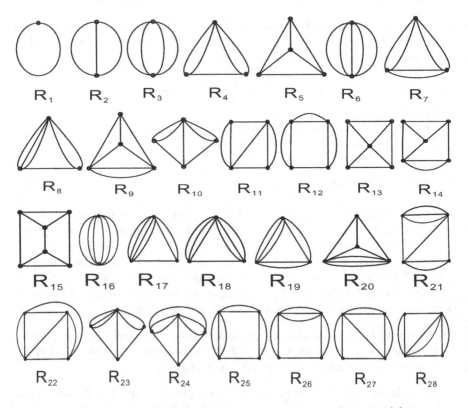

**Fig. 7.** Plane 3-edge-connected graph without cut-vertices and vertex of degree two

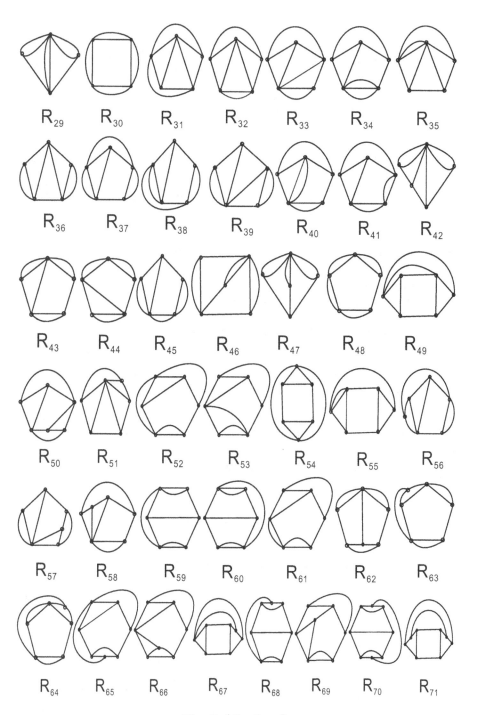

**Fig. 7.** (*Continued*)

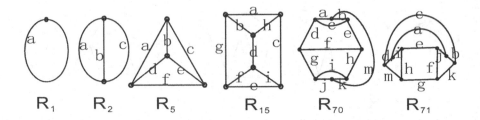

**Fig. 8.** Minimal set of universal graph with cyclomatic numbers $1 - 5$

cut-vertices are produced. Repeating this approach, we get a diagram of link $L$ whose corresponding signed plane graph is 3-edge-connected without cut-vertices.                                                                                        □

As the result of [2, Section 3], we can easily see that the set of universal graph are reduced to be 3-edge-connected plane graph without cut-vertices and vertex of degree 2.

A careful check of the graphs in [2,4] show that there are 1, 1, 3, 10 and 56 plane 3-edge-connected graphs without cut-vertices (see Fig. 7). In Fig. 7, we list all the 3-edge-connected plane graph without cut-vertices.

**Theorem 3.** *All the 71 graphs in Fig. 7 can be obtained by contracting some edges of 6 graphs in Fig. 8. Precisely, the minimal number of universal graph with cyclomatic number $1 - 4$ is 1, and cyclomatic number 5 is 2.*

*Proof.* From table 1, we can see how the 71 3-edge-connected graph without cut-vertices are produced from $R_1$, $R_2$, $R_3$, $R_5$, $R_{15}$, $R_{70}$ and $R_{71}$ by contracting some edges.

For example, $R_{20}$ can be obtained by contracting $m$, $g$, $k$, $f$ of $R_{70}$. Clearly, $R_1$, $R_2$, $R_5$, $R_{15}$ are the minimal set of universal graphs with cyclomatic number 1, 2, 3, and 4 resp. $R_{70}$ and $R_{71}$ are the minimal set of universal graph with cyclomatic number 5, since they have the same number of edges, any one of them can not be obtained from the other by contracting some edges.     □

## 4   Computing Kauffman Bracket Polynomials

Using theorem 3 we get the minimal set of universal graphs up to cyclomatic number 5. The Kauffman bracket polynomial of all the prime knots and links can be computed from chain polynomial of universal graphs by theorem 1 and the following lemma.

**Lemma 1.** ([5]) *Let $a$ be an edge of $K$. Let $M$ be the graph obtained from $K$ by contracting the edge $a$. Then $Ch[M]$ is obtained from $Ch[K]$ by putting $a = 1$.*

We compute the chain polynomial of $R_1$, $R_2$, $R_3$, $R_5$, $R_{15}$, $R_{70}$ and $R_{71}$ as following:

**Table 1.** How to obtain 71 graphs from $R_1$, $R_2$, $R_5$, $R_{15}$, $R_{70}$, $R_{71}$

| Graph | Universal graph | Contracted edges | Graph | Universal graph | Contracted edges |
|---|---|---|---|---|---|
| $R_3$. | $R_5$. | $a, e$. | $R_{38}$. | $R_{70}$. | $d, b, j$. |
| $R_4$. | $R_5$. | $a$. | $R_{39}$. | $R_{70}$. | $f, e, k$. |
| $R_6$. | $R_{15}$. | $g, e, d, h$. | $R_{40}$. | $R_{70}$. | $a, g, m$. |
| $R_7$. | $R_{15}$. | $a, e, d$. | $R_{41}$. | $R_{70}$. | $e, f, m$. |
| $R_8$. | $R_{15}$. | $e, d, h$. | $R_{42}$. | $R_{70}$. | $d, g, f$. |
| $R_9$. | $R_{15}$. | $e, d$. | $R_{43}$. | $R_{70}$. | $c, i, d$. |
| $R_{10}$. | $R_{15}$. | $a, c$. | $R_{44}$. | $R_{70}$. | $c, f, k$. |
| $R_{11}$. | $R_{15}$. | $e, h$. | $R_{45}$. | $R_{70}$. | $c, f, h$. |
| $R_{12}$. | $R_{15}$. | $a, e$. | $R_{46}$. | $R_{70}$. | $c, k, d$. |
| $R_{13}$. | $R_{15}$. | $c$. | $R_{47}$. | $R_{70}$. | $a, j, m$. |
| $R_{14}$. | $R_{15}$. | $e$. | $R_{48}$. | $R_{70}$. | $c, i, f$. |
| $R_{16}$. | $R_{70}$. | $a, e, m, g, k, f$. | $R_{49}$. | $R_{70}$. | $i, l$. |
| $R_{17}$. | $R_{70}$. | $e, m, g, k, f$. | $R_{50}$. | $R_{70}$. | $m, f$. |
| $R_{18}$. | $R_{70}$. | $h, e, c, i, d$. | $R_{51}$. | $R_{70}$. | $d, g$. |
| $R_{19}$. | $R_{70}$. | $e, m, g, k, f$. | $R_{52}$. | $R_{70}$. | $d, i$. |
| $R_{20}$. | $R_{70}$. | $m, g, k, f$. | $R_{53}$. | $R_{70}$. | $f, k$. |
| $R_{21}$. | $R_{70}$. | $e, c, i, d$. | $R_{54}$. | $R_{70}$. | $c, f$. |
| $R_{22}$. | $R_{70}$. | $h, d, b, j$. | $R_{55}$. | $R_{70}$. | $b, d$. |
| $R_{23}$. | $R_{70}$. | $b, d, e, m$. | $R_{56}$. | $R_{70}$. | $a, h$. |
| $R_{24}$. | $R_{70}$. | $h, b, g, d$. | $R_{57}$. | $R_{70}$. | $f, d$. |
| $R_{25}$. | $R_{70}$. | $b, d, e, k$. | $R_{58}$. | $R_{70}$. | $d, k$. |
| $R_{26}$. | $R_{70}$. | $h, c, i, d$. | $R_{59}$. | $R_{70}$. | $j, a$. |
| $R_{27}$. | $R_{70}$. | $m, c, i, d$. | $R_{60}$. | $R_{70}$. | $j, b$. |
| $R_{28}$. | $R_{70}$. | $m, c, k, d$. | $R_{61}$. | $R_{70}$. | $f, i$. |
| $R_{29}$. | $R_{70}$. | $f, d, g, m$. | $R_{62}$. | $R_{70}$. | $c, e$. |
| $R_{30}$. | $R_{70}$. | $m, c, i, f$. | $R_{63}$. | $R_{70}$. | $c, i$. |
| $R_{31}$. | $R_{70}$. | $k, i, m$. | $R_{64}$. | $R_{70}$. | $c, j$. |
| $R_{32}$. | $R_{70}$. | $h, b, d$. | $R_{65}$. | $R_{70}$. | $d$. |
| $R_{33}$. | $R_{70}$. | $h, m, f$. | $R_{66}$. | $R_{70}$. | $f$. |
| $R_{34}$. | $R_{70}$. | $g, k, f$. | $R_{67}$. | $R_{70}$. | $m$. |
| $R_{35}$. | $R_{70}$. | $b, g, d$. | $R_{68}$. | $R_{70}$. | $j$. |
| $R_{36}$. | $R_{70}$. | $f, a, j$. | $R_{69}$. | $R_{70}$. | $i$. |
| $R_{37}$. | $R_{70}$. | $e, k, d$. | | | |

$Ch[R_1] = a - w$

$Ch[R_2] = abc - w(a + b + c) + w^2 + w$

$Ch[R_5] = abfecd - w(abc + adf + bde + cef + ae + cd + bf) + w(1 + w)(a + b + c + d + e + f) - (2w + 3w^2 + w^3)$

$Ch[R_{15}] = (-2w - 3w^2 - w^3)(a + b + e + f + h + i) + w(1 + w)(ai + cf + af + ae + ad + bi + bf + gh + fh + eh + cb + be + gi + ih + de + cei + gef + agb + bdh + dfi + ach) - (1 + w)^2 w(d + c + g) + w^2 cgd - w(cbf + ade + gih + gefh + acfh + adfi + agbi + cbei + bdeh + agbe + bdfih + aceih + abcdhg + cgdefi) + w^4 + 5w^3 + 8w^2 + 4w + abcdefghi$

$Ch[R_{70}] = (4w + 8w^2 + 5w^3 + w^4)(c + b + f + i + a + j + k) - (1 + w)^2 w(ag + ek + hc + id + eg + bg + ej + mi + jd + ei + mc + bh + hd + ha + kd + mf + cg + ehf + bam +$

$fdg+jkm)+(-2w-3w^2-w^3)(jig+ebc+acd+hki+kg+jf+af+cf+ja+ea+$
$kf+jc+hj+bd+ak+ci+fi+ai+bf+bj+bk+kc+bi)+(1+w)w(bhd+bkf+$
$kcf+mcf+akf+hkid+akcd+hjkmig+hacd+bkd+bid+bjig+haki+ebhcf+$
$eag+bfdg+acfdg+jcf+jaig+mfi+jkmf+kfdg+bhj+ebci+hjd+ehaf+$
$ejig+ebamcd+hja+ebkc+mci+eai+afi+jfidg+bfi+bjd+bamf+cfi+$
$acid+hkci+akg+ehkfi+ebjc+hjc+jaf+ebcg+jkmc+eak+ehjf+bjakm+$
$ekg+bjf+bami+jcig+jacd+eja+bkg+kcg+bhki)-w(bkfdg+ebamcid+$
$jkmcf+bamfi+ebhkcfi+bhjd+akcfdg+hjkmcig+jacfidg+ehjkmfidg+$
$ebhjcf+ehakfi+hjacd+hakcid+bjakmf+bhjakmig+ejaig+ebkcg+eakg+$
$mcfi+ebjcig+ebhamcfdg+ebjakmcd+bhkid+bjfidg+ehjaf)+w^2(ejkmd+$
$emid+bhamg+hmcg+ehmfdg)+(2w+5w^2+4w^3+w^4)(h+d+e+g)-(1+$
$w)w^2(emd+hmg)+(1+w)^3wm-8w-20w^2-18w^3-7w^4-w^5+ebhjakmcfidg$
$Ch[R_{71}]=(-2w-3w^2-w^3)(bg+ic+mf+if+kfg+ak+jh+dmc+jg+be+mhg+$
$dj+ieh+bm+dg+me+abj+hc+ib+adi+am+af+jc+dk+ah+ke+kh+ig+jef+$
$de+bkc+cf)-(1+w)^2w(ij+jk+ik+ec+ae+bh+dh+cg+hf+df+km+eg+jm+$
$bf+im+ag+ac+bd)+(4w+8w^2+5w^3+w^4)(j+a+m+f+c+k+h+g+d+i+e+$
$b)+(1+w)w(ibkc+djk+dbkmc+abjef+abjef+djef+bkhc+bmf+adibj+jcg+$
$icf+jecf+djmc+deg+adik+dbe+abjg+dbg+bkcfg+jkefg+ibm+jmef+$
$dmhcg+imehg+dke+bme+ahc+adimc+acf+ame+ikeh+adif+ikfg+abjm+$
$kme+ahf+adieh+adig+imf+jhc+djg+kmhfg+bkec+akh+jmhg+abjkc+$
$amhg+ijehf+dkh+ibeh+jkh+akfg+ijg+ijc+ake+iehc+bmhg+hcf+abjh+$
$icg+amf+dkfg+akm+djh+beg+dmcf+dmec+ibg)-w(djkefg+ijehcf+$
$akmhfg+abjkhc+abjmef+ijcg+ibkcfg+adimcf+ibmehg+adikfg+djmhcg+$
$adibjg+ibkehc+ibmf+ijkmehfg+akme+djmecf+adimehcg+adibjehf+$
$dbeg+adibjkmc+ahcf+dbkmec+dbkmhcfg+abjkecfg+abjmhg+djkh+$
$adikeh)+w^2(ijkm+bdhf+aecg)-11w-25w^2-20w^3-7w^4-w^5+adibjkmehcfg$

In order to compute the Kauffman bracket polynomials of knots and links with crossing number no more than 11, we classify all of them according to their reduced 3-edge-connected signed plane graphs in table 2, 3. In table 2, we use the Rolfsen knot table of $1-10$ crossing knots and the Hoste-Thistlethwaite table of 11 crossing knots and where $k_i$ denotes the $i$-th knot with $k$ crossings. $11_{ai}$ denotes the i-th alternated knot with 11 crossings. $11_{ni}$ denotes the i-th non-alternated knot with 11 crossings. In table 3 we use the Thistlethwaite link table where $k_{ai}$ denotes the i-th alternated knot with $k$ crossings. $k_{ni}$ denotes the i-th non-alternated knot with $k$ crossings.

By table 1, if we know the chain polynomial of six universal graphs, all the Kauffman bracket polynomials of the links with no more than 11 crossings can be obtain by a special parametrization.

**Table 2.** The classification of knots with crossing number no more than 11

| Family | Members |
|---|---|
| $R_1$. | $0_1, 3_1, 5_1, 7_1, 9_1, 11_1, 11_{a367}$. |

$$11_{a182} \qquad R_{10}$$

**Fig. 9.** knot $11_{a182}$ and its correspond plane graph with $n_b = 1$, $n_d = 1$, $n_e = 1$, $n_f = 1$, $n_g = 5$, $n_h = 1$, $n_i = 1$

**Example 3.** The Kauffman bracket polynomial of $11_{a182}$ knot.

From table 1, the Kauffman bracket polynomial of $11_{a182}$ can be computed by the chain polynomial of $R_{10}$. Furthermore from table 2, $R_{10}$ is in the family of $R_{15}$. Firstly, we let $a = c = 1$ in $Ch[R_{15}]$, to get $Ch[R_{10}]$ . Secondly, we replace $\omega$ by $-A^4 - 1 - A^{-4}$, replace $b$, $d$, $e$, $f$, $h$, $i$ by $-A^{-4}$ for chain $b$, $d$, $e$, $f$, $h$, $i$, and replace $g$ by $\left(-A^{-4}\right)^5$ for chain $g$, resp. Then

$$
\begin{aligned}
\langle D \rangle &= \frac{A^{11}}{(-A^2 - A^{-2})^4} Ch[R_{10}] \\
&= \frac{A^{27} + A^{23} - 3A^{19} - 6A^{15} - 16A^{11} - 5A^7 + 6A^3}{(-A^2 - A^{-2})^4} \\
&+ \frac{4A^{-1} + 6A^{-9} + 6A^{-13} + 3A^{-17} - A^{-21} - A^{-25} - A^{-29} - A^{-33}}{(-A^2 - A^{-2})^4}.
\end{aligned}
$$

# References

1. Kauffman, L.H.: A Tutte Polynomial for Signed Graphs. Discrete Appl. Math. **25** (1989) 105–127
2. Jin, Xian'an, Zhang, F.-J: The Kauffman Brackets for Equivalence Classes of Links. Advances in Appl. Math. **34** (2005) 47–64
3. Heap, B.R.: The Enumeration of Homeomorphically Irreducible Star Graphs. J. Math. Phys. **7** (1996) 1852–1857
4. Yang, W.-L., Jin, X.: The Construction of 2-Connected Plane Graph with Cyclomatic Number 5. Journal of Mathematical Study **37** (2004) 83–95
5. Read, R.C., Whitehead, E.G.: Chromatic Polynomials of Homeomorphism Classes of Graphs. Discrete Math. **204** (1999) 337–356
6. Bllobas, B.: Modern Graph Theory. Springer (1998)

| Family | Members |
|---|---|
| $R_2$. | $4_1, 5_2, 6_1, 6_2, 7_2, 7_4, 8_1, 8_2, 8_4, 8_5, 8_{13}, 9_2, 9_5, 9_{35}, 9_{46}, 10_1, 10_2, 10_4, 10_8, 10_{46}$. |
| $R_2$. | $10_{61}, 10_{124}, 10_{125}, 10_{126}, 10_{140}, 10_{142}, 11_{a247}, 11_{a343}, 11_{a362}, 11_{a363}$. |
| $R_3$. | $7_3, 7_5, 9_3, 9_6, 9_9, 9_{16}, 11_{a234}, 11_{a240}, 11_{a263}, 11_{a338}, 11_{a364}, 11_{a238}, 11_{a334}$. |
| $R_3$. | $11_{a335}, 11_{a355}$. |
| $R_4$. | $6_3, 7_6, 7_7, 8_7, 8_8, 8_9, 8_{10}, 8_{12}, 8_{15}, 9_8, 9_{11}, 9_{14}, 9_{17}, 9_{20}, 9_{22}, 9_{36}, 9_{42}, 9_{45}, 10_5, 10_{10}$. |
| $R_4$. | $10_{15}, 10_{19}, 10_{28}, 10_{34}, 10_{47}, 10_{52}, 10_{54}, 10_{62}, 10_{68}, 10_{130}, 10_{131}, 11_{a9}, 11_{a59}, 11_{a62}$. |
| $R_4$. | $11_{a74}, 11_{a97}, 11_{a140}, 11_{a142}, 11_{a161}, 11_{a179}, 11_{a188}, 11_{a195}, 11_{a203}, 11_{a206}, 11_{a223}$. |
| $R_4$. | $11_{a225}, 11_{a250}, 11_{a258}, 11_{a259}, 11_{a308}, 11_{a333}, 11_{n19}$. |
| $R_5$. | $8_{16}, 9_{29}, 9_{41}, 10_{85}, 10_{93}, 10_{100}, 10_{108}, 10_{155}, 11_{a279}, 11_{a293}, 11_{a313}, 11_{a323}, 11_{a330}$. |
| $R_5$. | $11_{a346}, 11_{n169}, 11_{n181}$. |
| $R_6$. | $8_3, 8_6, 9_4, 9_{10}, 9_{18}, 10_3, 10_6, 10_{11}, 10_{76}, 11_{a342}, 11_{a359}, 11_{a366}$. |
| $R_7$. | $10_{49}, 10_{66}, 10_{127}, 10_{153}$. |
| $R_8$. | $8_{11}, 9_7, 9_{12}, 9_{13}, 9_{15}, 9_{21}, 9_{24}, 10_7, 10_9, 10_{13}, 10_{14}, 10_{16}, 10_{18}, 10_{20}, 10_{21}, 10_{22}, 10_{24}$. |
| $R_8$. | $10_{25}, 10_{26}, 10_{29}, 10_{35}, 10_{39}, 10_{50}, 10_{56}, 10_{64}, 10_{70}, 10_{72}, 10_{74}, 11_{a55}, 11_{a57}, 11_{a65}$. |
| $R_8$. | $11_{a145}, 11_{a153}, 11_{a165}, 11_{a211}, 11_{a219}, 11_{a222}, 11_{a229}, 11_{a230}, 11_{a231}, 11_{a237}, 11_{a246}$. |
| $R_8$. | $11_{a311}, 11_{a341}, 11_{a360}$. |
| $R_9$. | $8_{17}, 8_{19}, 8_{20}, 8_{21}, 9_{33}, 9_{38}, 9_{39}, 9_{44}, 9_{48}, 9_{49}, 10_{82}, 10_{86}, 10_{90}, 10_{92}, 10_{94}, 10_{96}$. |
| $R_9$. | $10_{98}, 10_{102}, 10_{105}, 10_{106}, 10_{110}, 10_{111}, 10_{156}, 11_{a67}, 11_{a68}, 11_{a123}, 11_{a152}$. |
| $R_9$. | $11_{a163}, 11_{a169}, 11_{a252}, 11_{a254}, 11_{a265}, 11_{a292}, 11_{a296}, 11_{a299}, 11_{a317}, 11_{a320}$. |
| $R_9$. | $11_{a324}, 11_{a345}, 11_{a354}, 11_{n83}, 11_{n129}, 11_{n162}, 11_{n163}$. |
| $R_{10}$. | $8_{14}, 9_{19}, 9_{23}, 9_{25}, 9_{26}, 9_{27}, 9_{30}, 9_{31}, 9_{37}, 9_{43}, 10_{30}, 10_{32}, 10_{36}, 10_{41}, 10_{53}, 10_{55}$. |
| $R_{10}$. | $10_{58}, 10_{59}, 10_{63}, 10_{67}, 10_{146}, 11_{a13}, 11_{a21}, 11_{a39}, 11_{a50}, 11_{a61}, 11_{a82}, 11_{a83}$. |
| $R_{10}$. | $11_{a90}, 11_{a93}, 11_{a95}, 11_{a103}, 11_{a108}, 11_{a111}, 11_{a119}, 11_{a133}, 11_{a174}, 11_{a177}$. |
| $R_{10}$. | $11_{a182}, 11_{a185}, 11_{a190}, 11_{a192}, 11_{a194}, 11_{a198}, 11_{a199}, 11_{a200}, 11_{a205}, 11_{a306}$. |
| $R_{10}$. | $11_{n38}, 11_{n70}, 11_{n100}, 11_{n101}, 11_{n102}, 11_{n104}, 11_{n137}, 11_{n138}$. |
| $R_{11}$. | $10_{38}, 10_{60}, 10_{137}, 10_{138}, 10_{144}, 11_{a4}, 11_{a15}, 11_{a33}, 11_{a37}, 11_{a86}, 11_{a214}, 11_{n12}, 11_{n16}$. |
| $R_{11}$. | $11_{n17}, 11_{n18}, 11_{n24}, 11_{n25}, 11_{n26}, 11_{n27}, 11_{n57}, 11_{n62}, 11_{n63}, 11_{n64}, 11_{n88}, 11_{n89}$. |
| $R_{12}$. | $9_{28}, 10_{75}, 10_{78}, 10_{80}, 11_{a22}, 11_{a40}, 11_{a46}, 11_{a107}$. |
| $R_{13}$. | $8_{18}, 9_{34}, 9_{47}, 10_{112}, 10_{114}, 10_{116}, 10_{119}, 10_{120}, 10_{141}, 10_{159}, 10_{163}, 10_{165}, 11_{a249}$. |
| $R_{13}$. | $11_{a256}, 11_{a268}, 11_{a278}, 11_{a281}, 11_{a282}, 11_{a286}, 11_{a290}, 11_{a301}, 11_{a303}, 11_{a305}, 11_{n150}$. |
| $R_{13}$. | $11_{n155}, 11_{n158}, 11_{n159}, 11_{n161}, 11_{n172}, 11_{n173}$. |
| $R_{14}$. | $9_{32}, 10_{87}, 10_{97}, 10_{101}, 10_{134}, 11_{a8}, 11_{a29}, 11_{a53}, 11_{a81}, 11_{a99}, 11_{a113}, 11_{a115}$. |
| $R_{14}$. | $11_{a127}, 11_{a129}, 11_{a137}, 11_{a141}, 11_{a148}, 11_{a158}, 11_{a215}, 11_{a216}, 11_{a232}, 11_{a261}, 11_{n49}$. |
| $R_{14}$. | $11_{n92}, 11_{n109}, 11_{n110}, 11_{n111}, 11_{n144}$. |
| $R_{15}$. | $9_{40}, 10_{122}, 10_{160}, 10_{164}, 11_{a269}, 11_{a302}, 11_{a348}, 11_{a352}, 11_{n160}$. |
| $R_{16}$. | $11_{a242}, 11_{a245}, 11_{a339}, 11_{a358}$. |
| $R_{17}$. | $10_{12}, 10_{17}, 10_{37}, 10_{48}, 10_{77}, 11_{a12}, 11_{a64}, 11_{a75}, 11_{a98}, 11_{a118}, 11_{a144}, 11_{a154}$. |

| Family | Members |
|---|---|
| $R_{17}$. | $11_{a166}, 11_{a210}, 11_{a220}, 11_{a224}, 11_{a260}, 11_{a310}$. |
| $R_{18}$. | $11_{a94}, 11_{a186}, 11_{a191}, 11_{a235}, 11_{a241}, 11_{a336}, 11_{a340}, 11_{a357}, 11_{a365}, 11_{a356}$. |
| $R_{19}$. | $10_{79}, 10_{149}, 10_{152}, 11_{a20}, 11_{a45}, 11_{a49}, 11_{a60}, 11_{a102}, 11_{a105}, 11_{a201}, 11_{a202}$. |
| $R_{19}$. | $11_{n37}, 11_{n79}, 11_{n80}, 11_{n97}, 11_{n98}, 11_{n99}, 11_{n139}, 11_{n140}, 11_{n141}$. |
| $R_{20}$. | $10_{104}, 10_{128}, 10_{129}, 10_{139}, 10_{148}, 10_{158}, 10_{161}, 11_{a149}, 11_{a173}, 11_{a228}, 11_{a280}$. |
| $R_{20}$. | $11_{a331}, 11_{a347}, 11_{n170}, 11_{n171}$. |
| $R_{21}$. | $10_{27}, 10_{33}, 10_{65}, 11_{a96}, 11_{a117}$. |
| $R_{22}$. | $11_{a124}, 11_{a212}, 11_{a213}, 11_{a227}, 11_{a244}, 11_{a291}, 11_{a298}, 11_{a318}, 11_{a319}, 11_{a353}$. |
| $R_{23}$. | $10_{23}, 10_{31}, 10_{40}, 10_{51}, 10_{57}, 10_{135}, 11_{a1}, 11_{a10}, 11_{a48}, 11_{a63}, 11_{a88}, 11_{a120}$. |
| $R_{23}$. | $11_{a143}, 11_{a180}, 11_{a183}, 11_{a193}, 11_{a204}, 11_{a207}, 11_{a208}, 11_{a226}, 11_{a257}, 11_{a307}$. |
| $R_{23}$. | $11_{a309}, 11_{n1}, 11_{n2}, 11_{n3}, 11_{n20}$. |
| $R_{24}$. | $10_{83}, 10_{89}, 10_{91}, 10_{95}, 10_{99}, 10_{103}, 10_{109}, 10_{143}, 10_{157}, 11_{a28}, 11_{a54}, 11_{a70}$. |
| $R_{24}$. | $11_{a112}, 11_{a116}, 11_{a122}, 11_{a130}, 11_{a132}, 11_{a146}, 11_{a157}, 11_{a172}, 11_{a209}, 11_{a218}$. |
| $R_{24}$. | $11_{a251}, 11_{a253}, 11_{a262}, 11_{a275}, 11_{a295}, 11_{a304}, 11_{a321}, 11_{a325}, 11_{a344}, 11_{n48}$. |
| $R_{24}$. | $11_{n84}, 11_{n108}, 11_{n112}, 11_{n113}, 11_{n114}, 11_{n115}, 11_{n116}, 11_{n117}, 11_{n121}, 11_{n122}$. |
| $R_{24}$. | $11_{n134}, 11_{n142}, 11_{n143}, 11_{n151}, 11_{n152}, 11_{n167}, 11_{n179}$. |
| $R_{25}$. | $11_{a42}, 11_{a56}, 11_{a58}, 11_{a139}, 11_{a181}, 11_{a221}$. |
| $R_{26}$. | $11_{a78}, 11_{a156}$. |
| $R_{27}$. | $10_{81}, 10_{151}, 10_{154}, 11_{a2}, 11_{a3}, 11_{a7}, 11_{a16}, 11_{a17}, 11_{a31}, 11_{a34}, 11_{a38}, 11_{a87}$. |
| $R_{27}$. | $11_{a100}, 11_{a128}, 11_{n4}, 11_{n5}, 11_{n6}, 11_{n7}, 11_{n8}, 11_{n9}, 11_{n10}, 11_{n11}, 11_{n13}, 11_{n14}$. |
| $R_{27}$. | $11_{n15}, 11_{n28}, 11_{n29}, 11_{n30}, 11_{n31}, 11_{n32}, 11_{n33}, 11_{n51}, 11_{n52}, 11_{n53}, 11_{n58}$. |
| $R_{27}$. | $11_{n59}, 11_{n60}, 11_{n65}, 11_{n66}, 11_{n67}, 11_{n68}, 11_{n69}, 11_{n90}, 11_{n91}$. |
| $R_{28}$. | $10_{43}, 10_{44}, 11_{a5}, 11_{a47}, 11_{a84}, 11_{a92}, 11_{a110}, 11_{a176}, 11_{a184}$. |
| $R_{29}$. | $10_{71}, 10_{73}, 11_{a6}, 11_{a23}, 11_{a41}, 11_{a44}, 11_{a51}, 11_{a89}, 11_{a106}, 11_{a159}, 11_{a187}$. |
| $R_{29}$. | $11_{a329}, 11_{n76}, 11_{n77}, 11_{n78}$. |
| $R_{30}$. | $11_{a43}, 11_{n71}, 11_{n72}, 11_{n73}, 11_{n74}, 11_{n75}$. |
| $R_{31}$. | $10_{121}, 11_{a271}, 11_{a287}, 11_{a297}, 11_{a332}, 11_{a349}, 11_{a351}, 11_{n176}, 11_{n177}, 11_{n178}$. |
| $R_{31}$. | $11_{n183}, 11_{n184}, 11_{n185}$. |
| $R_{32}$. | $10_{145}, 11_{a125}, 11_{a164}, 11_{a248}, 11_{a255}, 11_{a270}, 11_{a272}, 11_{a273}, 11_{a274}, 11_{a276}$. |
| $R_{32}$. | $11_{a277}, 11_{a300}, 11_{a315}, 11_{a322}, 11_{a328}, 11_{n118}, 11_{n119}, 11_{n120}, 11_{n130}, 11_{n149}$. |
| $R_{32}$. | $11_{n153}, 11_{n154}, 11_{n164}, 11_{n165}, 11_{n166}, 11_{n168}$. |
| $R_{33}$. | $10_{115}, 10_{117}, 11_{a24}, 11_{a26}, 11_{a71}, 11_{a79}, 11_{a160}, 11_{a196}, 11_{n39}, 11_{n40}, 11_{n41}$. |
| $R_{33}$. | $11_{n45}, 11_{n46}, 11_{n47}, 11_{n145}, 11_{n146}, 11_{n147}$. |
| $R_{34}$. | $11_{a294}, 11_{a312}, 11_{a316}, 11_{n180}$. |
| $R_{35}$. | $10_{118}, 11_{a136}, 11_{a155}, 11_{a283}, 11_{a289}, 11_{n124}, 11_{n126}$. |
| $R_{36}$. | $11_{a14}, 11_{n21}, 11_{n22}, 11_{n23}$. |
| $R_{37}$. | $10_{88}, 11_{a30}, 11_{a131}, 11_{a147}, 11_{a167}, 11_{n50}$. |

| Family | Members |
|--------|---------|
| $R_{38}$. | $11_{a76}, 11_{a150}, 11_{n87}$. |
| $R_{39}$. | $10_{42}, 11_{a35}, 11_{a91}, 11_{a121}, 11_{n61}$. |
| $R_{40}$. | $10_{84}, 10_{107}, 11_{a27}, 11_{a66}, 11_{a69}, 11_{a80}, 11_{a114}, 11_{a151}, 11_{a168}, 11_{a189}, 11_{a217}$. |
| $R_{40}$. | $11_{a264}, 11_{n82}, 11_{n125}, 11_{n127}, 11_{n131}, 11_{n132}, 11_{n135}$. |
| $R_{41}$. | $11_{a19}, 11_{a25}, 11_{n34}, 11_{n35}, 11_{n36}, 11_{n42}, 11_{n43}, 11_{n44}$. |
| $R_{42}$. | $10_{45}, 11_{a32}, 11_{a77}, 11_{a85}, 11_{a109}, 11_{a175}, 11_{a178}, 11_{n54}, 11_{n55}, 11_{n56}, 11_{n105}$. |
| $R_{42}$. | $11_{n106}, 11_{n107}$. |
| $R_{43}$. | $11_{a11}, 11_{a18}$. |
| $R_{44}$. | $11_{a36}$. |
| $R_{46}$. | $11_{a104}, 11_{a134}, 11_{n103}$. |
| $R_{47}$. | $10_{69}$. |
| $R_{49}$. | $11_{a267}, 11_{a288}, 11_{a350}$. |
| $R_{51}$. | $10_{123}, 11_{a327}$. |
| $R_{52}$. | $10_{113}, 10_{132}, 10_{133}, 10_{136}, 10_{147}, 10_{150}, 11_{a73}, 11_{a101}, 11_{a135}, 11_{a138}, 11_{n85}$. |
| $R_{52}$. | $11_{n86}, 11_{n93}, 11_{n94}, 11_{n95}, 11_{n96}, 11_{n123}$. |
| $R_{54}$. | $11_{a239}, 11_{a284}, 11_{a314}, 11_{n148}, 11_{n174}, 11_{n182}$. |
| $R_{55}$. | $11_{a285}, 11_{a326}, 11_{n175}$. |
| $R_{57}$. | $11_{a72}$. |
| $R_{58}$. | $11_{a162}, 11_{a171}, 11_{a233}, 11_{n128}$. |
| $R_{60}$. | $11_{a126}$. |
| $R_{62}$. | $11_{a52}, 11_{a197}, 11_{n136}$. |
| $R_{65}$. | $11_{a266}, 11_{n156}, 11_{n157}$. |
| $R_{68}$. | $11_{a170}, 11_{n133}$. |

**Table 3.** The classification of links with crossing number no more than 11

| Family | Members |
|--------|---------|
| $R_1$. | $2_{a1}, 4_{a1}, 6_{a3}, 8_{a14}, 10_{a118}$. |
| $R_2$. | $5_{a1}, 6_{a5}, 7_{a3}, 7_{a6}, 8_{a18}, 9_{a14}, 9_{a29}, 9_{a36}, 9_{a40}, 10_{a145}, 10_{a161}, 11_{a110}$. |
| $R_2$. | $11_{a202}, 11_{a277}, 11_{a318}, 11_{a360}, 11_{a372}$. |
| $R_3$. | $6_{a2}, 7_{a7}, 8_{a5}, 8_{a11}, 8_{a12}, 8_{a13}, 8_{a21}, 8_{n7}, 8_{n8}, 9_{a44}, 9_{a48}, 10_{a44}, 10_{a46}, 10_{a74}$. |
| $R_3$. | $10_{a98}, 10_{a102}, 10_{a110}, 10_{a114}, 10_{a117}, 10_{a119}, 10_{a167}, 11_{a393}, 11_{a414}, 11_{a442}$. |
| $R_3$. | $11_{a504}, 11_{a518}, 11_{n306}$. |
| $R_4$. | $6_{a1}, 7_{a2}, 7_{a4}, 7_{a5}, 8_{a2}, 8_{a3}, 8_{a6}, 8_{a8}, 8_{a10}, 9_{a7}, 9_{a13}, 9_{a18}, 9_{a25}, 9_{a28}, 9_{a35}, 9_{a39}$. |
| $R_4$. | $9_{a41}, 9_{a50}, 9_{a54}, 9_{n1}, 9_{n2}, 9_{n3}, 9_{n18}, 10_{a9}, 10_{a27}, 10_{a31}, 10_{a33}, 10_{a48}, 10_{a62}, 10_{a66}$. |

| Family | Members |
|---|---|
| $R_4$. | $10_{a67}, 10_{a73}, 10_{a89}, 10_{a97}, 10_{a99}, 10_{a101}, 10_{a105}, 10_{a172}, 10_{n17}, 10_{n18}, 10_{n24}, 11_{a37}.$ |
| $R_4$. | $11_{a76}, 11_{a116}, 11_{a120}, 11_{a123}, 11_{a132}, 11_{a165}, 11_{a192}, 11_{a201}, 11_{a275}, 11_{a276}, 11_{a298}.$ |
| $R_4$. | $11_{a304}, 11_{a324}, 11_{a358}, 11_{a367}, 11_{a374}, 11_{a378}, 11_{a410}, 11_{a448}, 11_{a452}, 11_{a474}, 11_{a496}.$ |
| $R_4$. | $11_{a510}, 11_{n44}, 11_{n45}, 11_{n46}, 11_{n254}, 11_{n367}.$ |
| $R_5$. | $6_{a4}, 6_{n1}, 7_{a1}, 7_{n1}, 7_{n2}, 8_{a7}, 8_{a16}, 8_{n3}, 8_{n4}, 8_{n5}, 9_{a2}, 9_{a10}, 9_{a22}, 9_{a32}, 9_{n4}, 9_{n5}, 9_{n6}.$ |
| $R_5$. | $9_{n17}, 10_{a50}, 10_{a78}, 10_{a80}, 10_{a113}, 10_{a138}, 10_{a141}, 10_{a157}, 10_{a163}, 10_{n77}, 10_{n78}.$ |
| $R_5$. | $10_{n79}, 10_{n95}, 11_{a10}, 11_{a16}, 11_{a52}, 11_{a103}, 11_{a147}, 11_{a175}, 11_{a236}, 11_{a254}, 11_{a286}.$ |
| $R_5$. | $11_{a293}, 11_{a319}, 11_{a325}, 11_{a331}, 11_{a528}, 11_{n223}.$ |
| $R_6$. | $9_{a17}, 9_{a30}, 9_{a55}, 10_{a125}, 10_{a133}, 10_{a174}, 10_{n112}, 10_{n113}, 11_{a125}, 11_{a130}, 11_{a205}.$ |
| $R_6$. | $11_{a278}, 11_{a347}, 11_{a532}, 11_{a535}.$ |
| $R_7$. | $8_{a17}, 8_{a20}, 8_{n6}, 9_{a5}, 9_{a6}, 9_{a12}, 9_{a23}, 9_{a33}, 9_{n9}, 9_{n10}, 9_{n11}, 9_{n12}, 10_{a134}, 10_{a142}, 10_{a146}.$ |
| $R_7$. | $10_{a152}, 10_{a153}, 10_{a154}, 10_{a159}, 10_{n71}, 10_{n72}, 10_{n73}, 10_{n82}, 10_{n84}, 10_{n85}, 10_{n86}, 10_{n87}.$ |
| $R_7$. | $10_{n88}, 10_{n89}, 11_{a25}, 11_{a31}, 11_{a33}, 11_{a71}, 11_{a72}, 11_{a107}, 11_{a153}, 11_{a163}, 11_{a186}, 11_{a190}.$ |
| $R_7$. | $11_{a240}, 11_{a273}, 11_{a295}, 11_{a315}, 11_{a332}, 11_{a546}, 11_{n62}, 11_{n63}, 11_{n64}, 11_{n68}, 11_{n69}, 11_{n70}.$ |
| $R_7$. | $11_{n74}, 11_{n95}, 11_{n96}, 11_{n97}, 11_{n98}, 11_{n165}, 11_{n166}, 11_{n167}, 11_{n168}, 11_{n197}, 11_{n228}.$ |
| $R_7$. | $11_{n456}, 11_{n457}.$ |
| $R_8$. | $8_{a9}, 9_{a4}, 9_{a8}, 9_{a11}, 9_{a15}, 9_{a24}, 9_{a26}, 9_{a34}, 9_{a37}, 9_{a38}, 9_{a43}, 9_{n20}, 9_{n21}, 9_{n22}, 10_{a19}, 10_{a36}.$ |
| $R_8$. | $10_{a39}, 10_{a68}, 10_{a109}, 10_{a122}, 10_{a123}, 10_{a128}, 10_{a131}, 10_{a143}, 10_{a144}, 10_{a150}, 10_{a160}.$ |
| $R_8$. | $10_{n39}, 10_{n62}, 10_{n65}, 10_{n66}, 10_{n67}, 11_{a21}, 11_{a43}, 11_{a56}, 11_{a59}, 11_{a64}, 11_{a90}, 11_{a93}.$ |
| $R_8$. | $11_{a105}, 11_{a111}, 11_{a112}, 11_{a115}, 11_{a121}, 11_{a162}, 11_{a166}, 11_{a188}, 11_{a193}, 11_{a199}, 11_{a203}.$ |
| $R_8$. | $11_{a218}, 11_{a221}, 11_{a259}, 11_{a263}, 11_{a270}, 11_{a299}, 11_{a301}, 11_{a312}, 11_{a345}, 11_{a346}, 11_{a355}.$ |
| $R_8$. | $11_{a361}, 11_{a362}, 11_{a364}, 11_{a366}, 11_{a369}, 11_{a370}, 11_{a385}, 11_{a388}, 11_{a392}, 11_{a420}, 11_{a503}.$ |
| $R_8$. | $11_{a537}, 11_{a539}, 11_{a544}, 11_{n271}, 11_{n272}, 11_{n273}, 11_{n448}, 11_{n449}.$ |
| $R_9$. | $8_{a19}, 9_{a1}, 9_{a9}, 9_{a19}, 9_{a21}, 9_{a31}, 9_{a42}, 9_{a53}, 9_{n19}, 9_{n27}, 9_{n28}, 10_{a3}, 10_{a5}, 10_{a20}, 10_{a24}, 10_{a55}.$ |
| $R_9$. | $10_{a79}, 10_{a147}, 10_{a148}, 10_{a155}, 10_{a162}, 10_{a164}, 10_{n7}, 10_{n8}, 10_{n9}, 10_{n13}, 10_{n14}, 10_{n15}.$ |
| $R_9$. | $10_{n16}, 10_{n30}, 10_{n31}, 10_{n53}, 10_{n90}, 10_{n91}, 11_{a5}, 11_{a9}, 11_{a44}, 11_{a47}, 11_{a80}, 11_{a98}, 11_{a100}.$ |
| $R_9$. | $11_{a127}, 11_{a136}, 11_{a137}, 11_{a144}, 11_{a172}, 11_{a208}, 11_{a212}, 11_{a213}, 11_{a215}, 11_{a224}, 11_{a238}.$ |
| $R_9$. | $11_{a239}, 11_{a253}, 11_{a279}, 11_{a283}, 11_{a287}, 11_{a292}, 11_{a308}, 11_{a313}, 11_{a316}, 11_{a339}, 11_{a350}.$ |
| $R_9$. | $11_{a354}, 11_{a377}, 11_{a379}, 11_{a380}, 11_{a421}, 11_{a472}, 11_{a479}, 11_{a495}, 11_{n13}, 11_{n14}, 11_{n15}.$ |
| $R_9$. | $11_{n26}, 11_{n27}, 11_{n28}, 11_{n77}, 11_{n78}, 11_{n105}, 11_{n106}, 11_{n124}, 11_{n125}, 11_{n153}, 11_{n186}.$ |
| $R_9$. | $11_{n187}, 11_{n213}, 11_{n224}, 11_{n251}, 11_{n307}, 11_{n308}, 11_{n309}, 11_{n363}, 11_{n372}, 11_{n406}, 11_{n407}.$ |
| $R_{10}$. | $9_{a47}, 10_{a8}, 10_{a26}, 10_{a29}, 10_{a60}, 10_{a64}, 10_{a65}, 10_{a93}, 10_{a103}, 10_{a130}, 10_{a135}, 10_{n21}.$ |
| $R_{10}$. | $10_{n22}, 10_{n23}, 10_{n54}, 11_{a18}, 11_{a19}, 11_{a63}, 11_{a66}, 11_{a74}, 11_{a156}, 11_{a158}, 11_{a196}.$ |
| $R_{10}$. | $11_{a219}, 11_{a264}, 11_{a289}, 11_{a300}, 11_{a310}, 11_{a406}, 11_{a409}, 11_{a441}, 11_{a447}, 11_{a453}, 11_{a498}.$ |
| $R_{10}$. | $11_{n50}, 11_{n51}, 11_{n52}, 11_{n53}, 11_{n227}, 11_{n345}.$ |
| $R_{11}$. | $9_{a27}, 10_{a63}, 10_{a88}, 10_{a90}, 11_{a280}, 11_{a515}.$ |
| $R_{12}$. | $8_{a4}, 8_{a15}, 9_{a16}, 9_{a45}, 9_{a49}, 9_{a52}, 9_{n26}, 10_{a13}, 10_{a18}, 10_{a35}, 10_{a41}, 10_{a59}, 10_{a126}, 10_{a132}.$ |

| Family | Members |
|--------|---------|
| $R_{12}$. | $10_{a139}, 10_{n80}, 11_{a35}, 11_{a57}, 11_{a85}, 11_{a129}, 11_{a267}, 11_{a397}, 11_{a418}, 11_{a437}, 11_{a451}.$ |
| $R_{12}$. | $11_{a455}, 11_{a469}, 11_{a477}, 11_{a489}, 11_{a514}, 11_{n343}, 11_{n358}.$ |
| $R_{13}$. | $9_{a20}, 9_{n13}, 9_{n14}, 9_{n15}, 9_{n16}, 10_{a53}, 10_{a112}, 10_{n40}, 10_{n44}, 10_{n45}, 10_{n46}, 10_{n47}, 10_{n48}.$ |
| $R_{13}$. | $10_{n49}, 10_{n50}, 11_{a142}, 11_{a143}, 11_{a168}, 11_{a210}, 11_{a250}, 11_{a252}, 11_{a322}, 11_{a330}, 11_{a353}.$ |
| $R_{13}$. | $11_{a516}, 11_{n138}, 11_{n139}, 11_{n140}, 11_{n141}, 11_{n142}, 11_{n143}, 11_{n144}, 11_{n145}, 11_{n170}, 11_{n207}.$ |
| $R_{13}$. | $11_{n208}, 11_{n209}, 11_{n210}, 11_{n212}.$ |
| $R_{14}$. | $8_{a1}, 8_{n1}, 8_{n2}, 9_{a3}, 9_{a46}, 9_{a51}, 9_{n7}, 9_{n8}, 9_{n23}, 9_{n24}, 9_{n25}, 10_{a2}, 10_{a4}, 10_{a7}, 10_{a22}, 10_{a23}.$ |
| $R_{14}$. | $10_{a52}, 10_{a54}, 10_{a84}, 10_{a106}, 10_{a170}, 10_{n4}, 10_{n5}, 10_{n6}, 10_{n10}, 10_{n11}, 10_{n12}, 10_{n19}, 10_{n20}.$ |
| $R_{14}$. | $10_{n51}, 10_{n56}, 10_{n108}, 10_{n109}, 11_{a15}, 11_{a53}, 11_{a134}, 11_{a148}, 11_{a198}, 11_{a234}, 11_{a321}.$ |
| $R_{14}$. | $11_{a328}, 11_{a400}, 11_{a403}, 11_{a427}, 11_{a431}, 11_{a433}, 11_{a439}, 11_{a461}, 11_{a462}, 11_{a484}, 11_{a508}.$ |
| $R_{14}$. | $11_{a521}, 11_{a525}, 11_{n29}, 11_{n30}, 11_{n31}, 11_{n42}, 11_{n43}, 11_{n81}, 11_{n156}, 11_{n288}, 11_{n289}.$ |
| $R_{14}$. | $11_{n290}, 11_{n323}, 11_{n324}, 11_{n325}, 11_{n329}, 11_{n330}, 11_{n331}, 11_{n385}, 11_{n386}.$ |
| $R_{15}$. | $10_{a70}, 10_{n52}, 11_{a181}, 11_{a291}, 11_{a506}, 11_{a529}, 11_{n182}, 11_{n424}, 11_{n425}, 11_{n426}.$ |
| $R_{16}$. | $10_{a120}, 11_{a396}, 11_{a417}, 11_{a548}.$ |
| $R_{17}$. | $10_{a75}, 11_{a75}, 11_{a86}, 11_{a108}, 11_{a124}, 11_{a128}, 11_{a200}, 11_{a204}, 11_{a206}, 11_{a260}, 11_{a302}.$ |
| $R_{17}$. | $11_{a344}, 11_{a356}, 11_{a357}, 11_{a365}, 11_{a371}, 11_{a383}, 11_{a389}, 11_{a390}, 11_{a411}, 11_{a419}, 11_{a465}.$ |
| $R_{17}$. | $11_{a531}, 11_{n438}, 11_{n439}, 11_{n440}, 11_{n441}.$ |
| $R_{18}$. | $10_{a94}, 10_{a100}, 10_{a115}, 10_{a116}, 11_{a394}, 11_{a407}, 11_{a415}, 11_{a443}, 11_{a444}, 11_{a481}.$ |
| $R_{18}$. | $11_{a502}, 11_{a513}, 11_{a534}, 11_{n255}, 11_{n256}, 11_{n257}, 11_{n417}.$ |
| $R_{19}$. | $10_{a32}, 10_{a40}, 10_{a82}, 10_{n32}, 10_{n33}, 10_{n34}, 11_{a34}, 11_{a73}, 11_{a114}, 11_{a117}, 11_{a164}.$ |
| $R_{19}$. | $11_{a189}, 11_{a274}, 11_{a296}, 11_{a327}, 11_{a386}, 11_{a424}, 11_{a425}, 11_{a436}, 11_{a468}, 11_{a475}.$ |
| $R_{19}$. | $11_{a478}, 11_{a491}, 11_{a499}, 11_{n75}, 11_{n99}, 11_{n100}, 11_{n101}, 11_{n169}, 11_{n233}, 11_{n258}.$ |
| $R_{19}$. | $11_{n259}, 11_{n260}, 11_{n261}, 11_{n262}, 11_{n263}, 11_{n264}, 11_{n314}, 11_{n315}, 11_{n316}, 11_{n317}.$ |
| $R_{19}$. | $11_{n342}, 11_{n357}, 11_{n368}, 11_{n369}, 11_{n370}, 11_{n403}, 11_{n408}, 11_{n409}.$ |
| $R_{20}$. | $10_{a140}, 10_{a158}, 10_{n81}, 11_{a14}, 11_{a17}, 11_{a51}, 11_{a54}, 11_{a87}, 11_{a101}, 11_{a146}, 11_{a149}.$ |
| $R_{20}$. | $11_{a173}, 11_{a174}, 11_{a231}, 11_{a235}, 11_{a255}, 11_{a256}, 11_{a285}, 11_{a288}, 11_{a482}, 11_{a541}.$ |
| $R_{20}$. | $11_{n40}, 11_{n41}, 11_{n47}, 11_{n48}, 11_{n49}, 11_{n80}, 11_{n82}, 11_{n83}, 11_{n154}, 11_{n155}.$ |
| $R_{20}$. | $11_{n157}, 11_{n196}, 11_{n214}, 11_{n222}, 11_{n225}, 11_{n374}, 11_{n450}, 11_{n451}, 11_{n452}.$ |
| $R_{21}$. | $10_{a83}, 11_{a258}, 11_{a262}, 11_{a294}, 11_{a368}.$ |
| $R_{22}$. | $10_{a77}, 10_{a107}, 10_{a121}, 10_{n57}, 11_{a423}, 11_{a428}, 11_{a460}, 11_{a463}, 11_{a500}, 11_{a523}.$ |
| $R_{22}$. | $11_{a530}, 11_{n312}, 11_{n313}, 11_{n410}, 11_{n411}, 11_{n437}.$ |
| $R_{23}$. | $10_{a57}, 10_{a58}, 10_{a96}, 11_{a58}, 11_{a62}, 11_{a78}, 11_{a109}, 11_{a113}, 11_{a154}, 11_{a187}, 11_{a195}, 11_{a230}.$ |
| $R_{23}$. | $11_{a269}, 11_{a271}, 11_{a284}, 11_{a305}, 11_{a320}, 11_{a359}, 11_{a363}, 11_{a376}, 11_{a408}, 11_{a449}, 11_{a473}.$ |
| $R_{23}$. | $11_{a511}, 11_{n84}, 11_{n85}, 11_{n86}, 11_{n229}, 11_{n364}, 11_{n365}, 11_{n366}, 11_{n442}, 11_{n443}, 11_{n444}.$ |
| $R_{24}$. | $10_{a81}, 11_{a41}, 11_{a81}, 11_{a118}, 11_{a209}, 11_{a225}, 11_{a229}, 11_{a237}, 11_{a261}, 11_{a282}, 11_{a303}.$ |
| $R_{24}$. | $11_{a329}, 11_{a340}, 11_{a373}, 11_{a375}, 11_{a382}, 11_{a454}, 11_{a476}, 11_{a527}, 11_{n76}, 11_{n107}, 11_{n108}.$ |
| $R_{24}$. | $11_{n118}, 11_{n119}, 11_{a340}, 11_{n120}, 11_{n121}, 11_{n122}, 11_{n123}, 11_{n195}, 11_{n216}, 11_{n237}, 11_{n252}.$ |

| Family | Members |
|---|---|
| $R_{24}$. | $11_{n253}, 11_{n346}, 11_{n347}, 11_{n348}, 11_{n349}, 11_{n371}$. |
| $R_{25}$. | $11_{a126}, 11_{a155}, 11_{a450}, 11_{a466}$. |
| $R_{26}$. | $10_{a15}, 10_{a37}, 10_{a45}, 10_{a47}, 10_{a72}, 10_{a95}, 10_{a151}, 11_{a27}, 11_{a36}, 11_{a40}, 11_{a42}, 11_{a94}$. |
| $R_{26}$. | $11_{a96}, 11_{a131}, 11_{a191}, 11_{a538}, 11_{n445}, 11_{n446}, 11_{n447}$. |
| $R_{27}$. | $10_{a10}, 10_{a11}, 10_{a12}, 10_{a25}, 10_{a28}, 10_{a61}, 10_{a171}, 10_{n25}, 10_{n26}, 10_{n27}, 10_{n28}, 11_{a22}$. |
| $R_{27}$. | $11_{a20}, 11_{a23}, 11_{a24}, 11_{a26}, 11_{a32}, 11_{a65}, 11_{a68}, 11_{a69}, 11_{a88}, 11_{a151}, 11_{a152}, 11_{a220}$. |
| $R_{27}$. | $11_{a404}, 11_{a405}, 11_{a440}, 11_{a446}, 11_{a456}, 11_{a467}, 11_{a470}, 11_{a488}, 11_{a497}, 11_{a512}, 11_{n54}$. |
| $R_{27}$. | $11_{n55}, 11_{n56}, 11_{n57}, 11_{n58}, 11_{n59}, 11_{n60}, 11_{n61}, 11_{n65}, 11_{n71}, 11_{n72}, 11_{n73}, 11_{n91}$. |
| $R_{27}$. | $11_{n92}, 11_{n93}, 11_{n94}, 11_{n109}, 11_{n162}, 11_{n163}, 11_{n164}, 11_{n189}, 11_{n190}, 11_{n191}, 11_{n350}$. |
| $R_{27}$. | $11_{n354}, 11_{n355}, 11_{n356}, 11_{n359}, 11_{n397}$. |
| $R_{28}$. | $10_{a87}, 10_{a92}, 11_{a89}, 11_{a222}, 11_{a272}, 11_{a297}$. |
| $R_{29}$. | $11_{a20}, 11_{a61}, 11_{a91}, 11_{a92}, 11_{a160}, 11_{a194}, 11_{a265}, 11_{a341}, 11_{a342}, 11_{a445}, 11_{n87}, 11_{n110}$. |
| $R_{29}$. | $11_{n111}, 11_{n112}, 11_{n113}, 11_{n114}, 11_{n115}, 11_{n238}, 11_{n239}, 11_{n240}, 11_{n241}, 11_{n242}, 11_{n243}$. |
| $R_{30}$. | $10_{a30}, 10_{a38}, 10_{a108}, 10_{a166}, 10_{a168}, 10_{n35}, 10_{n36}, 10_{n37}, 10_{n38}, 10_{n58}, 10_{n59}, 10_{n60}$. |
| $R_{30}$. | $10_{n61}, 10_{n96}, 10_{n97}, 10_{n98}, 10_{n99}, 10_{n100}, 10_{n101}, 10_{n102}, 10_{n103}, 11_{a387}, 11_{a391}$. |
| $R_{30}$. | $11_{a413}, 11_{a435}, 11_{a490}, 11_{a501}, 11_{n265}, 11_{n266}, 11_{n267}, 11_{n268}, 11_{n269}, 11_{n270}$. |
| $R_{30}$. | $11_{n301}, 11_{n302}, 11_{n303}, 11_{n304}, 11_{n305}, 11_{n337}, 11_{n338}, 11_{n339}, 11_{n340}, 11_{n341}$. |
| $R_{30}$. | $11_{n398}, 11_{n399}, 11_{n400}, 11_{n401}, 11_{n402}, 11_{n412}, 11_{n413}, 11_{n414}, 11_{n415}, 11_{n416}$. |
| $R_{31}$. | $11_{a317}, 11_{a351}, 11_{n249}, 11_{n250}$. |
| $R_{32}$. | $10_{a71}, 10_{a86}, 11_{a180}, 11_{a226}, 11_{a233}, 11_{a309}, 11_{a326}, 11_{a349}, 11_{a381}, 11_{a412}$. |
| $R_{32}$. | $11_{a522}, 11_{a524}, 11_{a526}, 11_{n180}, 11_{n181}, 11_{n193}, 11_{n226}, 11_{n246}, 11_{n247}, 11_{n248}$. |
| $R_{32}$. | $11_{n298}, 11_{n299}, 11_{n300}$. |
| $R_{33}$. | $10_{a51}, 10_{n40}, 10_{n41}, 10_{n42}, 10_{n43}, 11_{a3}, 11_{a11}, 11_{a13}, 11_{a45}, 11_{a97}, 11_{a183}$. |
| $R_{33}$. | $11_{a197}, 11_{a399}, 11_{a432}, 11_{a434}, 11_{a438}, 11_{a483}, 11_{a507}, 11_{n7}, 11_{n8}, 11_{n9}, 11_{n32}$. |
| $R_{33}$. | $11_{n33}, 11_{n34}, 11_{n35}, 11_{n38}, 11_{n39}, 11_{n116}, 11_{n117}, 11_{n285}, 11_{n286}, 11_{n287}$. |
| $R_{33}$. | $11_{n326}, 11_{n327}, 11_{n328}, 11_{n332}, 11_{n333}, 11_{n334}, 11_{n335}, 11_{n336}, 11_{n344}, 11_{n375}$. |
| $R_{33}$. | $11_{n376}, 11_{n377}, 11_{n378}, 11_{n379}, 11_{n380}, 11_{n381}, 11_{n382}, 11_{n383}, 11_{n384}$. |
| $R_{34}$. | $10_{a49}, 10_{a127}, 10_{a137}, 10_{n68}, 10_{n69}, 10_{n70}, 10_{n74}, 10_{n75}, 10_{n76}, 11_{a4}, 11_{a6}, 11_{a7}$. |
| $R_{34}$. | $11_{a48}, 11_{a50}, 11_{a102}, 11_{a145}, 11_{a182}, 11_{a257}, 11_{a306}, 11_{a542}, 11_{n10}, 11_{n11}, 11_{n12}$. |
| $R_{34}$. | $11_{n16}, 11_{n17}, 11_{n18}, 11_{n19}, 11_{n20}, 11_{n21}, 11_{n183}, 11_{n184}, 11_{n185}, 11_{n453}, 11_{n454}$. |
| $R_{35}$. | $10_{a76}, 10_{a156}, 10_{n92}, 10_{n93}, 10_{n94}, 11_{a67}, 11_{a141}, 11_{a169}, 11_{a176}, 11_{a177}, 11_{a178}, 11_{a179}$. |
| $R_{35}$. | $11_{a246}, 11_{a251}, 11_{a290}, 11_{a314}, 11_{a333}, 11_{a335}, 11_{a352}, 11_{n88}, 11_{n89}, 11_{n90}, 11_{n130}$. |
| $R_{35}$. | $11_{n131}, 11_{n132}, 11_{n133}, 11_{n134}, 11_{n135}, 11_{n136}, 11_{n137}, 11_{n146}, 11_{n147}, 11_{n148}, 11_{n149}$. |
| $R_{35}$. | $11_{n150}, 11_{n151}, 11_{n152}, 11_{n171}, 11_{n178}, 11_{n179}, 11_{n198}, 11_{n199}, 11_{n200}, 11_{n201}, 11_{n211}$. |
| $R_{37}$. | $10_{a85}, 10_{a104}, 10_{n55}, 11_{a214}, 11_{a223}, 11_{a493}, 11_{n192}, 11_{n230}, 11_{n231}$. |
| $R_{38}$. | $10_{a21}, 10_{a42}, 10_{a136}, 11_{a49}, 11_{a99}, 11_{a135}, 11_{a138}, 11_{a211}, 11_{a243}, 11_{a337}, 11_{a338}$. |
| $R_{38}$. | $11_{a384}, 11_{a429}, 11_{a459}, 11_{a492}, 11_{a547}, 11_{n79}, 11_{n236}, 11_{n458}, 11_{n459}$. |

| Family | Members |
|---|---|
| $R_{40}$. | $10_{a69}, 11_{a28}, 11_{a140}, 11_{a167}, 11_{a217}, 11_{a228}, 11_{a323}, 11_{a334}, 11_{a402}, 11_{a519}, 11_{n66}$. |
| $R_{40}$. | $11_{n188}, 11_{n194}, 11_{n232}, 11_{n234}, 11_{n235}, 11_{n294}, 11_{n295}, 11_{n296}, 11_{n297}$. |
| $R_{41}$. | $10_{a6}, 10_{a14}, 10_{a173}, 10_{n29}, 10_{n110}, 10_{n111}, 11_{a8}, 11_{a29}, 11_{a79}, 11_{a422}, 11_{a426}$. |
| $R_{41}$. | $11_{a457}, 11_{a471}, 11_{a480}, 11_{a486}, 11_{a494}, 11_{n22}, 11_{n23}, 11_{n24}, 11_{n25}, 11_{n67}, 11_{n102}$. |
| $R_{41}$. | $11_{n103}, 11_{n104}, 11_{n310}, 11_{n318}, 11_{n319}, 11_{n351}, 11_{n352}, 11_{n360}, 11_{n361}, 11_{n362}$. |
| $R_{41}$. | $11_{n373}, 11_{n389}, 11_{n390}, 11_{n391}, 11_{n392}, 11_{n393}, 11_{n404}, 11_{n405}$. |
| $R_{42}$. | $10_{a91}, 11_{a55}, 11_{a157}, 11_{a159}, 11_{a216}, 11_{a247}, 11_{a248}, 11_{a266}$. |
| $R_{43}$. | $10_{a17}, 10_{a129}, 11_{a30}, 11_{a38}, 11_{a39}, 11_{a60}, 11_{a77}, 11_{a84}, 11_{a106}, 11_{a185}, 11_{a268}, 11_{a543}$. |
| $R_{44}$. | $11_{a416}$. |
| $R_{45}$. | $10_{a34}, 11_{a83}, 11_{a119}, 11_{a122}, 11_{a161}$. |
| $R_{47}$. | $11_{a82}, 11_{a241}, 11_{a242}, 11_{a307}, 11_{a311}$. |
| $R_{48}$. | $10_{a124}, 10_{a165}, 11_{a95}, 11_{a343}, 11_{a395}, 11_{a533}, 11_{a536}, 11_{a540}$. |
| $R_{49}$. | $10_{a56}, 11_{a150}, 11_{a249}, 11_{a348}, 11_{a509}, 11_{n158}, 11_{n159}, 11_{n160}, 11_{n161}, 11_{n428}, 11_{n429}$. |
| $R_{49}$. | $11_{n430} 11_{n431}$. |
| $R_{50}$. | $10_{a169}, 10_{n104}, 10_{n105}, 10_{n106}, 10_{n107}, 11_{a401}, 11_{a487}, 11_{n291}, 11_{n292}, 11_{n293}, 11_{n394}$. |
| $R_{50}$. | $11_{n395} 11_{n396}$. |
| $R_{51}$. | $11_{n202}, 11_{n203}, 11_{n204}, 11_{n205}, 11_{n206}$. |
| $R_{52}$. | $11_{a207}, 11_{a232}$. |
| $R_{54}$. | $10_{a111}, 10_{n63}, 10_{n64}, 11_{a517}, 11_{n432}, 11_{n433}, 11_{n434}, 11_{n435}$. |
| $R_{55}$. | $11_{a170}, 11_{a171}, 11_{a245}, 11_{a281}, 11_{n172}, 11_{n173}, 11_{n174}, 11_{n175}, 11_{n176}, 11_{n177}$. |
| $R_{55}$. | $11_{n217}, 11_{n218}, 11_{n219}, 11_{n220}, 11_{n221}$. |
| $R_{56}$. | $11_{a505}, 11_{n418}, 11_{n419}, 11_{n420}, 11_{n421}, 11_{n422}, 11_{n423}$. |
| $R_{57}$. | $10_{a1}, 10_{n1}, 10_{n2}, 10_{n3}, 11_{a1}, 11_{a2}, 11_{a133}, 11_{a398}, 11_{a430}, 11_{n1}, 11_{n2}, 11_{n3}, 11_{n4}$. |
| $R_{57}$. | $11_{n5}, 11_{n6}, 11_{n274}, 11_{n275}, 11_{n276}, 11_{n277}, 11_{n278}, 11_{n279}, 11_{n280}, 11_{n281}, 11_{n282}$. |
| $R_{57}$. | $11_{n283}, 11_{n284}, 11_{n320}, 11_{n321}, 11_{n322}, 11_{n427}$. |
| $R_{58}$. | $11_{a184}, 11_{a244}$. |
| $R_{60}$. | $10_{a43}, 11_{a46}, 11_{a104}$. |
| $R_{62}$. | $11_{a139}, 11_{a336}, 11_{n126}, 11_{n127}, 11_{n128}, 11_{n129}$. |
| $R_{63}$. | $10_{a149}, 10_{n83}, 11_{a12}, 11_{a70}, 11_{a464}, 11_{a545}, 11_{n36}, 11_{n37}, 11_{n353}, 11_{n455}$. |
| $R_{64}$. | $11_{a227}$. |
| $R_{66}$. | $11_{a485}, 11_{n387}, 11_{n388}$. |
| $R_{67}$. | $11_{a520}, 11_{n436}$. |
| $R_{69}$. | $11_{n244}, 11_{n245}$. |

# Fractional Vertex Arboricity of Graphs[*]

Qinglin Yu[1,2] and Liancui Zuo[1,**]

[1] Center for Combinatorics, LPMC, Nankai University, Tianjin, 300071, China
[2] Department of Mathematics and Statistics, Thompson Rivers University,
Kamloops, BC, Canada
lczuo@cfc.nankai.edu.cn

**Abstract.** The *vertex arboricity* $va(G)$ of a graph $G$ is the minimum number of subsets into which the vertex set $V(G)$ can be partitioned so that each subset induces an acyclic subgraph. The fractional version of vertex arboricity is introduced in this paper. We determine fractional vertex arboricity for several classes of graphs, e.g., complete multipartite graphs, cycles, integer distance graphs, prisms and Peterson graph.

**Keywords:** vertex arboricity; tree coloring; fractional vertex arboricity; fractional tree coloring.

## 1 Introduction

In this paper, we use $\mathbb{Z}$ to denote the set of all integers and $|S|$ for the cardinality of a set $S$ ($|S| = +\infty$ means that $S$ is an infinite set).

A *k-coloring* of a graph $G$ is a mapping $g$ from $V(G)$ to $\{1, 2, \ldots, k\}$. With respect to a given $k$-coloring, $V_i$ denotes the set of all vertices of $G$ colored with $i$, and $\langle V_i \rangle$ denotes the subgraph induced by $V_i$ in G. If $V_i$ induces a subgraph whose connected components are trees, then $g$ is called a *k-tree coloring*. The *vertex arboricity* of a graph $G$, denoted by $va(G)$, is the minimum integer $k$ for which $G$ has a $k$-tree coloring. In other words, the vertex arboricity $va(G)$ of $G$ is the minimum number of subsets into which the vertex set $V(G)$ can be partitioned so that each subset induces an acyclic subgraph (i.e., a forest).

In fact, if $V_i$ is an independent set for each $i$ ($1 \leq i \leq k$), then $g$ is called a *proper k-coloring* and the chromatic number $\chi(G)$ of a graph $G$ is the minimum integer $k$ of colors for which $G$ has a proper $k$-coloring. So the proper coloring is a special case of the tree coloring.

Kronk and Mitchem [4] proved that $va(G) \leq \lceil \frac{\Delta(G)+1}{2} \rceil$ for any graph $G$. Chartrand etc. [2] showed $va(K(p_1, p_2, \ldots, p_n)) = n - \max\{k \mid \sum_{i=0}^{k} p_i \leq n-k\}$ for the complete $n$-partite graph $K(p_1, p_2, \ldots, p_n)$, where $p_0 = 0$, $1 \leq p_1 \leq p_2 \leq \cdots \leq p_n$.

In this paper, we introduce the fractional version of vertex arboricity and to determine fractional vertex arboricity for several families of graphs. This is

---

[*] This work is supported by Nankai University Overseas Scholar Grant, RFDP of Higher Education of China and Discovery Grant of NSERC of Canada.
[**] Corresponding author.

J. Akiyama et al. (Eds.): CJCDGCGT 2005, LNCS 4381, pp. 245–252, 2007.

the first paper in a series of investigations on fractional vertex arboricity, its relationship with other graphic parameters.

## 2  Fractional Vertex Arboricity of Graphs

Let $S$ be a set of subsets of a set $V$. A *covering* of $V$ is a collection of elements $L_1, L_2, \ldots, L_j$ of $S$ such that $V \subseteq L_1 \cup \cdots \cup L_j$.

For any graph $G$, let $\mathcal{F}(G)$ be the set of all subsets of $V(G)$ that induce forests of $G$.

We now define the fractional vertex arboricity $va_f(G)$ of a graph $G$ as follows.

**Definition 1.** *A fractional tree coloring of a graph $G$ is a mapping $g$ from $\mathcal{F}(G)$ to the interval $[0, 1]$ such that*

$$\sum_{L \text{ contains } x} g(L) \geq 1, \quad \text{for any } x \in V(G).$$

*The* weight *of a fractional tree coloring is the sum of its values, and the* fractional vertex arboricity *of a graph $G$ is the minimum possible weight of a fractional coloring, that is,*

$$va_f(G) = \min \left\{ \sum_{L \in \mathcal{F}(G)} g(L) \mid g \text{ is a fractional tree coloring of } G \right\}.$$

Clearly, we have $va_f(H) \leq va_f(G)$ for any subgraph $H$ of $G$.

If we restrict the range of a mapping $g$ to $\{0, 1\}$ instead of $[0, 1]$, then $va_f(G)$ is the usual vertex arboricity, $va(G)$.

If $g$ is a $va(G)$-tree coloring of $G$ and $V_i = \{v \mid v \in V(G), g(v) = i\}$ ($1 \leq i \leq va(G)$), then we can define a mapping $h \colon \mathcal{F}(G) \longrightarrow [0, 1]$ by

$$h(L) = \begin{cases} 1, & \text{for } L = V_i, 1 \leq i \leq va(G), \\ 0, & \text{otherwise.} \end{cases}$$

such that $h$ is a fractional tree coloring of $G$ which has the weight $va(G)$. Therefore, it follows immediately that $va_f(G) \leq va(G)$.

Conversely, if $G$ has a $(0, 1)$-valued fractional tree coloring $g$ of weight $k$. Then the support of $g$ consists of $k$ forests $V_1, V_2, \ldots, V_k$ whose union is $V(G)$. If we color any vertex $v$ with the smallest $i$ such that $v \in V_i$, then we have a $k$-tree coloring of $G$. Thus the vertex arboricity of $G$ is the minimum weight of a $(0, 1)$-valued fractional tree coloring.

*Remark 1.* Vertex arboricity of a finite graph $G$ can be seen as an optimal solution of an integer programming and its fractional version can be viewed as an optimal solution of its relaxed problem, i.e., a linear programming problem.

To each set $L_i \in \mathcal{F}(G)$ we associate a $(0, 1)$-variable $x_i$ with it. The vector $\mathbf{x} = \{x_i\}$ is an indicator of the sets we have selected for the covering. Let $M$ be the vertex-forest incident matrix of $G$, i.e., the $(0, 1)$-matrix whose rows are indexed by $V(G)$, columns are indexed by $\mathcal{F}(G)$ and $(i, j)$-entry is 1 only when $v_i \in L_j$. The condition that the indicator vector $\mathbf{x}$ corresponds to a covering is simply $M\mathbf{x} \geq 1$ (that is, every coordinate of $M\mathbf{x}$ is at least 1). Hence the vertex arboricity of $G$ is precisely the optimal value of the integer programming

$$
\begin{aligned}
&Min \qquad \sum_i x_i \\
&Subject\ to \\
&\qquad\qquad M\mathbf{x} \geq 1 \\
&\qquad\qquad x_i = 0\ or\ 1\ (1 \leq i \leq |\mathcal{F}(G)|).
\end{aligned}
\tag{1}
$$

The relaxation of the integer programming (1) is the following linear programming

$$
\begin{aligned}
&Min \qquad \sum_i x_i \\
&Subject\ to \\
&\qquad\qquad M\mathbf{x} \geq 1 \\
&\qquad\qquad 0 \leq x_i \leq 1\ (1 \leq i \leq |\mathcal{F}(G)|)
\end{aligned}
\tag{2}
$$

and the optimal value of (2) is the fractional vertex arboricity of $G$.

Using Weak Duality Theorem for dual problems, we can derive the lower bound for $va_f(G)$.

**Lemma 1.** Let $G$ be a finite graph, $t = \max\{|L| \mid L \in \mathcal{F}(G)\}$, then $va_f(G) \geq \frac{|V(G)|}{t}$.

*Proof.* The dual linear programming of (2) is the following

$$
\begin{aligned}
&Max \qquad \sum_j y_j \\
&Subject\ to \\
&\qquad\qquad M^T \mathbf{y} \leq 1 \\
&\qquad\qquad 0 \leq y_j \leq 1\ (1 \leq i \leq |V|).
\end{aligned}
\tag{3}
$$

Thus, if we define $f$ to take the value $f(v)$ on each vertex of $V(G)$ with $0 \leq f(v) \leq 1$ and $M^T \mathbf{y} \leq 1$ for $\mathbf{y} = (f(v_1), \ldots, f(v_n))^T$ with $n = |V|$, then $\mathbf{y}$ is a feasible solution of (3).

Let $\omega$ be the objective value of (3) for some feasible solution $\mathbf{y}$. Since (2) and (3) are a pair of dual problems, from Weak Duality Theorem (see [1]), we have $\omega \leq va_f(G)$.

If we assign each vertex of $G$ with a weight $\frac{1}{t}$, then we have a feasible solution of (3). Thus $va_f(G) \geq \frac{|V(G)|}{t}$. □

Therefore, $va_f(G) \geq 1$ for any nonempty graph $G$. Clearly, $va_f(G) = 1$ if a graph $G$ is a forest.

For a complete $n$-partite graph $G = K(m_1, m_2, \ldots, m_n)$, we denote the vertices of $n$-partite of $V(G)$ by

$$X_1 = \{v_{11}, v_{12}, \ldots, v_{1m_1}\}$$

$$X_2 = \{v_{21}, v_{22}, \ldots, v_{2m_2}\}$$

$$\cdots$$

$$X_n = \{v_{n1}, v_{n2}, \ldots, v_{nm_n}\},$$

where $|X_i| = m_i$ for $1 \leq i \leq n$.

**Theorem 1.** *Let $n \geq 2$. For a complete $n$-partite graph $G = K(m_1, m_2, \ldots, m_n)$,*

$$va_f(G) = n - \frac{n}{m+1}, \quad for \quad m_1 = m_2 = \cdots = m_n = m,$$

*and*

$$n - \frac{m}{m+1} \leq va_f(G) \leq n - \frac{m(n+1)}{(m+1)^2},$$

*for $m_1 = m_2 = \cdots = m_{n-1} = m > m_n = n$.*

*Proof.* (1) For $m \geq 3$, it is easy to see that $t = \max\{|X| \mid X \in \mathcal{F}(G)\} = m+1$. So $va_f(G) \geq \frac{mn}{m+1} = n - \frac{n}{m+1}$ by Lemma 1. Define a mapping $h_1 \colon \mathcal{F}(G) \longrightarrow [0, 1]$ by

$$h_1(X) = \begin{cases} \frac{1}{(m+1)(n-1)}, & \text{for } X = X_i \cup \{v_{kj}\}, \ 1 \leq i, j, k \leq n, \ i \neq k, \\ 0, & \text{otherwise.} \end{cases}$$

Since there are exactly $(m+1)(n-1)$ forests that have nonzero weights containing vertex $v_{ij}$ for $1 \leq i, j \leq m$, $h_1$ is a fractional tree coloring of $G$. The number of $(m+1)$-forests that contain $m$ elements in $X_i$ is $\binom{n-1}{1}\binom{m}{1} = m(n-1)$. So there are $nm(n-1)$ elements in $\mathcal{F}$ that have nonzero values or $va_f(G) \leq \frac{nm(n-1)}{(m+1)(n-1)} = n - \frac{n}{m+1}$. Therefore $va_f(G) = n - \frac{n}{m+1}$.

(2) For $m = 2$, it is straight forward to verify that $t = \max\{|X| \mid X \in \mathcal{F}(G)\} = 3$. So $va_f(G) \geq \frac{2n}{3}$. Define a mapping $h_2 \colon \mathcal{F}(G) \to [0, 1]$ by

$$h_2(X) = \begin{cases} \frac{1}{3(n-1)}, & \text{for } |X| = 3 \text{ and there exist } i < j \text{ such that } X \subseteq X_i \cup X_j, \\ 0, & \text{otherwise.} \end{cases}$$

The number of all 3-forests that contain two elements in $X_1$ is $2(n-1)$ and the number of all 3-forests that contain one element in $X_1$ is also $2(n-1)$. So there are $4(n-1) + 4(n-2) + \cdots + 8 + 4 = 2(n-1)n$ elements in $\mathcal{F}$ that have nonzero values. Then $h_2$ is a fractional tree coloring of $G$ which has weight $\frac{1}{3(n-1)}2(n-1)n = \frac{2n}{3}$ or $va_f(G) \leq \frac{2n}{3}$. Therefore $va_f(G) = \frac{2n}{3}$.

(3) For $m = 1$, define a mapping $h_3 \colon \mathcal{F}(G) \to [0, 1]$ by

$$h_3(X) = \begin{cases} \frac{1}{n-1}, & \text{if } |X| = 2, \\ 0, & \text{otherwise.} \end{cases}$$

Then $h_3$ is a fractional tree coloring of $G$ which has weight $\frac{n}{2}$. Thus $va_f(G) \leq \frac{n}{2}$. It is easy to see that $t = \max\{|X| \mid X \in \mathcal{F}(G)\} = 2$, so $va_f(G) \geq \frac{|V(G)|}{t} = \frac{n}{2}$. Hence, $va_f(G) = \frac{n}{2}$.

(4) For $m_1 = m_2 = \cdots = m_{n-1} = m > n$ and $m_n = n$, define a mapping $h_4 \colon \mathcal{F}(G) \to [0, 1]$ by

$$h_4(X) = \begin{cases} \frac{1}{(n-1)(m+1)}, & \text{if } X = X_i \cup \{v_{nj}\} \text{ for } i < n \\ & \text{or } X = X_n \cup \{v_{kj}\} \text{ for } k < n, \\ \frac{nm-m-2}{(n-1)(m+1)^2(n-2)}, & \text{if } X = X_i \cup \{v_{kj}\} \text{ for } i, k < n, \\ 0, & \text{otherwise.} \end{cases}$$

It is not hard to verify that $h_4$ is a fractional tree coloring. Moreover, there are $n(n-1) + (n-1)m$ forests that contain elements of $V_n$ and have nonzero values, $\binom{n-1}{1}\binom{n-2}{1}\binom{m}{1}$ forests that do not contain any element of $V_n$ and have nonzero values. Hence, $h_4$ has the weight

$$\frac{n+m}{m+1} + (n-1)(n-2)m\frac{nm-m-2}{(n-1)(m+1)^2(n-2)}$$

$$= \frac{n+m}{m+1} + m\frac{nm-m-2}{(m+1)^2} = n - \frac{m(n+1)}{(m+1)^2}.$$

So $va_f(G) \leq n - \frac{m(n+1)}{(m+1)^2}$.

Since $t = \max\{|X| \mid X \in \mathcal{F}(G)\} = m + 1$, so $va_f(G) \geq \frac{|V(G)|}{t} = \frac{n+(n-1)m}{m+1} = n - 1 + \frac{1}{m+1} = n - \frac{m}{m+1}$. $\square$

Next, we determine fractional vertex arboricities of several familiar graphs: cycles, prism of cycles and Petersen graph.

**Theorem 2. (1)** *For an $n$-cycle $C_n$, $va_f(C_n) = \frac{n}{n-1}$.*
**(2)** *Let $L_h$ be the prism of two $h$-cycles $(h \geq 3)$. Then $\frac{2h}{h+1} \leq va_f(L_h) \leq 2$.*
**(3)** *For Petersen graph $P(5, 2)$, we have $va_f(P(5, 2)) = \frac{10}{7}$.*

*Proof.* (1) Suppose that $C_n = a_0a_1 \cdots a_{n-1}a_0$. Let $P_i = a_ia_{i+1} \cdots a_{i+n-2}$, where the subscripts are taken with modulo $n$ and $0 \leq i \leq n - 1$. It is obvious that every $a_i$ is contained in exactly $n - 1$ paths $P_0, \ldots, P_i, P_{i+2}, \ldots, P_{n-1}$. Define a mapping $g \colon \mathcal{F} \to [0, 1]$ by

$$g(X) = \begin{cases} \frac{1}{n-1}, & \text{if } X = P_i \ (i = 0, 1, \ldots, n - 1), \\ 0, & \text{otherwise.} \end{cases}$$

Then $g$ is a fractional tree coloring of $C_n$ which has weight $\sum_{X \in \mathcal{F}(C_n)} g(X) = \frac{n}{n-1}$, so $va_f(C_n) \leq \frac{n}{n-1}$. Clearly, $t = \max\{|X| \mid X \in \mathcal{F}(C_n)\} = n - 1$, hence $va_f(C_n) \geq \frac{n}{n-1}$. Therefore $va_f(C_n) = \frac{n}{n-1}$.

(2) Denote $2h$ vertices of the prism $L_h$ by $u_1, u_2, \ldots, u_h$ and $v_1, v_2, \ldots, v_h$. Then the edges of $L_h$ are $u_i u_{i+1}$, $v_i v_{i+1}$ and $u_i v_i$ $(1 \leq i \leq n)$. Clearly, $t = \max\{|X| \mid X \in \mathcal{F}\} = h + 1$ and thus $va_f(L_h) \geq \frac{2h}{h+1}$. If we color the vertices $u_1, u_2, \ldots, u_{h-1}, v_{h-1}, v_1$ by 0 and the vertices $v_2, v_3, \ldots, v_{h-2}, v_h, u_h$ by 1, then it yields a tree coloring. Thus $va_f(L_h) \leq va(L_h) \leq 2$.

(3) Denote the vertex set of Petersen graph $P(5, 2)$ by $\{a, b, c, d, e, a_1, b_1, c_1, d_1, e_1\}$ and then the edge set is $\{ab, bc, cd, de, ea, aa_1, bb_1, cc_1, dd_1, ee_1, a_1c_1, a_1d_1, b_1d_1, b_1e_1, c_1e_1\}$. Since any eight vertices of $P(5, 2)$ would induce a cycle, we see $\max\{|X| \mid X \in \mathcal{F}\} = 7$. Then $va_f(P(5, 2)) \geq \frac{10}{7}$ by Lemma 1.

Let

$$
\begin{aligned}
S_1 &= \{a, b, c, d, a_1, b_1, e_1\}, & S_2 &= \{a, b, c, d, d_1, c_1, e_1\}, \\
S_3 &= \{b, c, d, e, e_1, a_1, d_1\}, & S_4 &= \{b_1, b, c, d, e, c_1, a_1\}, \\
S_5 &= \{c_1, c, d, e, a, d_1, b_1\}, & S_6 &= \{c, d, e, a, a_1, e_1, b_1\}, \\
S_7 &= \{d, d_1, e, a, b, e_1, c_1\}, & S_8 &= \{d, e, a, b, b_1, a_1, c_1\}, \\
S_9 &= \{e, a, b, c, c_1, b_1, d_1\}, & S_{10} &= \{e_1, e, a, b, c, a_1, d_1\}, \\
S_{11} &= \{a, a_1, c_1, c, d, e_1, b_1\}, & S_{12} &= \{a, a_1, d_1, d, c, b_1, e_1\}, \\
S_{13} &= \{b, b_1, d_1, d, e, a_1, c_1\}, & S_{14} &= \{b, b_1, e_1, e, d, c_1, a_1\}, \\
S_{15} &= \{c, c_1, e_1, e, a, b_1, d_1\}, & S_{16} &= \{c, c_1, a_1, a, e, b_1, d_1\}, \\
S_{17} &= \{d, d_1, b_1, b, a, c_1, e_1\}, & S_{18} &= \{d, d_1, a_1, a, b, c_1, e_1\}, \\
S_{19} &= \{e, e_1, b_1, b, c, d_1, a_1\}, & S_{20} &= \{e, e_1, c_1, c, b, a_1, d_1\}.
\end{aligned}
$$

Clearly, each $S_i$ $(1 \leq i \leq 20)$ induces a forest and each vertex is contained in exactly fourteen such forests. Define a mapping $g$ by

$$
g(X) = \begin{cases} \frac{1}{14}, & \text{if } X = S_i \ (1 \leq i \leq 20), \\ 0, & \text{otherwise}, \end{cases}
$$

then $g$ is a fractional tree coloring with the weight $\frac{20}{14} = \frac{10}{7}$. Hence, $va_f(P(5, 2)) \leq \frac{10}{7}$ and thus $va_f(P(5, 2)) = \frac{10}{7}$. □

In general, it is rather difficult to determine the exact values of either $va(G)$ or $va_f(G)$ for an *infinite* graph $G$. In the following, we investigate a family of special infinite graphs, integer distance graphs, and are able to determine values of $va_f(G)$ for some special cases. For a set $D$ of positive integers, the *integer distance graph* $G(D)$ is a graph with vertex set $\mathbb{Z}$ and two vertices $x$ and $y$ are adjacent if and only if $|x - y| \in D$, where $D$ is called the *distance set*.

**Theorem 3.** (1) For $D = \{1, 2, \ldots, m\}$, $va_f(G(D)) = \frac{m+1}{2}$.
(2) Let $P$ be the set of all prime numbers, then $va_f(G(P)) = 2$.

*Proof.* (1) Let

$$S_0 = \{\ldots, 0, 1, m+1, m+2, 2(m+1), 2(m+1)+1, \ldots\},$$
$$S_1 = \{\ldots, 1, 2, m+2, m+3, 2(m+1)+1, 2(m+1)+2, \ldots\},$$
$$S_2 = \{\ldots, 2, 3, m+3, m+4, 2(m+1)+2, 2(m+1)+3, \ldots\},$$

$$\cdots$$

$$S_{m-1} = \{\ldots, -2, -1, m-1, m, 2m, 2m+1, 3m+1, 3m+2, \ldots\},$$
$$S_m = \{\ldots, -1, 0, m, m+1, 2m+1, 2m+2, 2(m+1)+m, 3(m+1), \ldots\}.$$

Then each of $S_0, S_1, \ldots, S_m$ induces a forest and each integer $i$ is contained in exactly two $S_j$ $(0 \le j \le m)$. Define a mapping $g \colon \mathcal{F} \to [0, 1]$ by

$$g(X) = \begin{cases} \frac{1}{2}, & \text{if } X = S_j \ (j = 0, 1, \ldots, m), \\ 0, & \text{otherwise.} \end{cases}$$

Then $g$ is a fractional tree coloring of $G(D)$ which has the weight $\Sigma_{X \in \mathcal{F}(G(D))} g(X) = \frac{m+1}{2}$, so $va_f(G(D)) \le \frac{m+1}{2}$.

On the other hand, let $H$ be a subgraph induced by vertices $0, 1, \ldots, m$. Then $H$ is a complete graph of order $m+1$ and thus $va_f(G(D)) \ge va_f(H) = \frac{m+1}{2}$ by Theorem 1. Therefore, $va_f(G(D)) = \frac{m+1}{2}$.

(2) Let $S_i = \{n \mid n \equiv i \pmod 2, n \in \mathbb{Z}\}$ $(i = 0, 1)$, then $S_i$ induces a forest. It is obvious that each integer is contained in exactly one of such forests. Define a mapping $g \colon \mathcal{F} \to [0, 1]$ by

$$g(X) = \begin{cases} 1, & \text{if } X = S_i \ (i = 0, 1), \\ 0, & \text{otherwise.} \end{cases}$$

Then $g$ is a fractional tree coloring which has the weight 2. So $va_f(G(P)) \le 2$. Suppose that $H$ is the subgraph induced by vertices $0, 1, 2, \ldots, 7$. It is easy to verify that $t = \max\{|X| \mid X \subseteq V(H) \text{ and } X \text{ induces a forest of } H\} = 4$ and the vertex subset $\{0, 1, 2, 3\}$ induces a tree. So $va_f(H) \ge \frac{8}{4} = 2$. Hence $va_f(G(P)) = 2$. $\square$

## Acknowledgements

The authors are indebted to the anonymous referees for their constructive suggestions.

## References

1. Chvátal, V.: Linear Programming. W. H. Freeman and Company, New York (1983)
2. Chartrand, G., Kronk, H.V., Wall, C.E.: The Point Arboricity of a Graph. Israel J. Math **6** (1968) 169–175
3. Goddard, W.: Acyclic Coloring of Planar Graphs. Discrete Math. **91** (1991) 91–94

4. Kronk H.V., Mitchem, J.: Critical Point-Arboritic Graphs. J. London Math. Soc. **9** (2) (1975) 459–466
5. Škrekovski, R.On the Critical Point-Arboricity Graphs. J. Graph Theory **39** (2002) 50–61
6. Scheinerman, E.R., Ullman, D.H.: Fractional Graph Theory. John Wiley and Sons. Inc. New York (1997)

# Fitting Triangles into Rectangles*

Liping Yuan[1], Yevgenya Movshovich[2], and Ren Ding[1]

[1] College of Mathematics and Information Science
Hebei Normal University, Shijiazhuang 050016, China
lpyuan@hebtu.edu.cn, rending@hebtu.edu.cn
[2] Department of Mathematics and Computer Science
Eastern Illinois University, Charleston, IL 61920, USA
ymovshovich@eiu.edu

**Abstract.** A *figure* $F$ and a *target set* $T$ are given in the plane. We say that the figure $F$ *fits* in the target $T$, or, equivalently, the target $T$ *covers* the figure $F$, when there is a rigid motion $\mu$ (an isometry of the plane) so that $\mu(F) \subseteq T$, i.e., if $T$ has a subset *congruent* to $F$. In this paper we find necessary and sufficient conditions for a triangle with sides $a$, $b$, $c$ to fit into a rectangle with sides $u$ and $v$.

## 1 Introduction

Questions about precisely when one shape fits into another are fundamental to the elementary study of shapes in geometry, and yet such questions have only rarely been considered in the literature (see [4]). A *figure* $F$ and a *target set* $T$ are given in the plane. We say that the figure $F$ *fits* in the target $T$, or, equivalently, the target $T$ *covers* the figure $F$, when there is a rigid motion $\mu$ (an isometry of the plane) so that $\mu(F) \subseteq T$, i.e., if $T$ has a subset *congruent* to $F$. In 1993 Post [3] gave necessary and sufficient conditions on the six sides of two triangles to fit one into another, and in [2] this result was refined for equilateral triangles in triangles. Recently the necessary and sufficient conditions for the the first shape to fit into the second were given for the following figures: rectangles in triangles [6], rectangles in rectangles [5], squares in triangles [7], triangles in squares [1]. At the same time, the smallest triangular cover [8], rectangular cover [9], and parallelogram cover [10] for the family of triangles of diameter 1 were also considered. In this paper given a triangle with sides $a$, $b$, $c$ we find necessary and sufficient conditions on the sides $u$ and $v$ of a rectangle so that the triangle fits in the rectangle.

Let $T$ denote a triangle $\triangle ABC$ with sides $a \geq b \geq c$, angles $\alpha \geq \beta \geq \gamma$, and altitudes $h_a \leq h_b \leq h_c$. Let $R$ denote a rectangle $u \times v$ with $u \leq v$. Let $s$ denote the side of the minimal square containing $T$. It is clear that $T$ fits in $R$ when $u \geq s$ for any $v \geq u$. It is also clear that $T$ does not fit in $R$ when $u < h_a$. Therefore for $u \in [h_a, s]$ we will find $v(u)$ to determine the minimal rectangle $u \times v(u)$ containing $T$. Consequently $T$ will fit in $R$ if and only if $v \geq v(u)$.

---

* This research was supported by National Natural Science Foundation of China 10571042, Hebei NSF A2005000144, Tianyuan Fund for Mathematics 10426013 and Fund of Hebei Province for Doctorate B2004114.

J. Akiyama et al. (Eds.): CJCDGCGT 2005, LNCS 4381, pp. 253–258, 2007.
© Springer-Verlag Berlin Heidelberg 2007

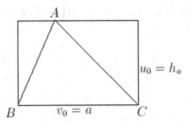

**Fig. 1.** The initial position of $T$ in $R_0$

We place $R = u \times v$ in a $u, v$-plane with the vertical independent axis $u$, so that the lower left corner of $R$ is at the origin. Now $T$ fits in a rectangle $R_0$ with sides $u_0 = h_a$ and $v_0 = a$ as shown on Fig. 1. We call it the *initial position* of $T$ in $R_0$, provided that the vertex $B$ is at the origin. Increasing $u$ gradually upwards, and rotating $T$ counter-clockwise around the origin, we decrease $v(u)$ accordingly to obtain minimal rectangles. We rotate $T$ as long as this continuous motion produces such rectangles. Then we flip or reflect the triangle to have a different vertex at the origin and perform the same type of rotation as long as it continuously produces minimal rectangles, etc. Theorems 1-3 state that the interval $[h_a, s]$ is split into several intervals of these continuous motions. One would expect the function $v(u)$ also to be piece-wise continuous. That is true for all but obtuse triangles, for which this function happens to be continuous (see Theorem 3). Lemmas 1 and 2 tell us which vertex of $T$ should be at the rotation corner on a given stretch of $u$. And finally, the side of the minimal square, which agrees with [1], will be determined by solving the equation $v(u) = u$ on the last interval of a continuous rotation.

**Definition 1.** *We say that a rectangle circumscribes $T$, if each side of the rectangle contains a vertex of $T$. When a vertex of $T$, say $C$, is at a corner of such a rectangle, the notation $v_C^b(u)$ corresponds to the position of $T$ with the side $b$ projected entirely on $v$.*

All minimal rectangles circumscribe the triangles with the exceptions described in Theorem 2.

From Fig. 2 we find that $v_C^b(u) = b \cos(\arcsin \frac{u}{a} - \gamma)$.

## 2    Main Results and Proofs

The following lemma is a reminiscence of Lemma 3 in [1]. The latter states that a circumscribing $T$ square is smaller if it is positioned around $T$ with a corner at the vertex with a larger angle.

**Lemma 1.** *If $u \in [h_a, h_b]$, then $v_B^a(u) \leq v_C^a(u)$. That is to say, a rotation of $T$ around the vertex with a larger angle produces smaller $v$ for the same $u$.*

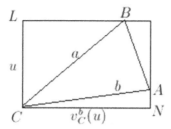

**Fig. 2.** Illustration of $v_C^b(u)$

*Proof.* When $u = h_a$, we have $\arcsin \frac{u}{c} - \beta = \arcsin \frac{u}{b} - \gamma = 0$. On the other hand, $\partial_u(\arcsin \frac{u}{c} - \arcsin \frac{u}{b}) > 0$. Thus, $\arcsin \frac{u}{c} - \beta \geq \arcsin \frac{u}{b} - \gamma$, therefore $v_B^a(u) = a\cos(\arcsin \frac{u}{c} - \beta) \leq v_C^a(u) = a\cos(\arcsin \frac{u}{b} - \gamma)$.

**Theorem 1.** *For $T$ with $\alpha \leq \frac{\pi}{2}$, $h_c \leq c$, the dimensions of minimal rectangles are:*

$u \times v_B^a(u), \quad u \in [h_a, h_b);$

$u \times v_A^b(u), \quad u \in [h_b, h_c);$

$u \times v_A^c(u), \quad u \in [h_c, s];$ *where the side of the minimal square, $s = \frac{bc\cos\alpha}{\sqrt{b^2+c^2-2bc\sin\alpha}}$, is found from equation $v_A^c(u) = u$.*

*Proof.* We place $T$ at its initial position in $R_0$. By Lemma 1, we rotate $T$ counter-clockwise around vertex $B$ by increasing $u$ so that $v = v_B^a(u)$. When $u$ has reached $h_b$ we flip $T$ so that its side $b$ rests on the lower side of the rectangle obtained with the vertex $A$ at the left corner. We reduce horizontal sides to $b$, because the law of sines yields

$$v_B^a(h_b) = a\cos(\alpha - \beta) = a\cos\alpha\cos\beta + b\sin^2\alpha = \cos\alpha\,[a\cos\beta - b\cos\alpha] + b > b.$$

Now in Lemma 1 we replace the interval by $[h_b, h_c]$ and side $a$ by $b$ to see that rotation around vertex $A$ produces smaller $v$. We keep rotating until $u = h_c$. Then we flip $T$ so that its side $c$ rests on the lower side of the rectangle, with the vertex $A$ at the left corner. We also reduce horizontal sides to $c$, because again, applying the law of sines, we see that

$$v_A^b(h_c) = b\cos\alpha\sqrt{1 - \frac{h_c^2}{c^2}} + b\frac{h_c}{c}\sin\alpha$$

$$= \sqrt{b^2 - h_c^2}\sqrt{1 - \frac{h_c^2}{c^2}} + \frac{h_c^2}{c}$$

$$\geq \sqrt{c^2 - h_c^2}\sqrt{1 - \frac{h_c^2}{c^2}} + \frac{h_c^2}{c}$$

$$= c = v_A^c(h_c).$$

This time in Lemma 1 we replace the interval by $[h_c, s]$ and side $a$ by $c$ to see again that rotation around vertex $A$ produces smaller $v$. And so we rotate $T$

around vertex $A$ until $v_A^c(u) = c \cos \alpha \sqrt{1 - \frac{u^2}{b^2}} + \frac{cu}{b} \sin \alpha = u$. We solve the equation for $u$ when we multiply it by $\frac{1}{u}$ and isolate

$$\frac{1}{u^2} = \frac{1}{b^2} + \frac{(b - c \sin \alpha)^2}{b^2 c^2 \cos^2 \alpha}.$$

**Lemma 2.** *Let a rectangle $R$ with sides $u > c$, $v = v_C^a(u)$ circumscribe $T$ so that vertex $C$ is at the lower left corner, while vertex $A$ is on the top side of $R$. Let a copy $T'$ of $T$ be placed in $R$ so that $C' = C$, while $B'$ is on the top side of $R$. Then vertex $A'$ is inside $R$.*

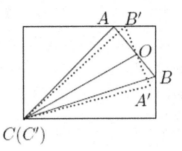

**Fig. 3.** Illustration of Lemma 2

*Proof.* We observe that $A'$ is above the bottom side of $R$, because $u > c$. It is also to the left of the right side of $R$, because $u \leq v$, and $\angle B'CA = \angle A'CB$. The angles are equal because when we connect vertex $C$ with the point of intersection $O$ of sides $c$ and $c'$, we see that

$$\triangle AOC \cong \triangle A'OC, \quad \text{and} \quad \triangle BOC \cong \triangle B'OC.$$

**Theorem 2.** *For $T$ with $\alpha \leq \frac{\pi}{2}$, $h_c > c$, the dimensions of minimal rectangles are:*

$u \times v_B^a(u), \quad u \in [h_a, h_b); \qquad u \times v_A^b(u), \qquad u \in [h_b, c];$
*and if $\gamma > \frac{\pi}{4}$, then*
$u \times h_c, \quad u \in (c, s], \qquad s = h_c$ *rectangles do not circumscribe $T$;*
*and if $\gamma \leq \frac{\pi}{4}$, then*
$u \times h_c, \quad u \in (c, a \cos(\alpha - \gamma))$, *rectangles do not circumscribe $T$;*
$u \times v_C^b(u), \quad u \in [a \cos(\alpha - \gamma), s], \quad s = \dfrac{ab \cos \gamma}{\sqrt{a^2 + b^2 - 2ab \sin \gamma}}$ *solves $v_C^b(u) = u$.*

*Proof.* For $u \in [h_a, c)$ we repeat the arguments from Theorem 1, replacing $h_c$ by $c$, arriving to the position of $T$ in $R = c \times h_c$ with side $c$ resting on the left side of $R$. For $u > c$ in some neighborhood of $c$ this would be the only position for $T$ to fit into $R = u \times h_c$. Eventually side $v = h_c$ can be decreased by rotating $T$ around vertex $C$, if $u$ is large enough for $T$ to fit into $R$ with $C$ at its corner. Apparently such $u < h_c$ exists only, when $\gamma < \frac{\pi}{4}$.

Indeed, let $u_1$, $u_2$ be so that $v_C^b(u_1) = v_C^a(u_2) = h_c$. Then from Fig. 2 with $v_C^b(u_1) = h_c$, we see that

$$a \cos(\angle BCL) = u_1 = a \cos(\alpha - \gamma),$$

because $\angle BCL = \frac{\pi}{2} - \gamma - \angle ACN$, and $\cos(\angle ACN) = \frac{h_c}{b} = \cos(\frac{\pi}{2} - \alpha)$. Similarly

$$u_2 = b \cos(\beta - \gamma).$$

Both $u_1$ and $u_2$ are strictly larger than $c$. Next we compute the differences:

(1)
$$h_c - u_2 = b \sin(\beta + \gamma) - b \cos(\beta - \gamma)$$

$$= b \cos \beta \cos \gamma (\tan \beta - 1)(1 - \tan \gamma),$$

and

(2)
$$h_c - u_1 = a \cos \alpha \cos \gamma (\tan \alpha - 1)(1 - \tan \gamma).$$

From (1) and (2) we see that $u_1 < u_2$, but this we knew already from Lemma 2. We can now observe the following:

When $\gamma > \frac{\pi}{4}$, the dimensions of minimal rectangles are $u \times h_c$ for all $u \in (c, h_c]$, and those rectangles do not circumscribe triangles $T$. Consequently, this will be the only type of triangles, whose minimal squares $\{h_c \times h_c\}$ do not circumscribe them (see also [1]).

When $\gamma \le \frac{\pi}{4}$, for $u \in (c, u_1)$ the triangle still fits into $R = u \times h_c$, only with $c$ resting on a side of $R$. On the other hand, when $u \ge u_1$, we see from (2) and Lemma 2 that a circumscribing $R = u \times v_C^b(u)$ is defined. Rotating $T$ around $C$ we reduce $v$ thus arriving to the minimal square, circumscribing $T$, with side $s = u$ that is found from the equation $v_C^b(u) = u$. We solve it as we solved $v_A^c(u) = u$ in Theorem 1.

**Theorem 3.** *For $T$ with $\alpha > \frac{\pi}{2}$, the dimensions of minimal rectangles are:*
*$u \times v_B^a(u), u \in [h_a, h_b]$;*
*and if $\beta \le \frac{\pi}{4}, u \times \sqrt{a^2 - u^2}, u \in [h_b, s]$, $s = \frac{a}{\sqrt{2}}$;*
*while if $\beta > \frac{\pi}{4}, u \times \sqrt{a^2 - u^2}, u \in [h_b, a \cos \beta]$;*
*$u \times v_C^b(u), u \in [a \cos \beta, s]$, $s = \frac{ab \cos \gamma}{\sqrt{a^2 + b^2 - 2ab \sin \gamma}}$.*

*Proof.* As in Theorems 1 and 2, we start from the initial position of $T$ in $R_0$. Keeping vertex $B$ at its place we rotate $T$ counter-clockwise with $v = v_B^a$ until $u = h_b$, so $b$ is on the top side of the rectangle obtained, and vertex $C$ is at the upper right corner. We continue to rotate $T$ about $B$ counter-clockwise keeping $C$ at the opposite corner with $v = \sqrt{a^2 - u^2}$.

If $\beta \le \frac{\pi}{4}$, we rotate $T$ until $u = a \sin \beta = h_c$ and then flip $T$ over the diagonal so that $c$ lies on the bottom of $R$ and continue rotating in the same fashion until $\sqrt{a^2 - u^2} = u$.

If $\beta > \frac{\pi}{4}$, we can rotate $T$ until side $c$ lies on the left side of $R$, and so $u = a \cos \beta$. Note that because $a^2 > b^2 + c^2$, by the law of cosines, $u > c$. By

Lemma 2, $v_C^b < v_C^a$, if both are defined. We can see that $v_C^b$ certainly is, when we flip $T$ so that $c$ lies on the right side of $R$. Now we rotate $T$ about $C$ counterclockwise with vertex $B$ moving along the top side of the rectangle, away from the corner, until $v_C^b(u) = u$.

# References

1. R. P. Jerrard and J. E. Wetzel, Triangles in squares, *Elem. Math.* 58 (2003), 65-72.
2. R. P. Jerrard and J. E. Wetzel, Equilateral triangles in triangles, *Amer. Math. Monthly* 109 (2002), 909-915.
3. K. A. Post, Triangle in a triangle: on a problem of Steinhaus, *Geom. Dedicata* 45 (1993), 115-120.
4. J. E. Wetzel, Fits and covers, *Math. Mag.* 76(5)(2003), 349-363.
5. J. E. Wetzel, Rectangles in rectangles, *Math. Mag.* 73(2000), 204-211.
6. J. E. Wetzel, Rectangles in triangles, *Journal of Geometry* 81(2004), No. 1-2, 180-191.
7. J. E. Wetzel, Squares in triangles, *Math. Gazzete* 86 (2002), 28-34.
8. L. Yuan, R. Ding, The Smallest Triangular Cover for Triangles of Diameter One, *Journal of Applied Mathematics and Computing*, Vol. 17, No. 1-2 (2005), 39-48.
9. L. Yuan, Y. Zhang, R. Ding, Box for Triangles of Diameter One, *The Mathematical Gazette*, to appear.
10. L. Yuan, Y. Movshovich, The Smallest Parallelogram Cover for Triangles of Diameter One, submitted.

# Regular Coronoids and Ear Decompositions of Plane Elementary Bipartite Graphs

Heping Zhang

School of Mathematics and Statistics, Lanzhou University, Lanzhou, Gansu 730000, P.R. China
zhanghp@lzu.edu.cn

**Abstract.** A connected bipartite graph is called *elementary* (or *normal*) if its every edge is contained in some perfect matching. In *rho* classification of coronoids due to Cyvin et al., normal coronoids are divided into two types: regular and half essentially disconnected. A coronoid is called *regular* if it can be generated from a single hexagon by a series of normal additions of hexagons (modes $L_1$, $L_3$ or $L_5$) plus corona condensations of hexagons of modes $L_2$ or $A_2$. Chen and Zhang (1997) gave a complete characterization: A coronoid is regular if and only if it has a perfect matching $M$ such that the boundaries of non-hexagon faces are all $M$-alternating cycles. In this article, a general concept for the regular addition of an allowed face is proposed and the above result is extended to a plane elementary bipartite graph some specified faces of which are forbidden by applying recently developed matching theory. As its corollary, we give an equivalent definition of regular coronoids as a special ear decomposition.

**Keywords:** Regular coronoid, normal coronoid, plane elementary bipartite graph, perfect matching, ear decomposition, regular decomposition.

## 1 Introduction

A benzenoid or hexagonal system is a finite connected plane graph with no cut-vertices in which every interior region is bounded by a regular hexagon of side length 1 [12]. A *coronoid* $G$ is a connected subgraph of a hexagonal system such that every edge lies in a hexagon of $G$ and $G$ contains at least one non-hexagon interior face (called corona hole), which should have a size of at least two hexagons. Further a coronoid is said to be *single* or *multiple* according as it contains a unique hole or at least two holes. For example, Fig. 1 illustrates a multiple coronoid with two holes. In general, coronoids can be regarded as benzenoids with holes. Since benzenoids and coronoids have counterparts in what called benzenoid and coronoid hydrocarbons, the studies of such systems are also of chemical relevance (see [2, 3, 5, 6, 7, 12, 13]).

A Kekulé structure $M$ (also called perfect matching or 1-factor in graph theory) of a graph $G$ is a set of pairwise disjoint edges of $G$ such that every vertex of $G$ is incident with an edge in $M$; a cycle $C$ of $G$ is called $M$-conjugated or $M$-alternating if the edges of $C$ alternate on and off $M$. For benzenoids or coronoids,

J. Akiyama et al. (Eds.): CJCDGCGT 2005, LNCS 4381, pp. 259–271, 2007.
© Springer-Verlag Berlin Heidelberg 2007

Kekuléan systems are those which possess a Kekulé structure. A Kekuléan system is called normal if every edge is contained in a Kekulé structure; essentially disconnected otherwise. In *neo* classification, these kinds of systems are classed into normal (n), essentially disconnected (e) and non-Kekuléan (o). Criteria for benzenoids and coronoids to be normal were given by Zhang and Chen [15], Zhang and Zheng [16].

**Theorem 1.** ([15, 16]) *A benzenoid or coronoid is normal if and only if the boundary of every non-hexagonal face is alternating with respect to a perfect matching.*

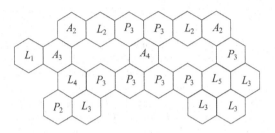

**Fig. 1.** Various modes of hexagons in a coronoid

Some modes of hexagons in a coronoid system are illustrated in Fig. 1. More complete and detailed definitions may be referred to Cyvin and Gutman [5, page 16]. In order to construct normal benzenoids and coronoids the following *addition* operations were introduced. A *normal addition* is to add a hexagon to a benzenoid or coronoid such that the added hexagon acquires the modes $L_1$, $L_3$ or $L_5$; A *corona condensation* is to add a hexagon to a benzenoid or coronoid such that the added hexagon acquires the modes $L_2$, $L_4$, $A_2$ or $A_3$ (see Fig. 1). For normal benzenoids the following construction was originally conjectured by Cyvin and Gutman [5], then proved by He and He (1990).

**Theorem 2.** ([8]) *Any normal benzenoid with $h + 1$ hexagons can be generated from a normal benzenoid with $h$ hexagons by a normal addition of one hexagon.*

In *rhe* classification of coronoids due to Cyvin et al., normal coronoids are divided into two types: Regular and half essentially disconnected (HED) [4]. Such a classification is useful for Kekulé structure count of HED coronoids.

**Definition 1.** ([2, 3]) *A coronoid $G$ is said to be* regular *if $G$ can be generated from a single hexagon by a series of normal additions plus corona condensations of modes $L_2$ or $A_2$.*

For example, the coronoid shown in Fig. 2 (a) is regular since it can be generated from a hexagon $h_1$ by normal additions of $h_2$, $h_3$, $h_4$, $h_5$, $h_6$ and $h_7$ (mode $L_1$), a corona condensation of $h_8$ (mode $A_2$) and normal additions of $h_9$ (mode $L_1$) and $h_{10}$ (mode $L_3$); whereas the coronoid shown in Fig. 2(b) is irregular, i.e. half essentially disconnected. Chen and Zhang gave the following characterization for a coronoid to be regular.

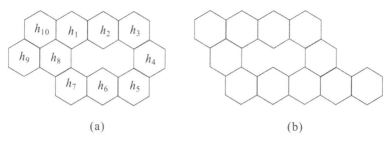

**Fig. 2.** (a) Regular coronoid, (b) irregular (HED) coronoid

**Theorem 3.** ([1,3]) *A coronoid is regular if and only if it has a perfect matching M such that the boundaries of non-hexagon faces are all M-alternating.*

A normal addition of a hexagon $h$ (modes $L_5$, $L_3$ or $L_1$) to a normal system has another explanation: A path $P$ of length 1, 3 or 5 is added so that such a path together with a part on the perimeter of a non-hexagon face forms a new hexagon $h$. Such a path of odd length is called an *ear* in matching theory of graphs [11]. Conversely, to remove such a hexagon $h$ means to delete the interior of $P$. The hexagon $h$ is called *removable* in the obtained larger system. Zheng [19] obtained a stronger result than Theorem 2: A normal benzenoid with at least two hexagons has two removable hexagons.

For a corona condensation of modes $L_4$ and $A_2$, two ears are added; for $A_3$ three ears are added. The corona condensation of mode $L_2$, however, is to add two paths of length 2, none of which is an ear. We wish to give an alternative construction of regular coronoids avoiding the corona condensation of mode $L_2$.

In this article the above discussion to regular coronoids is extended to plane elementary bipartite graphs. Corona holes are replaced by some specified faces called *forbidden faces*. A *regular addition* is defined in terms of ears and unifies normal additions and corona condensations of modes $L_4$, $A_2$ or $A_3$. The converse of regular addition is a *regular decomposition*. The main result (Theorem 10) in this paper extending Theorem 3 will be described in Section 2. Its proof will be given in Section 3 following a series of lemmas. As applications we give a novel characterization of regular coronoids as follows.

**Theorem 4.** *Let G be a coronoid. Then the following statements are equivalent:*

**(a)** *G is regular,*
**(b)** *G has a perfect matching M such that the boundary of every non-hexagon face is M-alternating cycle.*
**(c)** *G can be generated from a single hexagon by a series of normal additions plus corona condensations of modes $L_4$, $A_2$ or $A_3$.*

For regular single coronoids Theorem 4 (c) shows that in such a construction process the hexagon addition of mode $A_3$ does not appear; otherwise it would generate at least two holes. Thus we have the following result which is somewhat different from the corresponding result in [3, Chapter 8].

**Corollary 5.** *Let G be a single coronoid. Then G is regular if and only if G can be generated from a single hexagon by a series of normal additions plus only one corona condensation of mode $A_2$ or $L_4$.*

In general we propose a conjecture as follows.

**Conjecture 6.** *Let G be a coronoid. Then G is regular if and only if G can be generated from a single hexagon by a series of normal additions plus corona condensations of mode $A_2$.*

## 2    Regular Decomposition and Main Results

Let $G$ be a graph with the vertex set $V(G)$ and edge set $E(G)$. A connected bipartite graph is said to be *elementary* if every edge of $G$ is contained in a perfect matching. A plane elementary bipartite graph with more than two vertices is 2-connected and each face is bounded by a cycle. Various equivalent propositions for elementary bipartite graph were obtained. Here we only describe several useful in this paper. For convenience we always color all vertices of bipartite graphs considered with black and white such that any adjacent vertices receive different colors.

**Lemma 7.** *([9,11]) Let G be a connected bipartite graph. Then G is elementary if and only if for any vertices u and v of G with distinct colors, $G - u - v$ has a perfect matching.*

Let $G$ be a bipartite graph. An *ear* $P$ of $G$ is a path of odd length such that all internal vertices are of degree two. Let $G' := G - P$ denote the subgraph obtained from $G$ by removal of $P$, i.e. deleting all edges and internal vertices of $P$; whereas $G = G' + P$ denotes the supergraph obtained from $G'$ by adding an ear $P$: two end-vertices are in $G'$, but none of the internal vertices belongs to $G'$. Any elementary bipartite graphs admit a well-known "ear decomposition".

**Theorem 8.** *( [9, 11]) A bipartite graph G is elementary if and only if there exists a sequence $G_0 \subset G_1 \subset \cdots \subset G_r$ of subgraphs of G such that*

**(i)** $G_0 = K_2$, $G_r = G$, *and*
**(ii)** $G_i$ *is obtained from $G_{i-1}$ by adding an ear $P_i$ joining two vertices of $G_{i-1}$ with different colors for all $1 \leq i \leq r$.*

By Theorem 8 $G_0, \ldots, G_r(= G)$ is a sequence of elementary bipartite graphs. We call the sequence or $G = G_r = G_0 + P_1 + \cdots + P_r$ a (bipartite) *ear decomposition* of $G$ and $P_r$ a *removable ear* of $G$, where $r$ is equal to the cyclomatic number of $G$. An ear $P$ of $G$ is removable if and only if $G - P$ is elementary.

For a plane elementary bipartite graph $G$, an ear decomposition $G = G_0 + P_1 + \cdots + P_r$ is called a *reducible face decomposition* (RFD) of $G$ if every ear $P_{i+1}$ lies in the exterior face of $G_i$ for all $i \geq 1$; that is, every interior face of $G_i$ remains an interior face of $G$.

**Theorem 9.** ([18]) *Let $G$ be a plane elementary bipartite graph with at least one interior face. Then $G$ has an RFD.*

For an RFD $G = G_0 + P_1 + \cdots + P_r$, every $P_i$ and a path with odd length on the exterior face boundary of $G_{i-1}$ surround an interior face $s_i$. Hence an RFD of $G$ is also associated with a sequence $s_1, \ldots, s_r$ of all the interior faces of $G$; The last face $s_r$ is called a *reducible face* of $G$.

For any face $f$ of $G$, its boundary is a subgraph of $G$, denoted by $\partial f$; in particular, the exterior face boundary is called the boundary of $G$ and denoted by $\partial G$. Two faces of $G$ are said to be *disjoint* if their boundaries do not intersect; *adjacent* otherwise. The boundary of a face in $G$ is called a *facial cycle* if it is a cyle.

Let $F$ be a non-empty set of specified faces of $G$. The faces in $F$ are called *forbidden* if any two faces in $F$ are disjoint. The other faces of $G$ are called *allowed faces*, which form a multiple connected point set in the plane. We shall give a special ear decomposition of $G$ by using the $k := r + 1 - |F|$ allowed faces of $G$.

As a unified treatment for normal additions, corona condensations of modes $L_4$, $A_2$ or $A_3$ and reducible faces of a plane elementary bipartite graph, we now define a *regular addition* as follows.

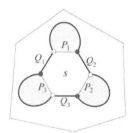

**Fig. 3.** A regular addition of allowed face $s$ in forbidden face $f$ (the thick cycle denotes the boundary of $f$)

Let $G'$ be a 2-connected plane bipartite graph with a forbidden face $f$. We embed an additive face $s$ with facial cycle $S$ into $f$: $G = G' + S$ is called a *regular addition* of $s$ in $f$ if the intersection of $S$ and $\partial f$ consists of paths $Q_1, Q_2, \ldots, Q_t (t \geq 1)$ with odd lengths, and $S - Q_1 - \cdots - Q_t$ consists of paths $P_1, P_2, \ldots, P_t$ with odd lengths (for example, see Fig. 3). Then the cycle $S$ alternates in the $P_i$'s and the $Q_i$'s. Hence such a regular addition can be written as $G = G' + S = G' + P_1 + \cdots + P_t$. In other words, a regular addition of a face $s$ in a forbidden face $f$ of $G'$ is to obtain graph $G$ by adding some $t \geq 1$ ears to $G'$ in $f$ so that such ears together with $t$ paths with odd lengths on the boundary of $f$ surround the face $s$. Then all allowed faces of $G'$ together with $s$ compose of all the allowed faces of $G$; $G$ inherits all forbidden faces of $G'$ except for $f$, and the $t$ connected components of the point set obtained from $f$ by deleting $s$ together with its boundary are the new forbidden faces of $G$.

We now define a regular decomposition as follows.

**Definition 2.** *A plane elementary bipartite graph $G$ with forbidden faces has a regular decomposition if it can be generated from a single allowed face by a series of regular additions of the other allowed faces of $G$; If such a sequence of the added allowed faces is denoted by $s_1, \ldots, s_k$, let $G_1 = S_1$ and $G_i = G_{i-1} + S_i$, $2 \leq i \leq k$, where $S_i := \partial s_i$. Then we call such a sequence $(s_1, \ldots, s_k)$ of the allowed faces or a sequence $G_1 \subset \cdots \subset G_k$ a regular decomposition (RD) of $G$ and $s_k$ a reducible face (RF) of $G$.*

For example, the antikekulene (see Fig. 4(a)) has an RD when the central face and the exterior face are forbidden and the other faces (hexagons and squares) are allowed; the polyomino in Fig. 4(b), however, has no RD when the non-square faces are viewed as forbidden faces.

As a main result of this article, a characterization for a plane elementary bipartite graphs with forbidden faces having an RD is obtained as follows.

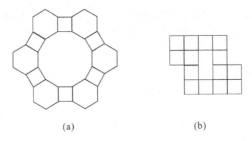

(a)                    (b)

**Fig. 4.** (a) Antikekulene, (b) a polyomino

**Theorem 10.** *Let $G$ be a plane elementary bipartite graph with forbidden faces. Then $G$ has a regular decomposition (RD) if and only if $G$ has a perfect matching $M$ such that the boundary of every forbidden face of $G$ is an $M$-alternating cycle.*

The proof will be given in next sections. We now describe some applications. First we can prove Theorem 4 by Theorem 10 whenever all non-hexagon faces of coronoids are always forbidden.

*Proof of Theorem 4.* The equivalence of (a) and (b) is described already by Theorem 3. So it is sufficient to prove that $(b) \Leftrightarrow (c)$.

All the non-hexagon faces of $G$ are specified as the forbidden faces and all the hexagons are allowed. Suppose that $G$ has a perfect matching $M$ such that the boundary of every non-hexagon face is $M$-alternating. So $G$ is normal by Theorem 1. Further, by Theorem 10 $G$ has an RD $(s_1, s_2, \ldots, s_k)$ which is a sequence of the hexagons. For each $i$ Definition 2 implies that a regular addition of hexagon $s_{i+1}$ to $G_i := s_1 + \cdots + s_i$ must be one in possible modes $L_1, L_3, L_5, L_4, A_2$ and $A_3$. Conversely, if $G$ can be generated from a single hexagon by a series of normal additions plus corona condensations of modes $L_4$, $A_2$ or $A_3$, $G$ is normal by Theorem 8 and has an RD by Definition 2. Further, Theorem 10 implies that

$G$ has a perfect matching $M$ such that the boundary of every non-hexagon face is $M$-alternating.    □

Then, Theorem 10 is a generalization of Theorem 9 since an RD of a plane elementary bipartite graph is actually an RFD whenever only the exterior face is forbidden. Finally, by Theorem 10 we can see that Theorem 9 is equivalent to the following theorem, considering that any face of a plane graph can become the exterior face under some stereographic projection.

**Theorem 11.** ([18]) *For any plane elementary bipartite graph with an interior face, the boundary of every face is an alternating cycle with respect to some perfect matching.*

## 3   Proof of Theorem 10

To prove Theorem 10, in this section we introduce some useful concepts and notations, and obtain some preliminary results.

Let $G$ be a plane bipartite graph. A subgraph $H \subseteq G$ is said to be *nice* if $G - V(H)$ has a perfect matching. For a cycle $C$ of $G$, let $I[C]$ (resp. $O[C]$) denote the subgraph of $G$ consisting of $C$ together with its interior (resp. exterior). Further, if $x$ and $y$ are vertices of $C$, $C[x, y]$ denotes the path lying in $C$ from $x$ to $y$ along the clockwise direction of $C$ and $C(x, y) := C[x, y] - x - y$.

Let $M$ be a perfect matching of $G$. A cycle (path) of $G$ is said to be $M$-*alternating* if its edges appear alternately in $M$ and $E(G) \backslash M$. Let $A$ and $B$ be finite sets. We define their *symmetric difference* as $A \oplus B := (A \cup B) \backslash (A \cap B)$. If $M$ and $M'$ are distinct perfect matchings of $G$, then $M \oplus M'$ forms a number of mutually disjoint $M$- and $M'$-alternating cycles; If $C$ is an $M$-alternating cycle, $M \oplus C$ is also a perfect matching of $G$, where $C$ is viewed as its edge set. An $M$-alternating face means an $M$-alternating cycle bounding a face.

From now on we are given a plane elementary bipartite graph $G$ with some forbidden faces and a perfect matching $M$.

**Lemma 12.** *Let $C$ be an $M$-alternating cycle of $G$. If $G$ has an edge $e$ that does not lie in $C$ but is incident with a vertex of $C$, then $G$ has an $M$-alternating path $P$ passing through $e$ such that only the end-vertices of $P$ lie in $C$.*

*Proof.* Let $e$ be an edge of $G$ not lying in $C$ and incident with a vertex $x$ of $C$. Since $G$ is elementary, it has a perfect matching $M'$ such that $e \in M'$. Then $M \oplus M'$ forms a number of disjoint $M$- and $M'$-alternating cycles; one (say $C'$), must contain $e$. Hence $M$-alternating cycles $C'$ and $C$ intersect. We choose a path $P$ of $C'$ only the end-vertices of which lie on the $C$ as required.    □

**Lemma 13.** ([18]) *If $C$ is an $M$-alternating cycle of $G$, then both $I[C]$ and $O[C]$ are elementary and nice subgraphs of $G$.*

Let $A$ be a region (simply or multiple connected open set) in the plane the boundary of which consists of mutually disjoint cycles of $G$. The *clockwise orientation* of $A$ means an orientation of its boundary such that $A$ always lies on the right side when one traverses the boundary along the direction.

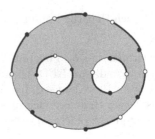

**Fig. 5.** A proper $M$-alternating region (shadow part), the perfect matching $M$ is indicated by thick edges

**Definition 3.** *The region $A$ is said to be $M$-alternating if its boundary consists of $M$-alternating cycles; Further an $M$-alternating region is called proper (improper) $M$-alternating if every edge in its boundary contained in (not in) $M$ goes from the white end-vertex to the black end-vertex along the clockwise orientation of the region (also see [10]). For example, Fig. 5 illustrates a proper $M$-alternating region. A usual proper or improper $M$-alternating cycle is with respect to its interior region.*

**Lemma 14.** ( [17, 18]) *Any two proper (resp. improper) $M$-alternating faces are disjoint.*

Let $R_G$ denote the point set in the plane by deleting all forbidden faces together with their boundaries. Since the forbidden faces are mutually disjoint, it is easily shown by Jordan Curve Theorem in the plane that $R_G$ is a region. So we call $R_G$ the *allowed region* of $G$. It is obvious that the boundary of $R_G$ is the union of all forbidden face boundaries. The following result is obvious.

**Lemma 15.** *The allowed region of $G$ is proper (improper) $M$-alternating if and only if the forbidden faces are all improper (proper) $M$-alternating.*

**Lemma 16.** *Every proper (improper) $M$-alternating cycle of $G$ contains a proper (improper) $M$-alternating face in its interior.*

*Proof.* We only consider the case that $C$ is a proper $M$-alternating cycle $C$ of $G$. We shall prove that $G$ has a proper $M$-alternating face lying in the interior of $C$ by induction on the number $k$ of interior faces of $C$. If $k = 1$, it is trivial. So we let $k \geq 2$. By Lemma 12 $G$ has an $M$-alternating path $P$ only end-vertices of which lie on the boundary $C$. Then $P$ partitions the interior of $C$ into two regions; the boundary of one must be a proper $M$-alternating cycle $C'$. By induction hypothesis $G$ has a proper $M$-alternating face in the interior of $C'$.    □

Let $G$ be a plane elementary bipartite graph with a set $F$ of forbidden faces $f_0, f_1, \ldots, f_t$ for $t \geq 0$. Recall that the $f_i$'s are mutually disjoint. An allowed face $s$ of $G$ is said to be *ideal* if $s$ is adjacent to some one(s) in $F$ and $\partial s \cap \partial f_i$ is either empty or a path of odd length for all $0 \leq i \leq t$.

Let $\mathcal{M}$ denote the set of perfect matchings of $G$ such that every face in $F$ is alternating; that is, for each $M \in \mathcal{M}$ the faces in $F$ are all $M$-alternating. To prove Theorem 10 we need the following crucial result.

**Lemma 17.** *Suppose that $\mathcal{M} \neq \emptyset$. If $G$ has at least **two** allowed faces, then there exists an $M$-alternating ideal face for some $M \in \mathcal{M}$.*

We first prove Theorem 10 by applying Lemma 17, then prove the lemma in next section.

*Proof of Theorem 10.* We proceed by induction on the number $k$ of allowed faces of $G(k \leq r)$. If $k = 1$, $G$ is exactly one cycle and has a unique forbidden face, the result is trivial. In what follows assume that $k \geq 2$. We suppose that for any plane elementary bipartite graph with less allowed faces the result holds.

*Necessity.* Suppose that $G$ has an RD $(s_1, \ldots, s_{k-1}, s_k)$. We want to prove that $\mathcal{M} \neq \emptyset$ for $G$. Let $G_i := S_1 + \cdots + S_i$, $1 \leq i \leq k$, where $S_i = \partial s_i$ are facial cycles. Then $s_1, \ldots, s_{k-1}$ are regarded as the allowed faces of $G_{k-1}$ and the others as forbidden faces. By Definition 2 $(s_1, \ldots, s_{k-1})$ is an RD of $G_{k-1}$. By induction hypothesis $G_{k-1}$ has a perfect matching $M$ such that all forbidden faces are $M$-alternating. $G = G_k$ is obtained from $G_{k-1}$ by the regular addition of the allowed face $s_k$ of $G$ in some forbidden face $f$ of $G_{k-1}$. Without loss of generality, suppose that $f$ is the exterior face of $G_{k-1}$. Put $C := \partial f$ and $S := S_k = \partial s_k$. Further, $C$ and $S$ intersect and $S \neq C$. Then the connected components of the point set $f \backslash (s_k \cup S) \neq \emptyset$ in the plane are forbidden faces of $G$ and bounded by cycles in $S \oplus C$.

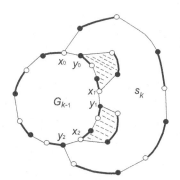

**Fig. 6.** Illustrations for the proof of Theorem 10 (thick lines represent edges in $M'$)

By the clockwise orientation of $C$ all the vertices with degree 3 in the graph $S \cup C$ are arranged as $x_0, y_0, x_1, y_1, \ldots, x_m, y_m (m \geq 0)$ such that all $C[x_i, y_i]$ are the components of $S \cap C$ (see Fig. 6). By the regular addition of $s_k$ the $C[x_i, y_i]$ are $M$-alternating paths of odd length and the $S[x_{i+1}, y_i]$ (the subscripts modulo $m + 1$) are ears of $G_k$. Hence the $x_i$ are of the same color, say white; and the $y_i$ are colored black. Without loss of generality, suppose that $C$ is an improper

$M$-alternating cycle (see Definition 3); Otherwise, let $M := M \oplus C$. Thus the $C[y_i, x_{i+1}]$ are $M$-alternating paths whose end-edges belong to $M$. The perfect matching $M$ of $G_{k-1}$, together with all second edges from end-edges for all ears $S(x_{i+1}, y_i)$, $0 \le i \le m$, compose a perfect matching $M'$ of $G$. Then the $C[y_i, x_{i+1}] + S[x_{i+1}, y_i]$ are $M'$-alternating cycles. The other forbidden faces of $G_{k-1}$ different from $f$ remain $M$-alternating. So every forbidden face of $G$ is $M'$-alternating.

*Sufficiency.* Suppose that $\mathcal{M} \ne \emptyset$. By Lemma 17 $G$ has an $M \in \mathcal{M}$ and an ideal face $s$ that is $M$-alternating. Let $S := \partial s$ and let $C_i := \partial f_i$, $0 \le i \le t$. The cycles in the $C_i (i \ge 1)$ intersecting $S$ are denoted by $C_{i_1}, \ldots, C_{i_\tau} (\tau \ge 1)$. Then each $P_{i_j} := S \cap C_{i_j}$ is an $M$-alternating path the end-edges of which belong to $M$, and is thus an ear of $G$; further, $S - P_{i_1} - \cdots - P_{i_\tau}$ consists of $\tau$ paths $Q_{i_j}$ of odd length none of the end-edges of which belong to $M$. Removal of all ears $P_{i_j}$ from $G$ results in graph $H$. The allowed face $s$ and forbidden faces $f_{i_1}, \ldots, f_{i_\tau}$ of $G$ are combined into one face $f$ of $H$, which can be viewed as a forbidden face of $H$. Hence $G$ can be obtained from $H$ by a regular addition of $s$ in $f$. It is obvious that $C := \partial f = S \oplus (\cup_{j=1}^\tau C_{i_j})$. Let $M' := M \oplus S$. Then all $C_{i_j} - P_{i_j}$ are $M'$-alternating paths none of the end-edges of which belong to $M'$ and all $Q_{i_j}$ are $M'$-alternating path the end-edges of which belong to $M'$. So $C$ is an $M'$-alternating cycle. By Lemma 13 $H$ (one of the subgraphs $I[C]$ and $O[C]$ in $G$) is elementary, and $H$ has a perfect matching $M'|_H$ such that all forbidden faces of $H$ are $M'$-alternating. Since $H$ has less allowed faces than $G$, by induction hypothesis $H$ has an RD. Hence $G = H + S$ has an RD as required.     □

## 4     Proof of Lemma 17

Let $C_i := \partial f_i$, $0 \le i \le t$. Without loss of generality, assume that $f_0$ is the exterior face of $G$. We choose an $M \in \mathcal{M}$ and additional cycles $C_{t+1}, \ldots, C_{t'} (t \le t')$ with the following properties:

**P1.** All the $C_i$, $0 \le i \le t'$ are mutually disjoint $M$-alternating cycles, and
**P2.** The interiors of all $C_i (1 \le i \le t')$ are mutually disjoint,

such that the region $R \subseteq R_G$ among the cycles $C_i$ is as minimal as possible in a sense of point set inclusion, where $R_G$ is the allowed region of $G$ and $R$ is a region in the plane obtained from $R_G$ by deleting all $C_i$ $(t+1 \le i \le t')$ together with their interior regions. It is obvious that $R$ contains at least two allowed faces of $G$.

Let $H$ be the subgraph of $G$ consisting of $\cup_{i=0}^{t'} C_i$ together with all edges and vertices lying in the region $R$. By Lemma 13 $H$ has the following properties:

**P3.** $H$ is a plane elementary bipartite graph, and
**P4.** $H$ is a nice subgraph of $G$.

Let $f_i$, $t+1 \le i \le t'$, denote the interior region of $C_i$ in the plane. Then $H$ is a plane elementary bipartite graph with forbidden faces $f_0, f_1, \ldots, f_{t'}$; the allowed region $R_H = R$.

Let $\mathcal{M}_0$ denote the set of perfect matchings $M_0$ of $H$ such that $R$ is proper or improper $M_0$-alternating region. Then $\mathcal{M}_0 \neq \emptyset$ follows from $\mathcal{M} \neq \emptyset$. We assert that the following result holds.

(∗) *There exist an* $M_0 \in \mathcal{M}_0$ *and an* $M_0$-*alternating ideal face of* $H$.

Let $M_0 \in \mathcal{M}_0$. Then $M_0' := M_0 \oplus (\cup_{i=0}^{t'} C_i) \in \mathcal{M}_0$. Without loss of generality, suppose that $R$ is a proper $M_0$- and thus improper $M_0'$-alternating region. Then there are neither improper $M_0$- nor proper $M_0'$-alternating face on $R$; otherwise it would be disjoint with all faces $f_0, \ldots, f_{t'}$ by Lemmas 14 and 15, contradicting the choice of $C_{t+1}, \ldots, C_{t'}$.

Let $F := \{f_i : 0 \leq i \leq t'\}$, $\Gamma := \{f : f$ is a proper $M_0$-alternating face on R$\}$, and $\Phi := \{f : f$ is an improper $M_0'$-alternating face on R$\}$. Then we have the following properties:

**P5.** $|\Gamma| \geq 1$, $|\Phi| \geq 1$ and $|F| \geq 1$. (By Lemmas 15 and 16.)
**P6.** The faces in each of $F$, $\Gamma$, $\Phi$ are mutually disjoint. (By Lemma 14.)
**P7.** Every face in $\Gamma \cup \Phi$ must be adjacent with some one in $F$. (By the minimality of region $R$.)
**P8.** $\Gamma \cap \Phi = \emptyset$. (Obvious.)

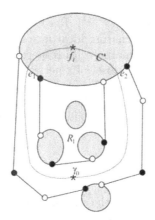

**Fig. 7.** Illustration for the proof of Lemma 17 (shadow regions represent forbidden faces)

Let $H^*$ denote a given *dual* of plane graph $H$ (cf. [14, page 72]). For $f \in F$ and $\gamma \in \Gamma \cup \Phi$, a pair of multiple edges between $f^*$ and $\gamma^*$ is called a *marked 2-cycle* of $H^*$. With each $\gamma \in \Gamma \cup \Phi$, we associate a *weight* $w(\gamma)$: the number of marked 2-cycles whose interior regions contain the point $\gamma^*$ in $H^*$. Then $w(\gamma) \geq 0$.

We choose a $\gamma_0 \in \Gamma \cup \Phi$ such that $w(\gamma_0)$ is as maximal as possible. Since $H^*$ is finite, such a $\gamma_0$ must exist. We assert that $\gamma_0$ is a desirable allowed face as in (∗). Without loss of generality, assume that $\gamma_0 \in \Gamma$; that is, $\gamma_0$ is a proper $M_0$-alternating face.

On the contrary, suppose that the subgraph $\partial \gamma_0 \cap \partial f_i \neq \emptyset$ is not connected for some $f_i \in F$; each component is an odd path since it is an $M_0$-alternating path whose end-edges belong to $M_0$. Let $e_1$ and $e_2$ be edges lying in different components of $\partial\gamma_0 \cap \partial f_i$. Then there exists an edge of $C_i$ in the interior of $C^*$ that bounds an allowed face $g$ of $H$, where $C^* = \{e_1^*, e_2^*\}$ is a marked 2-cycle in $H^*$ between $\gamma_0^*$ and $f_i^*$; otherwise both $e_1$ and $e_2$ would belong to one component of $\partial\gamma_0 \cap \partial f_i$, a contradiction. Let $R_1$ be the connected component of $R\backslash(\partial\gamma_0 \cup \gamma_0)$ lying in the interior of $C^*$, which contains $g$ (see Fig. 7). It follows that $R_1$ is an improper $M_0'$-alternating region by Lemma 15: the outer-boundary of $R_1$ is improper $M_0'$-alternating, each inner-boundary (if exists) remains improper $M_0$-alternating and is thus proper $M_0'$-alternating. By Lemma 16 there exists an improper $M_0'$-alternating face $\phi_0 \in \Phi$ on $R_1$. Then $\phi_0^*$ lies in the interior of the marked 2-cycle $C^*$ in $H^*$. On the other hand, any marked 2-cycle in $H^*$ containing $\gamma_0^*$ in its interior also contains $\phi_0^*$. Then $w(\phi_0) \geq w(\gamma_0) + 1$, which contradicts the maximality of $w(\gamma_0)$. Hence $\partial\gamma_0 \cap \partial f_i$ is a path of odd length whenever it is not empty. The assertion (∗) is proved.

Finally, we suppose that $R$ is a proper $M_0$-alternating region and $s$ is an $M_0$-alternating ideal face $s$ of $H$ for $M_0 \in \mathcal{M}_0$. Then $s$ is proper $M_0$-alternating. It is sufficient to prove that $s$ must be adjacent to some one of the $f_i$ for $0 \leq i \leq t$. Put $S := \partial s$.

Otherwise, let $C_{i_1}, \ldots, C_{i_\tau}$ denote all the cycles in the $C_i$'s that intersect $S$. Then $\emptyset \neq \{i_1, \ldots, i_\tau\} \subseteq \{t+1, \ldots, t'\}$. Recall that $M_0' := M_0 \oplus (\cup_{i=0}^{t'} C_i) \in \mathcal{M}_0$. Since $s$ is a proper $M_0$-alternating ideal face of $H$, $C := S \oplus (\cup_{j=1}^{\tau} C_{i_j})$ is a cycle that is proper $M_0'$-alternating with respect to its interior. Since $H$ is a nice subgraph of $G$ (P4), $M_0'$ is a subset of a perfect matching $M'$ of $G$ belonging to $\mathcal{M}$. For $M'$, the cycle $C$ and the other cycles in the $C_i$'s not intersecting $S$ satisfy the properties P1 and P2. However the interior of $C$ contains the ideal face $s$ and $f_{i_1}, \ldots, f_{i_\tau}$, contradicting the minimality of $R$. Therefore $s$ is an $M'$-alternating ideal face of $G$ for $M' \in \mathcal{M}$. The proof is completed.    □

# Acknowledgements

The article was completed when the author ever visited Center for Combinatorics and LPMC, Nankai University, Tianjin 300071, P. R. China. The author has been supported by Visiting Scholar Foundation of Key Lab. In University, NSFC grant 10471058 and TRAPOYT.

# References

1. Chen, R., Zhang, F.: Regular Coronoid Systems. Discrete Appl. Math. **74** (1997) 147–158
2. Cyvin, S.J., Brunvoll, J., Cyvin, B.N.: Theory of Coronoid Hydrocarbons. Lectures Notes in Chemistry **54**. Springer-Verlag, Berlin (1991)
3. Cyvin, S.J., Brunvoll, J., Chen, R., Zhang, F.: Theory of Coronoid Hydrocarbons II. Lectures Notes in Chemistry **62** Springer-Verlag, Berlin (1994)
4. Cyvin, S.J., Cyvin, B.N., Brunvoll, J.: Half Essentially Disconnected Coronoid Hydrocarbons. Chem. Phys. Letters **140** (1987) 124–129

5. Cyvin, S.J., Gutman, I.: Kekulé Structures in Benzenoid Hydrocarbons. Lectures Notes in Chemistry **46**. Springer-Verlag, Berlin (1988)
6. Gutman, I., Cyvin, S.J.: Introduction to the Theory of Benzenoid Hydrocarbons. Springer-Verlag, Berlin (1989)
7. Gutman, I., Cyvin, S.J.: Advances in the Theory of Benzenoid Hydrocarbons. Springer-Verlag, Berlin (1990)
8. He, W.C., He, W.J.: Some Topological Properties and Generation of Normal Benzenoids. Match **25** (1990) 225–236
9. Hetyei, G.: Rectangular Configurations which can be Covered by $2 \times 1$ Rectangles. Pécsi Tan. Főisk. Közl. **8** (1964) 351–367 (in Hungarian)
10. Lam, P.C.B., Zhang, H.: A Distributive Lattice on the Set of Perfect Matchings of a Plane Bipartite Graph. Order **20** (2003) 13–29
11. Lovász, L., Plummer, M.D.: Matching Theory. Annals of Discrete Mathematics **29** (1986)
12. Sachs, H.: Perfect Matchings in Hexagonal Systems. Combinatorica **4** (1) (1980) 89–99
13. Trinajastić, N.: Chemical Graph Theory. Second Edition. CRC Press, Boca Raton, Florida (1992)
14. Wilson, R.J.: Introduction to Graph Theory. Third Edition. Longman Inc., New York (1985)
15. Zhang, F., Chen, R.: When Each Hexagon of a Hexagonal System Covers it. Discrete Appl. Math. **30** (1991) 63–75
16. Zhang, F., Zheng, M.: Generalized Hexagonal Systems with Each Hexagon being Resonant. Discrete Appl. Math. **36** (1992) 67–73
17. Zhang, H., Zhang, F.: The Rotation Graphs of Perfect Matchings of Plane Bipartite Graphs. Discrete Appl. Math. **73** (1997) 5–12
18. Zhang, H., Zhang, F.: Plane Elementary Bipartite Graphs. Discrete Appl. Math. **105** (2000) 291–311
19. Zheng, M.: Perfect Matchings in Benzenoid Systems. Doctoral Thesis, Rutgers University (1992)

# On the Upper Chromatic Numbers of Mixed Interval Hypertrees*

Ping Zhao[1] and Kefeng Diao[2]

[1] Department of Math., Linyi Normal University, Linyi, Shandong, 276005,
P.R. China
zhaopingly@163.com
[2] Department of Math., Linyi Normal University, Linyi, Shandong, 276005,
P.R. China
kfdiao@lytu.edu.cn

**Abstract.** A mixed hypergraph consists of a finite set and two families
of subsets: C-edges and D-edges. In a coloring, every C-edge has at least
two vertices of the same color, and every D-edge has at least two vertices
colored differently. The maximum and minimum numbers of colors are
called the lower and upper chromatic numbers, respectively. A mixed
hypergraph $\mathcal{H}$ with vertex set $X$, C-edge set $\mathcal{C}$ and D-edge set $\mathcal{D}$ is
usually denoted by $\mathcal{H} = (X, \mathcal{C}, \mathcal{D})$. E. Bulgaru and V. Voloshin proved
that $\overline{\chi}(\mathcal{H}) = |X| - s(\mathcal{H})$ holds for any mixed interval hypergraph ([3]),
where $s(\mathcal{H})$ is the sieve number of $\mathcal{H}$. In this paper we prove that $\overline{\chi}(\mathcal{H}) =
|X| - s(\mathcal{H})$ also holds for any mixed interval hypertree.

**Keywords:** mixed hypergraph; upper chromatic number; mixed interval
hypergraph; mixed interval hypertree; sieve number.

**AMS Subject Classifications:** 05C65.

## 1   Introduction

Throughout the paper we basically use the terminology of [1] and [2].

The classical coloring theory of graphs and hypergraphs deals with the fundamental problem of partitioning a set of objects into classes according to certain
rules. They are based on the following fundamental restriction: the vertices of
every edge lie in at least two partite sets. The chromatic number of a hyper-
grap $\mathcal{H} = (X, \mathcal{E})$ is the minimum number $k$ for which there exists a partition
of $X$ into $k$ non-empty sets meeting these requirements. The maximum number
$k$ satisfying the above constrains is trivially equal to the number of vertices.
Therefore, the classical coloring theory is the minimum coloring theory.

It is very natural to think of coloring the vertices of a hypergraph such that in
some hyperedges at least two vertices have the same color. A hypergraph with
this kind of edges is called a mixed hypergraph. The notion of mixed hypergraphs
was introduced by Vitaly in 1995 ([2]).

---

* The work is supported by National Natural Science of China (No. 10471078).

J. Akiyama et al. (Eds.): CJCDGCGT 2005, LNCS 4381, pp. 272–277, 2007.
© Springer-Verlag Berlin Heidelberg 2007

Let $X = \{x_1, x_2, \ldots, x_n\}$ be a finite set and let $\mathcal{C} = \{C_1, C_2, \ldots, C_k\}$ and $\mathcal{D} = \{D_1, D_2, \ldots, D_m\}$ be families of subsets of $X$ where the size of every $C_i \in \mathcal{C}$ and every $D_j \in \mathcal{D}$ is not less than 2, in particular, $\mathcal{C}$ and /or $\mathcal{D}$ may be empty. Then the triple $\mathcal{H} = (X, \mathcal{C}, \mathcal{D})$ is called a mixed hypergraph on $X$. Every $C_i \in \mathcal{C}$ is called a C-edge and every $D_j \in \mathcal{D}$ is called a D-edge. It is convenient further to use the term C-hypergraph for a partial hypergraph $\mathcal{H}_C = (X, \mathcal{C})$ and use the term D-hypergraph for a partial hypergraph $\mathcal{H}_D = (X, \mathcal{D})$.

For any subset $Y \subseteq X$, the mixed hypergraph $\mathcal{H}' = (Y, \mathcal{C}', \mathcal{D}')$ is called a sub-hypergraph of $\mathcal{H}$ if $\mathcal{C}'$ is both a family of subsets of $Y$ and a subset of $\mathcal{C}$, and $\mathcal{D}'$ is both a family of subsets of $Y$ and a subset of $\mathcal{D}$. If further, $\mathcal{C}'$ consists of all those subsets in $\mathcal{C}$ that are contained in $Y$, and $\mathcal{D}'$ consists of all those subsets in $\mathcal{D}$ that are contained in $Y$, then we call it the induced sub-hypergraph of $\mathcal{H}$ on $Y$ and denot it by $\mathcal{H}[Y]$. The sub-hypergraph $\mathcal{H}'$ is called a partial hypergraph of $\mathcal{H}$ if $Y = X$.

The difference between D-edges and C-edges is in the coloring rules.

**Definition 1 (Voloshin [2]).** *A free coloring of a mixed hypergraph $\mathcal{H} = (X, \mathcal{C}, \mathcal{D})$ with $k$ colors is a maping $c\colon X \to \{1, 2, \ldots, k\}$ such that the following two conditions hold :*

**(1)** *Any C-edge $C_i \in \mathcal{C}$ has at least two vertices with the same color.*
**(2)** *Any D-edge $D_j \in \mathcal{D}$ has at least two vertices with different colors.*

A free coloring is called a strict coloring if it is onto. A mixed hypergraph is $k$-strict colorable if it has a strict coloring with $k$ colors.

In a mixed hypergraph it makes sense to ask for both the minimum and the maximum colors required for a strict coloring. Therefore, Vitaly I. Voloshin introduced the notions of upper and lower chromatic numbers of a mixed hypergraph.

**Definition 2 (Voloshin [2]).** *The maximum (minimum) $k$ for which there exists a strict coloring of a mixed hypergraph $\mathcal{H}$ with $k$ colors is called the upper (lower) chromatic number of $\mathcal{H}$ and is denoted by $\overline{\chi}(\mathcal{H})(\chi(\mathcal{H}))$.*

The classical coloring theory of hypergraphs becomes a special case, i.e., the case of mixed hypergraphs with only D-edges and lower chromatic numbers. We believe that the new coloring theory will have both impact on the classical coloring theory and its own direction of theoretic development. We also believe that the new coloring theory will has far more influence in practices and applications than the classical coloring theory.

**Definition 3 (Voloshin [2]).** *Let $\mathcal{H} = (X, \mathcal{C}, \mathcal{D})$ be a mixed hypergraph and $r$ be a positive integer no less than 2. If $|C_i| = r$ holds for every $C_i \in \mathcal{C}$ and $|D_j| = r$ holds for every $D_j \in \mathcal{D}$, then $\mathcal{H}$ is called an $r$-uniform mixed hypergraph.*

In this paper we concentrate on mixed interval hypergraphs. Bulgaru and Voloshin proved that $\overline{\chi}(\mathcal{H}) = |X| - s(\mathcal{H})$ holds for any mixed interval hypergraph ( [3]), where $s(\mathcal{H})$ is the sieve number of $\mathcal{H}$. In this paper we prove that $\overline{\chi}(\mathcal{H}) = |X| - s(\mathcal{H})$ also holds for any mixed interval hypertree.

## 2   Mixed Interval Hypergraphs

We first reproduce the concept of mixed interval hypergraph.

**Definition 4 (Bulgaru [3]).** *A mixed hypergraph $\mathcal{H}$ is a mixed interval hypergraph if there exists a linear ordering of the vertex set such that each $C$-edge represents an interval, and each $D$-edge also represents an interval in this ordering.*

The colorings of mixed interval hypergraphs can be applied in Molecule Biology. The colorings of mixed interval hypergraphs were investigated in paper [3] and the exact values of lower and upper chromatic numbers are found. Whether or not a mixed interval hypergraph is colorable depends upon wether or not it contains uncolorable $D$-edges which is defined as follows.

**Definition 5 (Bulgaru [3]).** *In a mixed hypergraph $\mathcal{H} = (X, \mathcal{C}, \mathcal{D})$, a $D$-edge $D$, $D \in \mathcal{D}$, is called uncolorable if for any pair of vertices $x$, $y \in D$, there exists a sequence $xC_1z_1C_2z_2 \cdots C_{l-1}z_{l-1}C_ly$ such that:*

**(1)** $z_1, z_2, \ldots, z_{l-1} \in D$;
**(2)** $C_i \in \mathcal{C}$,  $i = 1, 2, \ldots, l$;
**(3)** $C_1 = \{x, z_1\}$,  $C_2 = \{z_1, z_2\}, \ldots, C_l = \{z_{l-1}, y\}$.

E. Bulgaru and V. Voloshin proved the following result in [3].

**Theorem 1 (Bulgaru [3]).** *A mixed interval hypergraph $\mathcal{H} = (X, \mathcal{C}, \mathcal{D})$ is colorable if and only if it does not contain uncolorable $D$-edges. In this case we have $\chi(\mathcal{H}) \leq 2$.*

From the result of Theorem 1, we obviously have that: for a mixed interval hypergraph $\mathcal{H} = (X, \mathcal{C}, \mathcal{D})$, if $\mathcal{H}$ has no $D$-edges, then the lower chromatic number is 1; if $\mathcal{H}$ has at least one $D$-edge and contains no uncolorable $D$-edges, then the lower chromatic number is 2; if $\mathcal{H}$ contains uncolorable $D$-edges, then $\mathcal{H}$ is uncolorable.

Therefore, in the following we focus on the upper chromatic number. The upper chromatic number of a mixed interval hypergraph depends on its sieve number.

**Definition 6 (Bulgaru [3]).** *In a mixed hypergraph $\mathcal{H} = (X, \mathcal{C}, \mathcal{D})$ the subfamily $\mathcal{C}' \subset \mathcal{C}$ is called a sieve, if for any $x$, $y \in X$ and any $C_1, C_2 \in \mathcal{C}'$ the following implication holds:*

$$x, y \in C_1 \cap C_2 \Rightarrow (x, y) = D_j \in \mathcal{D} \text{ for some } j \in J.$$

*The maximum cardinality of a sieve of a mixed hypergraph $\mathcal{H}$ is called the sieve-number of $\mathcal{H}$ and is denoted by $s(\mathcal{H})$.*

E. Bulgaru and V. Voloshin also proved the following result in the same paper.

**Theorem 2 (Bulgaru [3]).** *If* $\mathcal{H} = (X, \mathcal{C}, \mathcal{D})$ *is a mixed interval hypergraph, then*

$$\overline{\chi}(\mathcal{H}) = |X| - s(\mathcal{H}).$$

We generalize this conclusion to mixed interval hypertrees.

**Definition 7.** *If the vertex set of a mixed hypergraph* $\mathcal{H}$ *can be ordered as an out-oriented tree, such that each C-edge contains consecutive vertices on one branch, and each D-edge contains consecutive vertices on one branch, then* $\mathcal{H}$ *is called a mixed interval hypertree.*

It is not difficult to notice from the definition that every branch of a mixed interval hypertree induces a mixed interval hypergraph.

**Definition 8.** *The out-oriented tree formed by the vertices of a mixed interval hypertree* $\mathcal{H}$ *is called the mother tree of* $\mathcal{H}$. *The vertex with in-degree 0 in the mother tree of* $\mathcal{H}$ *is called the root of* $\mathcal{H}$, *the vertices with out-degree being greater than one are called branch nods of* $\mathcal{H}$ *and the vertex with out-degree 0 is called a leaf of* $\mathcal{H}$.

In a mixed interval hypertree $\mathcal{H} = (X, \mathcal{C}, \mathcal{D})$, $x < y$ means that $x$ is left to $y$ if $x$ and $y$ are on the same branch, or $x$ is on a branch that lies under the branch containing $y$. About mixed interval hypertrees we have the following result.

**Theorem 3.** *If* $\mathcal{H} = (X, \mathcal{C}, \mathcal{D})$ *is a 3-uniform mixed interval hypertree, then*

$$\overline{\chi}(\mathcal{H}) = |X| - s(\mathcal{H}).$$

*Proof.* First we show that $\overline{\chi}(\mathcal{H}) \leq |X| - s(\mathcal{H})$. Let us suppose that $\overline{\chi}(\mathcal{H}) \geq |X| - s(\mathcal{H}) + 1$ and consider some strict coloring of $\mathcal{H}$ with $\overline{\chi}(\mathcal{H})$ colors. Let $\mathcal{C}_1 = \{C_1, C_2, \ldots, C_s\}$ be a sieve of maximum cardinality, $|\mathcal{C}_1| = s(\mathcal{H}) = s$, and let $x_i, y_i \in C_i$, $x_i < y_i$, $i = 1, 2, \ldots, s$ be the vertices colored with the same color $c_i$ in each $C_i \in \mathcal{C}_1$. Then if all $c_i$, $i = 1, 2, \ldots, s$ are different, the total number of used colors is at most

$$s + |X| - 2s = |X| - s,$$

which contradicts to the assumption. If there are two $c_i$s, say, $c_l$ and $c_j$, such that $c_l = c_j$, then since $\mathcal{C}_1$ is a sieve, $|\{x_i, y_i\} \cap \{x_j, y_j\}| \leq 1$, $i, j = 1, 2, \ldots, s$, $i \neq j$. Therefore $|\{x_l, y_l, x_j, y_j\}| \geq 3$. If even $|\{x_l, y_l, x_j, y_j\}| = 3$, and if even all the other $s - 2$ colors are all different and all the vertices of $X - \left( \bigcup_{i \neq l, j} C_i \right)$ are colored differently, the total number of used colors is

$$1 + (s - 2) + (|X| - 3 - 2(s - 2)) = |X| - s.$$

Hence, we again have a contradiction. Therefore

$$\overline{\chi}(\mathcal{H}) \leq |X| - s(\mathcal{H}).$$

Now we show that we can color $\mathcal{H}$ with $|X| - s$ colors, i.e., $\overline{\chi}(\mathcal{H}) \geq |X| - s(\mathcal{H})$ by induction on $s(\mathcal{H})$. Let $s(\mathcal{H}) = 1$. Then from the discussion of E. Bulgaru and V. Voloshin in [3] we have

$$\overline{\chi}(\mathcal{H}) = \alpha(\mathcal{H}_C) = |X| - 1 = |X| - s(\mathcal{H}).$$

Suppose that our assertion $\overline{\chi}(\mathcal{H}') \geq |X| - s(\mathcal{H}')$ is true for any 3-uniform mixed interval hypertree $\mathcal{H}'$ with $s(\mathcal{H}') < s(\mathcal{H})$.

Let $\mathcal{C}_1 = \{C_1, C_2, \ldots, C_s\}$ and $s = s(\mathcal{H})$ be again a maximum and ordered from left to right and lower to upper sieve of a 3-uniform mixed interval hypertree $\mathcal{H} = (X, \mathcal{C}, \mathcal{D})$, and let $C_1 = \{x, y, z\}$ be the first $C$-edge in this sieve.

If $C_1$ is not the very left and bottom $C$-edge of $\mathcal{H}$, then we can construct another maximum sieve of $\mathcal{H}$ replacing $C_1$ in $\mathcal{C}_1$ by the very left and bottom $C$-edge of $\mathcal{H}$. Therefore without loss of generality, let us assume without loss of generality that $C_1$ is the very left and bottom $C$-edge of $\mathcal{H}$. Let $x_1$ be the root of $\mathcal{H}$ and $x_m$ be the first branch nod, $x_{m+1}^i$, $x_{m+2}^i$, $m_{m+3}^i$ be the first three vertices of the $i$-th branch from the nod $x_m$ and $x_{m+1}^i < x_{m+2}^i < m_{m+3}^i$. Consider the following three possible cases.

*Case 1. $m \geq 3$.*

In this case we have that $C_1 = \{x_1, x_2, x_3\}$ and $\{x_1, x_3\}$ is not a $D$-edge . Then color $x_1$ and $x_2$ with the color 1 and 2, respectively, and color $x_3$ with the color 1. Consider the sub-hypergraph

$$\mathcal{H}_1 = \mathcal{H}/(X - \{x_1\}) = (X_1, \mathcal{C}^1, \mathcal{D}^1)$$

with

$$s(\mathcal{H}_1) = s(\mathcal{H}) - 1, \ |X_1| = |X| - 1.$$

By virtue of the induction hypothesis $\overline{\chi}(\mathcal{H}_1) = |X_1| - s(\mathcal{H}_1)$. Color $\mathcal{H}_1$ with $\overline{\chi}(\mathcal{H}_1)$ colors using the colors $1, 2, 3, \ldots, \overline{\chi}(\mathcal{H}_1)$ in such a way that the vertex $x_2, x_3$ are colored with the color 2 and 1 respectively. We obtain a coloring of the initial hypergraph $\mathcal{H}$, and

$$\overline{\chi}(\mathcal{H}) \geq \overline{\chi}(\mathcal{H}_1) = |X_1| - s(\mathcal{H}_1)$$
$$= |X| - 1 - (s(\mathcal{H}) - 1) = |X| - s(\mathcal{H}).$$

*Case 2. $m = 2$.*

In this case we have $C_1 = \{x_1, x_2, x_3^i\}$. If $\{x_1, x_2\} \notin \mathcal{D}$, then color $x_1$ and $x_2$ with the color 1. If $x_1 \in C$ for some $C \in \mathcal{C}$, then $x_2 \in C$. Hence any such $C$-edge is colored correctly. Consider the sub-hypergraph

$$\mathcal{H}_1 = \mathcal{H}/(X - \{x_1, \}) = (X_1, \mathcal{C}^1, \mathcal{D}^1)$$

with

$$s(\mathcal{H}_1) = s(\mathcal{H}) - 1, \ |X_1| = |X| - 1.$$

By virtue of the induction hypothesis $\overline{\chi}(\mathcal{H}_1) = |X_1| - s(\mathcal{H}_1)$. Color $\mathcal{H}_1$ with $\overline{\chi}(\mathcal{H}_1)$ colors using the colors $1, 2, 3, \ldots, \overline{\chi}(\mathcal{H}_1)$ in such a way that the vertex $x_2$, $x_3$ are colored with the color 2 and 1, respectively. We obtain a coloring of the initial hypergraph $\mathcal{H}$, and

$$\overline{\chi}(\mathcal{H}) \geq \overline{\chi}(\mathcal{H}_1) = |X_1| - s(\mathcal{H}_1)$$
$$= |X| - 1 - (s(\mathcal{H}) - 1) = |X| - s(\mathcal{H}).$$

*Case 3. $m = 1$.*

Then the vertices of the lowest branch including the root $x_1$, say $X_1$, induces a mixed interval hypergraph, say $\mathcal{H}_1$. Suppose that $\mathcal{H}_1$ has $n_1$ vertices and the sieve number of this branch is $s_1$, the induced sub-hypergraph of $\mathcal{H}$ on $X_2 = (X - X_1) \cup \{x_1\}$ is also a mixed interval hypertree, say $\mathcal{H}_2$, and the sieve number of $\mathcal{H}_2$ is $s_2 = s(\mathcal{H}) - s_1$. By virtue of the induction hypothesis $\overline{\chi}(\mathcal{H}_1) = |X_1| - s(\mathcal{H}_1)$ and $\overline{\chi}(\mathcal{H}_2) = |X_2| - s(\mathcal{H}_2)$. Assume that $|X_1| - s(\mathcal{H}_1) = m_1$ and $|X_2| - s(\mathcal{H}_2) = m_2$. Then color the vertices of $\mathcal{H}_1$ with the color $1, 2, \ldots, m_1$, respectively in such a way that the vertex $x_1$ is colored with the color 1, and color the vertices of $\mathcal{H}_2$ with the color $1, m_1 + 1, m_1 + 2, \ldots, m_1 + m_2 - 1$, respectively in such a way that the vertex $x_1$ is colored with the color 1. We obtain a coloring of the initial hypergraph $\mathcal{H}$, and

$$\overline{\chi}(\mathcal{H}) \geq m_1 + m_2 - 1 = (|X_1| - s_1) + (|X_2| - s_2) - 1$$
$$= (|X_1| + |X_2|) - (s_1 + s_2) - 1$$
$$= |X| + 1 - s - 1 = |X| - s(\mathcal{H}).$$

From the above discussion, we have that

$$\overline{\chi}(\mathcal{H}) = |X| - s(\mathcal{H}). \qquad \square$$

# References

1. Berge, C.: Graphs Hypergraphs. North-Holland, Amsterdam (1973)
2. Voloshin, V.I.: On the Upper Chromatic Number of a Hypergraph. Austr. J. Conbin. **11** (1995) 25–45
3. Bulgaru, E., Voloshin, V.I.: Mixed Interval Hypergraphs. Discrete Appl. Math. **77** (1997) 29–41
4. Diao, K.F., Zhao, P., Zhou, H.S.: About the Upper Chromatic Number of a Co-Hypergraph. Discrete Mathematics **220** (2000) 67–73
5. Milazzo, L., Tuza, Zs.: Upper Chromatic Number of Steiner Triple and Quadruple Systems. Discrete Math. **174** (1-3) (1997) 247–260

# Note on Characterization of Uniquely 3-List Colorable Complete Multipartite Graphs

Yongqiang Zhao[1], Wenjie He[2], Yufa Shen[3], and Yanning Wang[4]

[1] Department of Mathematics, Shijiazhuang College, Shijiazhuang 050035,
P.R. China
yqzhao1970@yahoo.com
[2] Applied Mathematics Institute, Hebei University of Technology, Tianjin 300130,
P.R. China
He_wenjie@yahoo.com
[3] Department of Mathematics and Physics, Hebei Normal University of Science and
Technology, Qinhuangdao 066004, P.R. China
[4] Department of Mathematics, Yanshan University, Qinhuangdao 066004, P.R. China

**Abstract.** Let $G$ be a graph and suppose that for each vertex $v$ of $G$, there exists a list of $k$ colors, $L(v)$, such that there is a unique proper coloring for $G$ from this collection of lists, then $G$ is called a uniquely k-list colorable graph. M. Ghebleh and E. S. Mahmoodian characterized uniquely 3-List colorable complete multipartite graphs except for nine graphs. Recently, except for graph $K_{2,3,4}$, the other eight graphs were shown not to be uniquely 3-list colorable by W. He and Y. Shen, etc. In this paper, it is proved that $K_{2,3,4}$ is not uniquely 3-list colorable, and then the uniquely 3-list colorable complete multipartite graphs are characterized completely.

**Keywords:** List coloring, Uniquely k-list colorable, Complete multipartite graph.

**AMS Subject Classifications:** 05C15.

## 1 Introduction and Preliminaries

We consider finite and undirected simple graphs. Let $G$ be a graph and $L(v)$ denote a list of colors available for vertex $v$ of $G$. A list coloring from the given collection of lists is a proper coloring $c$ such that $c(v)$ is chosen from $L(v)$. We will refer to such a coloring as an $L$-coloring. The idea of list colorings of graphs is due independently to Vizing [9] and to Erdős, Rubin and Taylor [1].

Suppose that for each vertex $v$ of $G$, there exists a list of $k$ colors, $L(v)$, such that there is a unique proper coloring for $G$ from this collection of lists, then $G$ is called a uniquely $k$-list colorable (U$k$LC) graph. We say that a graph $G$ has the property $M(k)$ if and only if it is not uniquely $k$-list colorable. We use the notation $K_{s*r}$ for a complete $r$-partite graph in which each part is of size $s$. Notations such as $K_{s*r,\,t}$, etc. are used similarly.

J. Akiyama et al. (Eds.): CJCDGCGT 2005, LNCS 4381, pp. 278–287, 2007.
© Springer-Verlag Berlin Heidelberg 2007

M. Mahdian and E. S. Mahmoodian characterized U2LC graphs as follows:

**Theorem 1 (Mahdian and Mahmoodian [6]).** *A connected graph $G$ has the property $M(2)$ if and only if every block of $G$ is either a cycle, a complete graph, or a complete bipartite graph.*

They extensively studied the unique list colorability about complete multipartite graphs, especially for unique 3-list colorability.

**Theorem 2 (Ghebleh and Mahmoodian [3]).** *The graphs $K_{3,3,3}$, $K_{2,4,4}$, $K_{2,3,5}$, $K_{2,2,9}$, $K_{1,2,2,2}$, $K_{1,1,2,3}$, $K_{1,1,1,2,2}$, $K_{1*4,6}$, $K_{1*5,5}$, and $K_{1*6,4}$ are U3LC.*

Except for nine graphs, they characterized U3LC complete multipartite graphs as follows:

**Theorem 3 (Ghebleh and Mahmoodian [3]).** *Let $G$ be a complete multipartite graph that is not $K_{2,2,r}$, for $r = 4, 5, \ldots, 8$, $K_{2,3,4}$, $K_{1*4,4}$, $K_{1*4,5}$, or $K_{1*5,4}$. Then $G$ is U3LC if and only if it has one of the graphs in Theorem 2 as an induced subgraph.*

They also provided the following problem for the excepted graphs.

*Problem 1 (Ghebleh and Mahmoodian [3]).* Verify the property $M(3)$ for the graphs $K_{2,2,r}$, for $r = 4, 5, \ldots, 8$, $K_{2,3,4}$, $K_{1*4,4}$, $K_{1*4,5}$, and $K_{1*5,4}$.

Now except $K_{2,3,4}$, all the graphs in the above problem were proved to have the property $M(3)$ by W. He and Y. Shen, etc..

**Theorem 4 (He et al. [4]).** *Graph $K_{2,2,r}$ has the property $M(3)$, where $r = 4, 5, \ldots, 8$.*

**Theorem 5 (Shen et al. [8]).** *Graph $K_{1*5,4}$ has the property $M(3)$.*

**Theorem 6 (He et al. [5]).** *Graphs $K_{1*4,4}$ and $K_{1*4,5}$ have the property $M(3)$.*

Aided by a computer program, $K_{2,3,3}$ was shown to have property $M(3)$ by M. Ghebleh and E. S. Mahmoodian [3]. In this paper we will show that $K_{2,3,4}$ has the property $M(3)$. Then it will be a direct corollary that $K_{2,3,3}$ has the property $M(3)$. Furthermore, we completely characterize the U3LC complete multipartite graphs. In Section 2, we provide some propositions and lemmas which are helpful in proving the main result. In Section 3, we prove that $K_{2,3,4}$ has the property $M(3)$. In Section 4, we completely characterize the U3LC complete multipartite graphs.

## 2    Propositions and Lemmas

Below are three theorems useful in our proofs.

**Proposition 1 (Ghebleh and Mahmoodian [3]).** *If $L$ is a $k$-list assignment to the vertices in the graph $G$, and $G$ has a unique $L$-coloring, then $|\bigcup_{v \in V(G)} L(v)| \geq k + 1$ and all these colors are used in the unique $L$-coloring of $G$.*

**Proposition 2 (Ganjalia et al. [2]).** *If $G$ is a complete tripartite U3LC graph, then all vertices in each part can not take the same color in the unique 3-list coloring of $G$.*

**Proposition 3 (Ghebleh and Mahmoodian [3]).** *If $G$ is a complete multipartite graph which has an induced U$k$LC subgraph, then $G$ is U$k$LC.*

In the following, let $\{v_1, v_2\}$, $\{v_3, v_4, v_5\}$ and $\{v_6, v_7, v_8, v_9\}$ be the three parts of $K_{2,3,4}$. Assign the list $\{c_{i1}, c_{i2}, c_{i3}\}$ to the vertex $v_i$, $i = 1, 2, \ldots, 9$. Now we suppose that there exists a unique $L$-coloring $c$ for $K_{2,3,4}$, say $c(v_i) = c_{i1}$, $i = 1, 2, \ldots, 9$. Then the following lemmas hold under this assumption. For short, denote $\{v_6, v_7, v_8, v_9\}$ by $S$, $\{c_{61}, c_{71}, c_{81}, c_{91}\}$ by $A$.

**Lemma 1.** $|\{c_{11}, c_{21}\}| = 2$, $|\{c_{31}, c_{41}, c_{51}\}| = 2$ and $|A| \geq 2$.

*Proof.* By Proposition 2, $|\{c_{11}, c_{21}\}| = 2$, $|\{c_{31}, c_{41}, c_{51}\}| \geq 2$ and $|A| \geq 2$. So in the following we just need to show that $|\{c_{31}, c_{41}, c_{51}\}| \neq 3$. By contradiction, suppose that $|\{c_{31}, c_{41}, c_{51}\}| = 3$, and consider the graph $G'$ obtained from the graph $K_{2,3,4}$ by adding edges $v_1 v_2$, $v_3 v_4$, $v_3 v_5$ and $v_4 v_5$ to it. It is easy to see that $G'$ is $K_{1*5,4}$, and $c$ is also a proper coloring for $G'$. By Theorem 5, there is another $L$-coloring $c'$ for $G'$. Since $G'$ is gotten from $K_{2,3,4}$ by adding some edges, $c'$ is also an $L$-coloring for $K_{2,3,4}$, which contradicts to that $c$ is a unique $L$-coloring for $K_{2,3,4}$. So we have $|\{c_{31}, c_{41}, c_{51}\}| \neq 3$. The proof is complete.  □

By Lemma 1, in the following, let $c_{i1} = i$ for $i = 1, 2, 3$, and $c_{41} = c_{51} = 4$.

**Lemma 2.** *If $v_i$, $v_j$ are two vertices of $K_{2,3,4}$ in the same part, then $c_{i1} \notin \{c_{j2}, c_{j3}\}$.*

*Proof.* If $c_{i1} = c_{j1}$, then the result is obvious. If $c_{i1} \neq c_{j1}$ and $c_{i1} \in \{c_{j2}, c_{j3}\}$, then let $c'(v_j) = c_{i1}$, and $c'(v_k) = c(v_k) = c_{k1}$ for $k \neq j$. It is easy to see that $c'$ is another $L$-coloring of $K_{2,3,4}$, a contradiction.  □

**Corollary 1.** $\{c_{j2}, c_{j3}\} \subseteq \{1, 2, 3, 4\}$, *where $j = 6, 7, 8, 9$.*

*Proof.* The result follows from Proposition 1 and Lemma 2.  □

**Lemma 3.** *For any $i \in \{1, 2\}$, if $i \in \{c_{32}, c_{33}\}$, then $3 \notin \{c_{i2}, c_{i3}\}$; if $4 \in \{c_{i2}, c_{i3}\}$, then $i \notin \{c_{42}, c_{43}\} \cap \{c_{52}, c_{53}\}$.*

*Proof.* For any $i \in \{1, 2\}$, if $i \in \{c_{32}, c_{33}\}$ and $3 \in \{c_{i2}, c_{i3}\}$, then let $c'(v_i) = 3$, $c'(v_3) = i$, and $c'(v_k) = c(v_k)$ for $k \neq i, 3$. If $4 \in \{c_{i2}, c_{i3}\}$ and $i \in \{c_{42}, c_{43}\} \cap \{c_{52}, c_{53}\}$, then let $c'(v_i) = 4$, $c'(v_4) = c'(v_5) = i$, and $c'(v_k) = c(v_k)$ for $k \neq i, 4, 5$. It is easy to check that $c'$ is another $L$-coloring of $K_{2,3,4}$ in each case, which is a contradiction to that $c$ is a unique $L$-coloring for $K_{2,3,4}$.  □

**Lemma 4.** *There exists at least one $i \in \{1, 2, 3, 4, 5\}$ such that $\{c_{i2}, c_{i3}\} \subseteq A$.*

*Proof.* Otherwise, for each $i \in \{1, 2, 3, 4, 5\}$, $|\{c_{i1}, c_{i2}, c_{i3}\} \setminus A| \geq 2$. Let $G'$ be the subgraph of $K_{2,3,4}$ induced by vertices $v_1, v_2, v_3, v_4$ and $v_5$. It is easy to see that $G'$ is $K_{2,3}$. By Theorem 1, $G'$ has the property $M(2)$. So there is another $L$-coloring $c'$ for vertices $v_1, v_2, v_3, v_4$ and $v_5$ without using any color in $A$ which, together with $c'(v_i) = c(v_i)$ for $i = 6, 7, 8, 9$, is another $L$-coloring of $K_{2,3,4}$, a contradiction. □

**Lemma 5.** *If the other parts are unconsidered, any three colors in $\{1, 2, 3, 4\}$ can be used to $L$-color $S$.*

*Proof.* Otherwise, without loss of generality, we may assume that $\{1, 2, 3\}$ can not be used to $L$-color $S$ when the other parts are unconsidered. Then $\{c_{i2}, c_{i3}\} \cap \{1, 2, 3\} = \phi$ for some $i \in \{6, 7, 8, 9\}$. Since $c_{i2} \neq c_{i3}$ and by Corollary 1, this is impossible. □

**Lemma 6.** *If the other parts are unconsidered, there exist at least two pairs of colors in $\{1, 2, 3, 4\}$, such that each pair of them can be used to $L$-color $S$.*

*Proof.* Otherwise, suppose to the contrary that at most one pair of colors in $\{1, 2, 3, 4\}$ can be used to $L$-color $S$ when the other parts are unconsidered, say $\{1, 2\}$ or none. Then each of $\{1, 3\}$, $\{1, 4\}$, $\{2, 3\}$, $\{2, 4\}$ and $\{3, 4\}$ can not be used to $L$-color $S$. Correspondingly, $\{2, 4\}$, $\{2, 3\}$, $\{1, 4\}$, $\{1, 3\}$ and $\{1, 2\}$ must be all in $\{\{c_{62}, c_{63}\}, \{c_{72}, c_{73}\}, \{c_{82}, c_{83}\}, \{c_{92}, c_{93}\}\}$, which is impossible. □

**Lemma 7.** *If $\{c_{12}, c_{13}\} = \{c_{22}, c_{23}\} = \{3, 4\}$, then there is another $L$-coloring of $K_{2,3,4}$.*

*Proof.* By Proposition 1, Lemmas 2 and 3, $\{c_{32}, c_{33}\} = \{a, b\} \subseteq A$. If $\{c_{42}, c_{43}\} \cap A \neq \phi$ and $\{c_{52}, c_{53}\} \cap A \neq \phi$, say $s \in \{c_{42}, c_{43}\} \cap A$ and $t \in \{c_{52}, c_{53}\} \cap A$. Let $c'(v_1) = c'(v_2) = 4$, $c'(v_3) = a$, $c'(v_4) = s$, $c'(v_5) = t$, and $c'$ $L$-colors $S$ with $\{1, 2, 3\}$. By Lemma 5, $c'$ is another $L$-coloring of $K_{2,3,4}$. Otherwise, by Proposition 1, Lemmas 2 and 3, without loss of generality, we may assume that $\{c_{42}, c_{43}\} = \{1, 2\}$ and $\{c_{52}, c_{53}\} = \{x, y\}$, where $x, y \in A$. At this time, if $S$ is unconsidered, it is easy to verify that we can $L$-color $v_1, v_2, v_3, v_4$ and $v_5$ without using $\{1, 2\}$, $\{1, 3\}$, $\{1, 4\}$, $\{2, 3\}$ or $\{2, 4\}$ respectively. But by Lemma 6, at least one of these can be used to $L$-color $S$. Thus we may get another $L$-coloring of $K_{2,3,4}$. □

## 3   Main Result

**Theorem 7.** *Graph $K_{2,3,4}$ has the property $M(3)$.*

*Proof.* By contradiction. Suppose that there exists a unique $L$-coloring $c$ for $K_{2,3,4}$, say $c(v_i) = c_{i1}$, $i = 1, 2, \ldots, 9$. By Lemma 1, let $c_{i1} = i$ for $i = 1, 2, 3$, and $c_{41} = c_{51} = 4$. By Lemma 6, we just need to consider the following nine cases, $\{1, 2\}$ and $\{1, 3\}$, $\{1, 2\}$ and $\{1, 4\}$, $\{1, 2\}$ and $\{3, 4\}$, $\{1, 3\}$ and $\{2, 3\}$, $\{1, 3\}$ and $\{1, 4\}$, $\{1, 3\}$ and $\{2, 4\}$, $\{1, 3\}$ and $\{3, 4\}$, $\{1, 4\}$ and $\{2, 4\}$, or $\{1, 4\}$ and $\{3, 4\}$ can respectively be used to color $S$ when the other parts are unconsidered, since each of the other cases is equivalent to one of the nine cases. Furthermore, if the following four claims hold, in each case we may get another $L$-coloring of $K_{2,3,4}$, which is contradicting to that $c$ is the unique $L$-coloring for $K_{2,3,4}$. Then the proof is complete.  □

In order to complete the proof of Theorem 7, we begin to prove the four claims. For short, throughout the proof of claims, denote Proposition 1 and Lemma 2 by PL12, Proposition 1, Lemmas 2 and 3 by PL123.

**Claim 1.** *If $\{1, 2\}$ can be used to $L$-color $S$ when the other parts are unconsidered, then there is another $L$-coloring of $K_{2,3,4}$.*

*Proof.* If $\{c_{12}, c_{13}\} \cap A \neq \phi$ and $\{c_{22}, c_{23}\} \cap A \neq \phi$, say $a \in \{c_{12}, c_{13}\} \cap A$ and $b \in \{c_{22}, c_{23}\} \cap A$. Then let $c'(v_1) = a$, $c'(v_2) = b$, $c'(v_3) = 3$, $c'(v_4) = c'(v_5) = 4$, and $c'$ $L$-colors $S$ with $\{1, 2\}$. Otherwise, by PL12, without loss of generality, we may assume that $\{c_{12}, c_{13}\} = \{3, 4\}$. By PL123, $\{c_{32}, c_{33}\} \cap A \neq \phi$, and we just need to consider the following two cases.

*Case 1.* $\{c_{32}, c_{33}\} = \{a, b\} \subseteq A$.

If $\{c_{22}, c_{23}\} = \{3, 4\}$, then by Lemma 7, another $L$-coloring of $K_{2,3,4}$ exists. Otherwise, by PL12, $\{c_{22}, c_{23}\} \cap A \neq \phi$, say $t \in \{c_{22}, c_{23}\} \cap A$. Let $c'(v_1) = 3$, $c'(v_2) = t$, $c'(v_4) = c'(v_5) = 4$, $c'$ $L$-colors $v_3$ with one of $a$ and $b$ that is different from $t$, and $c'$ $L$-colors $S$ with $\{1, 2\}$.

*Case 2.* $\{c_{32}, c_{33}\} = \{2, b\}$, where $b \in A$.

By PL123, $\{c_{22}, c_{23}\} \cap A \neq \phi$ and $3 \notin \{c_{22}, c_{23}\}$. If $\{c_{22}, c_{23}\} = \{x, y\} \subseteq A$, then we may get another $L$-coloring of $K_{2,3,4}$ by the similar way used in Case 1. Otherwise, $\{c_{22}, c_{23}\} = \{4, a\}, a \in A$. If $a \neq b$, then we may also get another $L$-coloring of $K_{2,3,4}$ by the similar way above. Now we consider the case $a = b$. If $\{c_{42}, c_{43}\} \cap A \neq \phi$ and $\{c_{52}, c_{53}\} \cap A \neq \phi$, then we may get another $L$-coloring of $K_{2,3,4}$ easily. Otherwise, by PL123, without loss of generality, we assume that $\{c_{42}, c_{43}\} = \{1, 2\}$ and $\{c_{52}, c_{53}\} = \{x, y\}$, where $x, y \in A$. At this time, it is not difficult to see that we can not $L$-color $v_1$, $v_2$, $v_3$, $v_4$ and $v_5$ without using $\{1, 2\}$ or $\{2, 4\}$, even if we do not consider $S$. But if we do not consider $S$, we can $L$-color $v_1$, $v_2$, $v_3$, $v_4$ and $v_5$ without using $\{1, 3\}$, $\{1, 4\}$, $\{2, 3\}$ or $\{3, 4\}$. So when $v_1$, $v_2$, $v_3$, $v_4$ and $v_5$ are unconsidered, if one of $\{1, 3\}$, $\{1, 4\}$, $\{2, 3\}$ and $\{3, 4\}$ can be used to $L$-color $S$, then we may get another $L$-coloring of $K_{2,3,4}$. Otherwise, by lemma 6, only $\{1, 2\}$ and $\{2, 4\}$ can be used to $L$-color $S$ when $v_1$, $v_2$, $v_3$, $v_4$ and $v_5$ are unconsidered. At this time, $\{\{c_{62}, c_{63}\}, \{c_{72}, c_{73}\}, \{c_{82}, c_{83}\}, \{c_{92}, c_{93}\}\} = \{\{2, 4\}, \{2, 3\}, \{1, 4\}, \{1, 2\}\}$. Then it is easy to see that $\{1, 3, w\}$, where $w \in A$, can be used to $L$-color $S$ when $v_1$, $v_2$, $v_3$, $v_4$ and $v_5$ are unconsidered. Let $c'(v_1) = c'(v_2) = 4$, $c'(v_3) = c'(v_5) = 2$, $c'$ $L$-colors $v_4$ with one of $x$ and $y$ that is different from $w$, and $c'$ $L$-colors $S$ with $\{1, 3, w\}$.

Combining Cases 1 and 2, it is easy to see that we may get another $L$-coloring $c'$ of $K_{2,3,4}$ in each case. The proof is complete.    □

**Claim 2.** *If $\{1, 3\}$ can be used to $L$-color $S$ when the other parts are unconsidered, then there is another $L$-coloring of $K_{2,3,4}$.*

*Proof.* By Lemma 4, we just need to consider the following four cases.

*Case 1.* $\{c_{12}, c_{13}\} = \{a, b\}$, where $a, b \in A$.

If $\{c_{32}, c_{33}\} \cap A \neq \phi$, then another $L$-coloring of $K_{2,3,4}$ exists obviously. Otherwise, by PL123, $\{c_{32}, c_{33}\} = \{1, 2\}$ and $\{c_{22}, c_{23}\} \cap A \neq \phi$, we may also get another $L$-coloring of $K_{2,3,4}$.

*Case 2.* $\{c_{22}, c_{23}\} = \{a, b\}$, where $a, b \in A$.

If $\{c_{12}, c_{13}\} \subseteq A$, it is Case 1. Otherwise, by PL12, we just need to consider the following two subcases.

*Subcase 2.1.* $\{c_{12}, c_{13}\} = \{3, 4\}$.

By PL123, $\{c_{32}, c_{33}\} \cap A \neq \phi$, say $s \in \{c_{32}, c_{33}\} \cap A$. If $\{c_{42}, c_{43}\} \cap A \neq \phi$ and $\{c_{52}, c_{53}\} \cap A \neq \phi$, then another $L$-coloring of $K_{2,3,4}$ exists. Otherwise, by PL123, without loss of generality, we may suppose that $\{c_{42}, c_{43}\} = \{1, 2\}$, and $\{c_{52}, c_{53}\} = \{2, x\}$ or $\{t, w\}$, where $x, t, w \in A$. It is easy to check that another $L$-coloring of $K_{2,3,4}$ exists in each case.

*Subcase 2.2.* $|\{c_{12}, c_{13}\} \cap A| = 1$.

Suppose $\{c_{12}, c_{13}\} \cap A = \{s\}$. If $2 \in \{c_{32}, c_{33}\}$, or $\{c_{32}, c_{33}\} \cap A$ includes a color $t$ such that $t \neq s$, then we may get another $L$-coloring of $K_{2,3,4}$. Otherwise, by PL123, $\{c_{12}, c_{13}\} = \{4, s\}$, $\{c_{32}, c_{33}\} = \{1, s\}$, and the discussion is similar to Subcase 2.1.

*Case 3.* $\{c_{32}, c_{33}\} = \{a, b\}$, where $a, b \in A$.

If $\{c_{12}, c_{13}\} \cap A \neq \phi$, then another $L$-coloring of $K_{2,3,4}$ exists obviously. Otherwise, by PL12, $\{c_{12}, c_{13}\} = \{3, 4\}$. If $\{c_{22}, c_{23}\} = \{3, 4\}$, by Lemma 7, another $L$-coloring of $K_{2,3,4}$ exists. So we just need to consider the case $\{c_{22}, c_{23}\} \cap A \neq \phi$, and the discussion is similar to Subcase 2.1.

*Case 4.* $\{c_{42}, c_{43}\} = \{a, b\}$, where $a, b \in A$.

If $\{c_{32}, c_{33}\} \subseteq A$, it is Case 2. Otherwise, by PL12, we just need to consider the following three subcases.

*Subcase 4.1.* $\{c_{32}, c_{33}\} = \{1, 2\}$.

By PL123, $\{c_{12}, c_{13}\} \cap A \neq \phi$ and $\{c_{22}, c_{23}\} \cap A \neq \phi$. At this time, we may get another $L$-coloring of $K_{2,3,4}$ easily.

*Subcase 4.2.* $\{c_{32}, c_{33}\} = \{1, s\}$, $s \in A$.

If $\{c_{12}, c_{13}\} \cap A$ includes a color $t$ such that $t \neq s$, then another $L$-coloring of $K_{2,3,4}$ exists obviously. Otherwise, by PL123, $\{c_{12}, c_{13}\} = \{4, s\}$. If $\{c_{52}, c_{53}\} \cap A \neq \phi$, we may get another $L$-coloring of $K_{2,3,4}$. Otherwise, by PL123, $\{c_{52}, c_{53}\} = \{1, 2\}$. If $4 \in \{c_{22}, c_{23}\}$, or $\{c_{22}, c_{23}\}$ includes a color $t \in A$ such that $t \neq s$, then we may also get another $L$-coloring of $K_{2,3,4}$. Otherwise, $\{c_{22}, c_{23}\} = \{3, s\}$. Let $c'(v_1) = c'(v_2) = s$, $c'(v_3) = c'(v_5) = 1$, $c'$ $L$-colors $v_4$ with one of $a$ and

$b$ that is different from $s$, and $c'$ $L$-colors $S$ with $\{2, 3, 4\}$. By Lemma 5, $c'$ is another $L$-coloring of $K_{2,3,4}$.

*Subcase 4.3.* $\{c_{32}, c_{33}\} = \{2, s\}$, $s \in A$.

If $\{c_{22}, c_{23}\} \subseteq A$, it is Case 2. Otherwise, by PL123, $\{c_{22}, c_{23}\} = \{4, t\}$, where $t \in A$. If $\{c_{12}, c_{13}\} \cap A \neq \phi$, then another $L$-coloring of $K_{2,3,4}$ exists. Otherwise, by PL12, $\{c_{12}, c_{13}\} = \{3, 4\}$. Now whether $\{c_{52}, c_{53}\} \cap A \neq \phi$ or $\{c_{52}, c_{53}\} \cap A = \phi$, we may always get another $L$-coloring of $K_{2,3,4}$.

Combining all the cases above, we get another $L$-coloring $c'$ of $K_{2,3,4}$ in each case. The claim is proved. □

**Claim 3.** *If $\{1, 4\}$ and $\{3, 4\}$ can respectively be used to $L$-color $S$ when the other parts are unconsidered, then there is another $L$-coloring for $K_{2,3,4}$.*

*Proof.* By Lemma 4, without loss of generality, we just need to consider the following four cases.

*Case 1.* $\{c_{42}, c_{43}\} = \{a, b\}$, where $a, b \in A$.

*Subcase 1.1.* $\{c_{52}, c_{53}\} \cap A \neq \phi$.

If $\{c_{32}, c_{33}\} \cap A \neq \phi$, then another $L$-coloring of $K_{2,3,4}$ exists obviously. Otherwise, by PL123, $\{c_{32}, c_{33}\} = \{1, 2\}$, $\{c_{12}, c_{13}\} \cap A \neq \phi$ and $\{c_{22}, c_{23}\} \cap A \neq \phi$. If we may select $w \in \{c_{12}, c_{13}\} \cap A$ and $s \in \{c_{52}, c_{53}\} \cap A$ such that $w \neq s$, then another $L$-coloring of $K_{2,3,4}$ exists. Otherwise, by PL123, $\{c_{12}, c_{13}\} = \{4, s\}$, $\{c_{52}, c_{53}\} = \{1, s\}$ or $\{2, s\}$, where $s \in A$. It is not difficult to check another $L$-coloring of $K_{2,3,4}$ exists in each case.

*Subcase 1.2.* $\{c_{52}, c_{53}\} \cap A = \phi$.

By PL12, $\{c_{52}, c_{53}\} = \{1, 2\}$. If $1 \in \{c_{32}, c_{33}\}$, by PL123, $\{c_{12}, c_{13}\} \cap A \neq \phi$, then another $L$-coloring of $K_{2,3,4}$ exists. If $2 \in \{c_{32}, c_{33}\}$, the discussion is similar to above. Otherwise, by PL12, $\{c_{32}, c_{33}\} = \{s, t\}$, where $s, t \in A$. If $\{c_{12}, c_{13}\} \cap A \neq \phi$ or $\{c_{22}, c_{23}\} \cap A \neq \phi$, we may always get another $L$-coloring of $K_{2,3,4}$. Otherwise, by PL12, $\{c_{12}, c_{13}\} = \{c_{22}, c_{23}\} = \{3, 4\}$. By Lemma 7, another $L$-coloring of $K_{2,3,4}$ exists.

*Case 2.* $\{c_{32}, c_{33}\} = \{a, b\}$, where $a, b \in A$.

If $\{c_{42}, c_{43}\} \cap A \neq \phi$ and $\{c_{52}, c_{53}\} \cap A \neq \phi$, then another $L$-coloring of $K_{2,3,4}$ exists obviously. Otherwise, by PL12, without loss of generality, we may assume that $\{c_{42}, c_{43}\} = \{1, 2\}$. If $1 \in \{c_{52}, c_{53}\}$, by PL123, $\{c_{12}, c_{13}\} \cap A \neq \phi$, and we may get another $L$-coloring of $K_{2,3,4}$. If $2 \in \{c_{52}, c_{53}\}$, the discussion is similar to above. Otherwise, by PL12, $\{c_{52}, c_{53}\} \subseteq A$, it is equivalent to Case 1.

*Case 3.* $\{c_{12}, c_{13}\} = \{a, b\}$, where $a, b \in A$.

If $\{c_{42}, c_{43}\} \subseteq A$ or $\{c_{52}, c_{53}\} \subseteq A$, it is Case 1. By PL12, we just need to consider the following three subcases.

*Subcase 3.1.* $1 \in \{c_{42}, c_{43}\}$ and $1 \in \{c_{52}, c_{53}\}$.

If $1 \in \{c_{32}, c_{33}\}$, then another $L$-coloring of $K_{2,3,4}$ exists obviously. If $2 \in \{c_{32}, c_{33}\}$, by PL123, $\{c_{22}, c_{23}\} \cap A \neq \phi$, we may also get another $L$-coloring of $K_{2,3,4}$. Otherwise, by PL12, $\{c_{32}, c_{33}\} \subseteq A$, it is Case 2.

*Subcase 3.2.* $2 \in \{c_{42}, c_{43}\}$ and $2 \in \{c_{52}, c_{53}\}$.

By PL123, $\{c_{22}, c_{23}\} \cap A \neq \phi$, and another $L$-coloring of $K_{2,3,4}$ exists obviously.

*Subcase 3.3.* $1 \in \{c_{42}, c_{43}\}$ and $2 \in \{c_{52}, c_{53}\}$.

Now $\{c_{42}, c_{43}\} = \{1, s\}$ and $\{c_{52}, c_{53}\} = \{2, t\}$, where $s$, $t \in A$. Otherwise, it will be Subcase 3.1 or 3.2. If $1 \in \{c_{32}, c_{33}\}$, then another $L$-coloring of $K_{2,3,4}$ exists obviously. The following discussion is similar to Subcase 3.1.

*Case 4.* $\{c_{22}, c_{23}\} = \{a, b\}$, where $a$, $b \in A$.

If $\{c_{32}, c_{33}\} \subseteq A$, then it is Case 2. Otherwise, By PL123, we just need to consider the following two subcases.

*Subcase 4.1.* $1 \in \{c_{32}, c_{33}\}$.

By PL123, $\{c_{12}, c_{13}\} \cap A \neq \phi$, say $t \in \{c_{12}, c_{13}\} \cap A$. If $\{c_{42}, c_{43}\} \subseteq A$ or $\{c_{52}, c_{53}\} \subseteq A$, it is Case 1. Otherwise, by PL12, 1 or $2 \in \{c_{42}, c_{43}\}$, and 1 or $2 \in \{c_{52}, c_{53}\}$. It is easy to check that another $L$-coloring of $K_{2,3,4}$ exists.

*Subcase 4.2.* $\{c_{32}, c_{33}\} = \{2, s\}$, $s \in A$.

The discussion is similar to Case 2.

Combining all the cases above, we get another $L$-coloring $c'$ of $K_{2,3,4}$ in each case. The proof is complete.  □

**Remark 1.** In $\{1, 2, 3, 4\}$, if only $\{1, 4\}$ and $\{2, 4\}$ can be used to $L$-color $S$ when $v_1$, $v_2$, $v_3$, $v_4$ and $v_5$ are unconsidered, then $\{\{c_{62}, c_{63}\}, \{c_{72}, c_{73}\}, \{c_{82}, c_{83}\}, \{c_{92}, c_{93}\}\} = \{\{3, 4\}, \{2, 4\}, \{1, 4\}, \{1, 2\}\}$. Furthermore, it is easy to see that any one in $\{\{1, 2, 3\}, \{1, 2, x\}, \{1, 3, y\}, \{2, 3, z\}, \{4, w\}\}$ can be used to color $S$ when $v_1$, $v_2$, $v_3$, $v_4$ and $v_5$ are unconsidered, where $x$, $y$, $z$, $w \in A$.

**Claim 4.** *In* $\{1, 2, 3, 4\}$, *if only* $\{1, 4\}$ *and* $\{2, 4\}$ *can respectively be used to $L$-color $S$ when the other parts are unconsidered, then there is another $L$-coloring for* $K_{2,3,4}$.

*Proof.* By Lemma 4, we just need to consider the following three cases.

*Case 1.* $\{c_{12}, c_{13}\} = \{a, b\}$, where $a$, $b \in A$.

*Subcase 1.1.* $\{c_{22}, c_{23}\} = \{s, t\}$, where $s$, $t \in A$.

By Remark 1, $\{1, 2, x\}$, where $x \in A$, can be used to $L$-color $S$ when $v_1$, $v_2$, $v_3$, $v_4$ and $v_5$ are unconsidered. So another $L$-coloring of $K_{2,3,4}$ exists obviously.

*Subcase 1.2.* $3 \in \{c_{22}, c_{23}\}$.

By PL123, $\{c_{32}, c_{33}\} \subseteq A$ or $\{c_{32}, c_{33}\} = \{1, s\}$, where $s \in A$. For each case, the discussion is similar to Case 2 in the proof of Claim 3.

*Subcase 1.3.* $\{c_{22}, c_{23}\} = \{4, s\}$, where $s \in A$.

If $\{c_{32}, c_{33}\} \cap A$ includes a color $t$ such that $t \neq s$, then by Remark 1, we may get another $L$-coloring of $K_{2,3,4}$. Otherwise, by PL12, $\{c_{32}, c_{33}\} = \{1, s\}$, $\{2, s\}$ or $\{1, 2\}$. Now we suppose $\{c_{32}, c_{33}\} = \{1, s\}$, and consider the following subcases.

*Subcase 1.3.1.* $\{c_{42}, c_{43}\} = \{t, p\}$, where $t, p \in A$.

If 1 or 2 $\in \{c_{52}, c_{53}\}$, we may get another $L$-coloring of $K_{2,3,4}$. Otherwise, by PL12, $\{c_{52}, c_{53}\} = \{q, r\}$, where $q, r \in A$, we may also get another $L$-coloring of $K_{2,3,4}$ by Remark 1.

*Subcase 1.3.2.* $\{c_{42}, c_{43}\} = \{1, 2\}$.

The discussion is similar to Subcase 1.3.1.

*Subcase 1.3.3.* $\{c_{42}, c_{43}\} = \{1, t\}$, where $t \in A$.

If $\{c_{52}, c_{53}\} \subseteq A$ or 1 $\in \{c_{52}, c_{53}\}$, the discussion is similar to Subcase 1.3.1. So we just need to consider the case $\{c_{52}, c_{53}\} = \{2, p\}$, where $p \in A$. If $t \neq s$ or $p \neq s$, we may get another $L$-coloring of $K_{2,3,4}$ respectively. Otherwise, $t = p = s$, we may also get another $L$-coloring of $K_{2,3,4}$ easily.

*Subcase 1.3.4.* $\{c_{42}, c_{43}\} = \{2, t\}$, where $t, \in A$.

The discussion is similar to Subcase 1.3.3.

When $\{c_{32}, c_{33}\} = \{2, s\}$ or $\{1, 2\}$, the proof is similar to case $\{c_{32}, c_{33}\} = \{1, s\}$.

*Case 2.* $\{c_{32}, c_{33}\} = \{a, b\}$, where $a, b \in A$.

If $\{c_{12}, c_{13}\} \subseteq A$, it is Case 1. If $\{c_{22}, c_{23}\} \subseteq A$, it is equivalent to Case 1. If $\{c_{12}, c_{13}\} = \{c_{22}, c_{23}\} = \{3, 4\}$, by Lemma 7, another $L$-coloring of $K_{2,3,4}$ exists. So we just need to consider the case 3 or 4 $\in \{c_{12}, c_{13}\} \cap \{c_{22}, c_{23}\}$, and the case $\{c_{12}, c_{13}\} = \{3, s\}$ and $\{c_{22}, c_{23}\} = \{4, t\}$, where $s, t \in A$. By Remark 1, for each case, the discussion is similar to Case 2 in the proof of Claim 3.

*Case 3.* $\{c_{42}, c_{43}\} = \{a, b\}$, where $a, b \in A$.

If $\{c_{32}, c_{33}\} \subseteq A$, it is Case 2. Otherwise, by PL12, we just need to consider the following three subcases.

*Subcase 3.1.* $\{c_{32}, c_{33}\} = \{1, s\}$, where $s \in A$.

By PL123, $\{c_{12}, c_{13}\} \subseteq A$ or $\{c_{12}, c_{13}\} = \{4, t\}$, $t \in A$. Since $\{c_{12}, c_{13}\} \subseteq A$ or $\{c_{22}, c_{23}\} \subseteq A$ is Case 1. So we just need to consider the case $\{c_{12}, c_{13}\} = \{4, t\}$, $t \in A$, and the following two subcases.

*Subcase 3.1.1.* 3 $\in \{c_{22}, c_{23}\}$.

If $\{c_{52}, c_{53}\} \cap A \neq \phi$, we may get another $L$-coloring of $K_{2,3,4}$. Otherwise, by PL123, $\{c_{52}, c_{53}\} = \{1, 2\}$, and we may also get another $L$-coloring of $K_{2,3,4}$.

*Subcase 3.1.2.* $\{c_{22}, c_{23}\} = \{4, p\}$, where $p \in A$.

If $\{c_{52}, c_{53}\} \cap A \neq \phi$, by Remark 1, we may get another $L$-coloring of $K_{2,3,4}$. Otherwise, by PL123, $\{c_{52}, c_{53}\} = \{1, 2\}$. By Remark 1, we may also get another $L$-coloring of $K_{2,3,4}$.

*Subcase 3.2.* $\{c_{32}, c_{33}\} = \{2, s\}$, where $s \in A$.

The proof is similar to Subcase 3.1.

*Subcase 3.3.* $\{c_{32}, c_{33}\} = \{1, 2\}$.

If $\{c_{12}, c_{13}\} \subseteq A$ or $\{c_{22}, c_{23}\} \subseteq A$, it is Case 1. Otherwise, by PL123, 4 $\in \{c_{12}, c_{13}\}$ and 4 $\in \{c_{22}, c_{23}\}$. The discussion is similar to Subcase 3.1.2.

Combining all the cases above, the proof is complete.    □

**Corollary 2.** *Graph $K_{2,3,3}$ has the property $M(3)$.*

*Proof.* Otherwise, $K_{2,3,3}$ is U3LC. Since $K_{2,3,3}$ is an induce subgraph of $K_{2,3,4}$, by Proposition 3, $K_{2,3,4}$ is U3LC, which contradicts Theorem 7.     □

# 4   Conclusion

All the graphs in Problem 1 now was proved to have the property $M(3)$. Therefore, we can completely characterize the U3LC complete multipartite graphs as follows.

**Theorem 8.** *Let $G$ be a complete multipartite graph, then $G$ is U3LC if and only if it has one of the graphs $K_{3,3,3}$, $K_{2,4,4}$, $K_{2,3,5}$, $K_{2,2,9}$, $K_{1,2,2,2}$, $K_{1,1,2,3}$, $K_{1,1,1,2,2}$, $K_{1*4,6}$, $K_{1*5,5}$, and $K_{1*6,4}$ as an induced subgraph.*

# Acknowledgements

The authors are grateful to the anonymous referees for their careful comments and valuable suggestions.

# References

1. Erdős, P., Rubin, A. L., Taylor, H.: Choosability in Graphs. Congr. Numer. **26** (1979) 125–157
2. Ganjali, Y.G., Ghebleh, M., Hajiabohassan, H., Mirzadeh, M., Sadjad, B.S.: Uniquely 2-List Colorable Graphs. Discrete Appl. Math. **119** (2002) 217–225
3. Ghebleh, M., Mahmoodian, E.S.: On Uniquely List Colorable graphs. Ars Combin. **59** (2001) 307–318
4. He, W., Shen, Y., Zhao, Y., Sun, S.: $K_{2*2,r}, r = 4, 5, 6, 7, 8$, Have the Property $M(3)$. Submitted
5. He, W., Shen, Y., Zhao, Y., Wang, Y., Ma, X.: On Property $M(3)$ of Some Complete Multipartite Graphs. Austral. J. Combin. **35** (2006) 211–220
6. Mahdian, M., Mahmoodian, E.S.: A Characterization of Uniquely 2-List Colorable Graphs. Ars Combin. **51** (1999) 295–305
7. Mahmoodian, E.S., Mahdian, M.: On the Uniquely List Colorable Graphs. In: Proceedings of the 28-th Annual Iranian Mathematics Conference, Part 1, number 377 in Tabriz Univ. Ser., Tabriz (1997) 319–326
8. Shen, Y., Wang, Y., He, W., Zhao, Y.: On Uniquely List Colorable Complete Multipartite Graphs. Ars Combin. (to appear)
9. Vizing, V.G.: Vertex Coloring with Given Colors (in Russian). Diskret. Analiz. **29** (1976) 3–10

# Author Index

Printing: Mercedes-Druck, Berlin
Binding: Stein+Lehmann, Berlin

# Lecture Notes in Computer Science

For information about Vols. 1–4284

please contact your bookseller or Springer

Vol. 4331: G. Min, B. Di Martino, L.T. Yang, M. Guo, G. Ruenger (Eds.), Frontiers of High Performance Computing and Networking – ISPA 2006 Workshops. XXXVII, 1141 pages. 2006.

Vol. 4330: M. Guo, L.T. Yang, B. Di Martino, H.P. Zima, J. Dongarra, F. Tang (Eds.), Parallel and Distributed Processing and Applications. XVIII, 953 pages. 2006.

Vol. 4329: R. Barua, T. Lange (Eds.), Progress in Cryptology - INDOCRYPT 2006. X, 454 pages. 2006.

Vol. 4328: D. Penkler, M. Reitenspiess, F. Tam (Eds.), Service Availability. X, 289 pages. 2006.

Vol. 4327: M. Baldoni, U. Endriss (Eds.), Declarative Agent Languages and Technologies IV. VIII, 257 pages. 2006. (Sublibrary LNAI).

Vol. 4326: S. Göbel, R. Malkewitz, I. Iurgel (Eds.), Technologies for Interactive Digital Storytelling and Entertainment. X, 384 pages. 2006.

Vol. 4325: J. Cao, I. Stojmenovic, X. Jia, S.K. Das (Eds.), Mobile Ad-hoc and Sensor Networks. XIX, 887 pages. 2006.

Vol. 4323: G. Doherty, A. Blandford (Eds.), Interactive Systems. XI, 269 pages. 2007.

Vol. 4320: R. Gotzhein, R. Reed (Eds.), System Analysis and Modeling: Language Profiles. X, 229 pages. 2006.

Vol. 4319: L.-W. Chang, W.-N. Lie (Eds.), Advances in Image and Video Technology. XXVI, 1347 pages. 2006.

Vol. 4318: H. Lipmaa, M. Yung, D. Lin (Eds.), Information Security and Cryptology. XI, 305 pages. 2006.

Vol. 4317: S.K. Madria, K.T. Claypool, R. Kannan, P. Uppuluri, M.M. Gore (Eds.), Distributed Computing and Internet Technology. XIX, 466 pages. 2006.

Vol. 4316: M.M. Dalkilic, S. Kim, J. Yang (Eds.), Data Mining and Bioinformatics. VIII, 197 pages. 2006. (Sublibrary LNBI).

Vol. 4314: C. Freksa, M. Kohlhase, K. Schill (Eds.), KI 2006: Advances in Artificial Intelligence. XII, 458 pages. 2007. (Sublibrary LNAI).

Vol. 4313: T. Margaria, B. Steffen (Eds.), Leveraging Applications of Formal Methods. IX, 197 pages. 2006.

Vol. 4312: S. Sugimoto, J. Hunter, A. Rauber, A. Morishima (Eds.), Digital Libraries: Achievements, Challenges and Opportunities. XVIII, 571 pages. 2006.

Vol. 4311: K. Cho, P. Jacquet (Eds.), Technologies for Advanced Heterogeneous Networks II. XI, 253 pages. 2006.

Vol. 4309: P. Inverardi, M. Jazayeri (Eds.), Software Engineering Education in the Modern Age. VIII, 207 pages. 2006.

Vol. 4308: S. Chaudhuri, S.R. Das, H.S. Paul, S. Tirthapura (Eds.), Distributed Computing and Networking. XIX, 608 pages. 2006.

Vol. 4307: P. Ning, S. Qing, N. Li (Eds.), Information and Communications Security. XIV, 558 pages. 2006.

Vol. 4306: Y. Avrithis, Y. Kompatsiaris, S. Staab, N.E. O'Connor (Eds.), Semantic Multimedia. XII, 241 pages. 2006.

Vol. 4305: A.A. Shvartsman (Ed.), Principles of Distributed Systems. XIII, 441 pages. 2006.

Vol. 4304: A. Sattar, B.-H. Kang (Eds.), AI 2006: Advances in Artificial Intelligence. XXVII, 1303 pages. 2006. (Sublibrary LNAI).

Vol. 4303: A. Hoffmann, B.-H. Kang, D. Richards, S. Tsumoto (Eds.), Advances in Knowledge Acquisition and Management. XI, 259 pages. 2006. (Sublibrary LNAI).

Vol. 4302: J. Domingo-Ferrer, L. Franconi (Eds.), Privacy in Statistical Databases. XI, 383 pages. 2006.

Vol. 4301: D. Pointcheval, Y. Mu, K. Chen (Eds.), Cryptology and Network Security. XIII, 381 pages. 2006.

Vol. 4300: Y.Q. Shi (Ed.), Transactions on Data Hiding and Multimedia Security I. IX, 139 pages. 2006.

Vol. 4299: S. Renals, S. Bengio, J.G. Fiscus (Eds.), Machine Learning for Multimodal Interaction. XII, 470 pages. 2006.

Vol. 4297: Y. Robert, M. Parashar, R. Badrinath, V.K. Prasanna (Eds.), High Performance Computing - HiPC 2006. XXIV, 642 pages. 2006.

Vol. 4296: M.S. Rhee, B. Lee (Eds.), Information Security and Cryptology – ICISC 2006. XIII, 358 pages. 2006.

Vol. 4295: J.D. Carswell, T. Tezuka (Eds.), Web and Wireless Geographical Information Systems. XI, 269 pages. 2006.

Vol. 4294: A. Dan, W. Lamersdorf (Eds.), Service-Oriented Computing – ICSOC 2006. XIX, 653 pages. 2006.

Vol. 4293: A. Gelbukh, C.A. Reyes-Garcia (Eds.), MICAI 2006: Advances in Artificial Intelligence. XXVIII, 1232 pages. 2006. (Sublibrary LNAI).

Vol. 4292: G. Bebis, R. Boyle, B. Parvin, D. Koracin, P. Remagnino, A. Nefian, G. Meenakshisundaram, V. Pascucci, J. Zara, J. Molineros, H. Theisel, T. Malzbender (Eds.), Advances in Visual Computing, Part II. XXXII, 906 pages. 2006.

Vol. 4291: G. Bebis, R. Boyle, B. Parvin, D. Koracin, P. Remagnino, A. Nefian, G. Meenakshisundaram, V. Pascucci, J. Zara, J. Molineros, H. Theisel, T. Malzbender (Eds.), Advances in Visual Computing, Part I. XXXI, 916 pages. 2006.

Vol. 4290: M. van Steen, M. Henning (Eds.), Middleware 2006. XIII, 425 pages. 2006.

Vol. 4289: M. Ackermann, B. Berendt, M. Grobelnik, A. Hotho, D. Mladenič, G. Semeraro, M. Spiliopoulou, G. Stumme, V. Svátek, M. van Someren (Eds.), Semantics, Web and Mining. X, 197 pages. 2006. (Sublibrary LNAI).

Vol. 4288: T. Asano (Ed.), Algorithms and Computation. XX, 766 pages. 2006.

Vol. 4287: C. Mao, T. Yokomori (Eds.), DNA Computing. XII, 440 pages. 2006.

Vol. 4286: P.G. Spirakis, M. Mavronicolas, S.C. Kontogiannis (Eds.), Internet and Network Economics. XI, 401 pages. 2006.

Vol. 4285: Y. Matsumoto, R.W. Sproat, K.-F. Wong, M. Zhang (Eds.), Computer Processing of Oriental Languages. XVII, 544 pages. 2006. (Sublibrary LNAI).